Briquetting in Metallurgy

Aitber Bizhanov
J.C. Steele & Sons
Representative in Russia&CIS

CRC Press
Taylor & Francis Group
Boca Raton London New York

CRC Press is an imprint of the
Taylor & Francis Group, an **informa** business

A SCIENCE PUBLISHERS BOOK

Cover image provided by the author, Aitber Bizhanov.

First edition published 2022
by CRC Press
6000 Broken Sound Parkway NW, Suite 300, Boca Raton, FL 33487-2742

and by CRC Press
4 Park Square, Milton Park, Abingdon, Oxon, OX14 4RN

Library of Congress Cataloging-in-Publication Data (applied for)

ISBN: 978-0-367-46230-7 (hbk)
ISBN: 978-1-032-34101-9 (pbk)
ISBN: 978-1-003-02764-5 (ebk)

DOI: 10.1201/9781003027645

Typeset in Times New Roman
by Radiant Productions

Preface

What time are we now living in? People tend to define different stages of life and eras in history. The modern era is also defined as the era of information technology, the era of artificial intelligence and digitalization. It is difficult to disagree with the fact that new information technologies play a significant role in the life of a modern person, but one should not forget that metallurgy has long been and remains the locomotive of scientific and technological progress since the role that steel plays in the life of modern society is incomparable with any other building and construction material. Therefore, in a sense, humanity is still living in the … Iron Age.

Metallurgy, especially ferrous metallurgy, will undoubtedly play a dominant role in industry in the foreseeable future. Along with this, it is metallurgy that is one of the main sources of harmful emissions into the atmosphere. The fact is that in the process of metallurgical production, a significant amount of fine-grained materials (dust and sludge) with a large content of iron and other important components is formed. Their direct reuse in metallurgical furnaces and reactors is impossible without preliminary agglomeration. And such agglomeration—the formation of a dense mass from finely dispersed raw materials during its high-temperature processing—is accompanied by emissions of substances extremely hazardous to human health.

How can one fail to recall the famous myth about Prometheus, who stole fire from Hephaestus, the patron saint of metallurgy, and gave it to people, for which, as you know, he was severely punished by the Gods. The acquisition of fire, of course, changed a person's life in the literal and figurative sense, but, in particular, it turned into many problems. Punishment, according to Greek mythology, came to people in the form of Pandora with her ill-fated box. Similarly, pyrometallurgy is one of the main sources of harmful emissions into the atmosphere that pose a serious threat to the very life of mankind. In the structure of these emissions, a significant share (up to 50%) are emissions from sinter plants. Therefore, the efforts of scientists and metallurgists to find ways of so-called cold agglomeration of raw materials for iron smelting do not stop.

This book is the result of more than ten years of research by the author aimed at finding an effective method for briquetting in ferrous metallurgy. Acquaintance with briquette took place when I was the commercial director of one of the largest Russian producers of commercial pig iron and ferromanganese. Located almost within the city limits, the enterprise did not have any reserves of territory and the possibility of building a sinter plant, which, together with the accumulated large volumes of anthropogenic materials, made the choice of briquetting uncontested. I am still grateful to the owners of the metallurgical plant for the support of my initiatives and the unique experience I have gained in briquetting. I will note, by the way, which, as often happens, the first pancake, as they say in Russia, came out lumpy. The briquette factory (based on vibropressing) did not work for long and was dismantled. Fortunately, I saw no other road to success for myself other than scientific research. Already in those years, I was greatly impressed by the scientific works of Professors Julian Yusfin and Viktor Dashevsky, whose approach I chose for myself in research on briquetting for blast furnace and ferroalloy production. It was necessary to thoroughly understand the mechanism of hot strength of the briquette, with the influence of the method of its production on the so-called metallurgical properties of such agglomerated product, to assess the prospects for using briquette in the charge of modern metallurgical furnaces and reactors. Even then, I observed the fact that metallurgical scientists paid attention mainly to high-temperature agglomeration-agglomeration and the production of pellets. As a rule, less notice was paid to briquetting, which, however, reflected the real ratio of the degree of commercialization of such technologies. The book contains the results of my research, carried out independently or in collaboration with famous metallurgical scientists (Professors Ivan Kurunov, Tatiana Malysheva, Alexander Pavlov and others). I hope this book will help fill the information gap in knowledge about briquetting that is still present in the metallurgical science literature.

A milestone in the search for the optimal method of briquetting was my acquaintance with the company J.S. Stele and Sons (NC)—the creator of the stiff extrusion technology for the production of ceramic bricks. I remember how, at a critical moment for my business, when my customer refused my proposed method of briquetting blast-furnace materials due to the inadequacy of the size of the briquette to the requirements of the technology, I asked myself—how else can a briquette can be made. And what is a briquette, if not a small brick, which follows from the very definition. How are bricks made? Soon, I turned to the representative of J.C. Steele & Sons in Russia, Mr. Roman Berman, who not only brought me to one of the top managers of the company, Rick Steele, but also told me about the company's experience in ... briquetting of blast-furnace materials (Bethlehem Steel project). "Truly I tell you, no prophet is accepted in his hometown". After some time in the office of the company in Statesville, I signed an agency agreement and soon, based on the previously obtained theoretical ideas, I was able to comprehensively study the metallurgical side of this now widely recognized briquette technology. The book provides an objective comparison of stiff extrusion with other traditional briquette technologies, each of which has its own advantages and disadvantages, areas of preferred use.

I would like to warn the reader that the book is not a reference book on briquetting and does not cover all the many aspects of the problem of briquetting. So, in particular, it does not consider the issues of briquetting in steel production and issues related to coal briquetting. The book contains exclusively scientific and technical results obtained by me personally or jointly with my colleagues in Russia, India, the USA, South Africa and other countries. It is a declaration of love for metallurgy, an amazing, amazingly interesting area of human knowledge and skills that largely determines the level of modern life.

Working on the book, I was greatly helped by direct communication with my colleagues at the Institute for Briquetting and Agglomeration (IBA)—Rick Komarek, Don Fosnacht, Tom Battle, Jim Torok and others. IBA brings together scientists and engineers—enthusiasts of agglomeration technologies and plays a significant role in uniting the efforts of specialists in the search for optimal ways to use fine natural and anthropogenic materials in metallurgy.

Modern briquette technology has reached the level of technical and logistical development, which has led to the appearance on the market of specialized companies providing briquetting services (Harsco Metals, TMS Tube City, Phoenix Services, etc.), which indicates the undoubted commercial prospects of briquette technologies—a safe way for nature to solve environmental problems in the areas of metallurgical production.

The publication is addressed to metallurgical engineers, personnel of metallurgical enterprises, researchers and can be useful to teachers of higher educational institutions of metallurgical or polytechnic profile, graduate students, undergraduates, students studying the specialty "Metallurgy".

With best wishes,   **Aitber Bizhanov**

Time to cast away stones
and a time to gather stones together...
(Ecclesiastes 3:5, English Standard version)

Contents

Introduction

The development of ferrous metallurgy, an increase in the capacity of Blast Furnaces (BF) and the volume of ore consumption required the expansion of a technology for agglomeration of fines formed during the extraction and processing of ore. The first industrial agglomeration technology, which appeared back in the 19th century, was the briquetting technology—obtaining lumps of a regular geometric shape from fine ore by pressing it under pressure using various binders or without them, followed by drying and strengthening firing of the resulting briquettes or with hardening aging on air. Industrial briquetting dates back to the first commercially successful project for the production of briquettes from fine magnetite iron ore, implemented in 1899 in Finland (the Gröndal project, [1]). Briquettes were obtained from moistened magnetite concentrate without using a binder on Sutcliffe type presses for the production of bricks with a pressing pressure of 30–50 MPa. Electricity consumption for the production of a ton of briquettes was 5 kWh. Raw briquettes ($150 \times 150 \times 75$ mm) were subjected to strengthening oxidative firing in a tunnel kiln at temperatures up to 1400°C. In the process of firing, sulfide sulfur was oxidized with its removal by 98%, and magnetite was oxidized to hematite. Despite their large size, these briquettes were successfully used in a low-shaft blast furnace with a capacity of 50–140 t/d at the plant in Pitkäranta (Finland, Former Russian Empire). Fired briquettes (porosity 40%) had high mechanical strength (10 MPa) and high reducibility, and their use led to a decrease in coke consumption and an increase in the productivity of the blast furnace. The success of the project contributed to the rapid distribution of such technology. In 1913, 38 similar briquetting lines were already in operation (16 in Sweden, 12 in England, 6 in the USA).

Later, lever, revolving, ring, conveyor and roller presses were used for the production of briquettes. Depending on the type of material to be pressed and the binder used (or its absence), briquettes were pressed with low (50–150 kg/cm²), medium (150–750 kg/cm²) or high (over 750 kg/cm²) pressure [2]. Low productivity of technological processes of briquetting and, often, insufficient strength properties of briquettes did not allow this technology to take a significant role in the agglomeration of raw materials for blast furnace production. This role is currently played by the technologies of sintering and production of pellets developed at the beginning of the 20th century. The first patents for the method of sintering iron ores were obtained in Germany in 1902 and 1905, but the date of the high-performance sintering process is 1906, when A. Dwight and R. Lloyd patented a conveyor sintering machine in the USA. The high adaptability of the process and the possibility of utilization of dispersed iron-containing wastes in it, which are inevitable in the smelting of iron and steel, stimulated the development of this process, and until now agglomeration by sintering is the main technological process for agglomeration of raw materials in ferrous metallurgy, which is also widely used in non-ferrous metallurgy. The main disadvantage of this agglomeration technology is its large negative impact on the environment. A sinter plant at an integrated metallurgical plant is the source of more than 50% of all dust and gas emissions from the plant.

The increase in the extraction of poor iron ores and the production of iron ore concentrate from them led to the development of a new technology for agglomeration of finely dispersed concentrate. This technology was the production of pellets—the pelletizing of iron ore concentrates with the formation of spherical granules with a size of 9–16 mm and their subsequent strengthening firing in shaft furnaces or on conveyor roasting machines. Andersen obtained a patent for this technology in

Sweden in 1916, and the industrial production of pellets began to develop rapidly from the second half of the 20th century.

The first two chapters of this book provide background information on these flagship technologies, without which, in my opinion, a discussion of the prospects for briquetting would be incomplete. The presentation is based on a book I wrote in collaboration with Professors Valentina Chizhikova, Viktor Dashevskii (Russia, NUST MISIS) and prof. Mikhail Gasik (National Academy of Metallurgy of Ukraine)—Agglomeration in Metallurgy (Springer, 2020) and Ferroalloys. Theory and Practice (Springer, 2020).

At the turn of the 20th and 21st centuries, briquetting technology again begun to be used in the ferrous metallurgy, primarily for the utilization of fine iron and iron-zinc containing waste–sludge and dusts. In Germany, a technology was developed and implemented for the production of briquettes from iron-zinc sludge with their subsequent melting in an Oxy-Cup cupola, operating on an oxygen-enriched blast [3]. Briquettes are produced by vibrocompression technology using Portland cement as a binder. In terms of their metallurgical properties, such briquettes meet the requirements for a blast furnace charge. Vibrocompression technology is used at metallurgical plants in Sweden, where blast furnaces operate on 100% pellets and briquetting is used as a utilization technology that recycles blast furnace dust and pellets screenings. Vibrocompression briquetting began to be used in the USA, Russia and other countries, not only for waste utilization, but also for agglomeration of iron ore concentrate.

In addition to the technology of vibrocompression, compaction in rolls is also used for the production of briquettes from metallurgical waste. The use of an organic binder ensures high cold strength of briquettes and allows them to be effectively used, but only in steelmaking processes. The use of such briquettes in BF is possible only in a very limited amount due to their destruction when heated. To ensure high hot strength of roller briquettes, it is required to use cement as a binder and in an amount much higher than in the manufacture of briquettes by vibrocompression.

At the beginning of the 21st century, a new briquetting technology, Stiff Vacuum Extrusion (SVE), appeared on the market for agglomeration of finely dispersed anthropogenic and natural mineral raw materials, which has been used for many decades for the production of bricks and since the beginning of the 21st century for the production of briquettes from fine nickel ore. The debut of screw extrusion in the field of preparing agglomerated raw materials for blast furnaces took place in the 90s of the last century, but did not become widespread. The metallurgical plant of the Bethlehem Steel (USA) company, where a line for the production of cement-bonded briquettes from sludge and dust for melting in blast furnaces, had been operating for three years, was closed. The experience of industrial production of extruded briquettes (2011–present) as a notable component of the charge in a blast furnace attracted the attention and increased the interest of specialists in the technology of extrusion briquetting. Such briquettes are called BREX in the technical literature [4].

References

[1] Pietsch, W. 2005. Agglomeration in Industry. Occurrence and Applications Wiley, 375 p.
[2] Yuzbashev, L. 1901. The method of obtaining various ores and fossil combustible artificial lump ore and artificial lump fuel from small things. Gornyi zhurnal, 2(6): 257–288 (in Russian).
[3] http://www.kuettner.com.
[4] BREX. Certificate of trademark (service mark) No. 498006, application No. 2012706053 of 02.03.2012. Right holder A.M. Bizhanov (in Russian).

CHAPTER 1
General Information on Briquetting of Natural and Anthropogenic Raw Materials

The desire to obtain the most valuable raw materials for smelting iron, requires deep enrichment of iron ore to produce concentrates so fine that their introduction into the mixture, for example, a blast furnace without preliminary agglomeration would simply disrupt its operation. It is well known that a decrease in the fine fraction (5–0 mm) in the iron ore charge of a blast furnace for every 1% leads to a decrease in coke consumption by 0.4–0.7% and to an increase in furnace productivity by 1% [1].

The question is whether it is possible in some way to deliver fine material to the furnace without agglomerating it at all, but, to pack it into a container made of a material that would ensure the integrity of such a container to temperatures sufficiently high to metallize the concentrate? Indeed, in this case, the need for sintering of the dispersed raw material would disappear. More recently, it was on this path that specialists of one of the ferroalloy enterprises intended to go. It was suggested to pack concentrates in ... tin cans. And even laboratory experiments seemed to confirm the fundamental possibility of introducing unfocused material into the charge of the ore-smelting furnace, but a simple calculation showed that for a similar recycling of the resulting dispersed material would require construction, next to the ferroalloy plant, a separate cannery… When experts tried to introduce a concentrate into the charge of such a furnace without agglomerating or packaging it, it led to a deep disorder of the technology, to a significant increase in the formation of dust from gas treatment plants, to a sharp increase in the accident rate of furnaces and their failure. And the most important, that the use of raw concentrate has led to a deterioration of technical and economic indicators, and, consequently, to an increase in production costs.

It is clear that for traditional metallurgical furnaces and reactors, agglomeration is an inevitable stage in the preparation of charge materials. Nevertheless, the search for methods of smelting metals using non-agglomerated raw materials in ferrous metallurgy practically does not stop.

In the monograph published in 2013 with the original name "The incongruities of metallurgy" [2], it was proposed, in particular, to metallize a dust-gas suspension from a fine or dust-like concentrate mixed with coal dust in a recuperator. According to the author, when "such a dust-gas suspension of concentrate and coal dust will come to a zone with a temperature above 720°C, an intensive reduction of iron oxides by carbon will begin. Gases (CO, CO_2), heated in a recuperator, will react with pulverized solid reagents in 5–10 seconds, during the residence time of the dust-gas slurry in the recuperator. At the outlet of the heat exchanger, a suspension of fine iron powder in metallization gases (CO) will be obtained. Metallization will take place due to cheap recuperative heat. Further, such a dust-gas suspension of iron powder in CO can be blown into the blast furnace." Unfortunately, the author of this concept confined himself to general considerations and did not confirm them with either a laboratory experiment or full-scale industrial test. In addition, he also suggested feed metallized material "to the charge of the steel-smelting unit in cans, bags or in the form of briquettes." But what about the desire to avoid agglomeration in general?

The author also did not mention the well-known experience of the Magnitogorsk Metallurgical Combine (MMC, USSR), where in 1953 they developed different methods for the direct production of iron and even patented a method of producing metal by reducing melted ores and a continuously operating unit for its implementation [3–4]. The essence of the method of MMC lies in the idea of a straight-through two-chamber unit, including vertical and horizontal cameras with the organization of the torch from top to bottom in a vertical chamber. Fine iron ore concentrate is fed into the torch. The principle of "flow through" allowed excluding the stages of agglomeration and pelletizing of the charge for the production of pellets.

The 30 heats carried out in 1962 showed that the metallization of iron ore concentrate took place in the conditions of an experimental furnace. The experience of operation of the furnace showed a low resistance of the lining, made of fireclay, carbon, zirconium blocks and magnesia brick. When switching to intensive cooling of working areas with coils, in refrigerators, they were covered with a layer of skull, the slag corresponded to the calculated one, but the heat loss from the water increased. Iron ore fines, obtained by grinding lumpy iron ore in an amount of 135–170 kg to a fraction of 0–0.5 mm, were used for each melt. Coke and coal fractions of 5–40 mm were mixed and, accordingly, 30 and 60 kg were consumed for smelting.

It should be noted that the use of larger fractions of coal (coke) slows down the process of FeO reduction in the melt, and smaller fractions are removed from the furnace by the gas stream. Experiments confirmed that the temperature in the vertical chamber on the lining with intensive cooling and combustion of natural gas is sufficient to melt the iron ore part of the charge, reheat the carbon of coal (coke) to 900–1000°C and reduce iron oxides by 80–85% to FeO. However, in the horizontal chamber on the lining with intensive cooling to create a skull and when burning natural gas through arch burners to complete the recovery process and heating the melt to 1400–1450°C heat was not enough. In some heats, due to large heat losses, the degree of FeO reduction did not exceed 70%. The identified problems did not allow at that time to go to the stage of practical implementation of the patented MMK method. This has not been implemented to date.

If one discusses the commercialized metallurgical processes in which the metallization of non-agglomerated raw materials is used, first fluidized bed apparatus should be mentioned [5]. In such units, in contrast to shaft furnaces, the charge particles randomly move in a certain volume and, with correctly chosen values of the gas flow rate, do not leave the working chamber. And if in the shaft furnace the charge particles are in direct contact with each other, which determines the specificity of heat and mass transfer processes in a dense layer, then in a fluidized bed condition, such processes occur individually for each particle. It should be noted that the practice of operating known installations of metallization in a fluidized bed has shown that there are restrictions on the grain size distribution of iron ores. In the well-known FINMET process [6], the proportion of ore particles with a size of less than 0.15 mm should not exceed 20%. The degree of metallization achieved on plants commissioned in 1999 and 2000 in Australia and Venezuela based on this process reached 92% with an average carbon content of 1.3%. Natural gas consumption was 13–16% higher than in shaft furnaces. In 2005, the plant in Venezuela ceased to exist, and the plant in Australia has not yet reached its design capacity. Another process, worthy of mention, is the metallization of iron in a fluidized bed CIRCORED, developed by Outokumpu (former Lurgi Metallurgie) [7]. However, this process has also used agglomeration. For effective metallization of fine materials (gas cleaning dust), their preliminary agglomeration was necessary, for which the company Outokumpu developed and patented a process for the production of micro granules. The only industrial installation operating under the CIRCORED process was commissioned in May 1999 in Trinidad. The actual annual production of the metallized product was 360 thousand tons, with the project capacity being 500 thousand tons. The plant was shut down in 2005 [8]. The CIRCORED process has yet to prove its worth as a process working on ore fines. At least, it is not yet possible to describe its economic attractiveness.

There are also attempts to avoid the agglomeration of raw materials in the smelting of ferrochrome in a fluidized bed reactor. What did it lead to? An experimental study showed that with the restoration of fines of chromic ore with a particle size of less than 75 microns in a laboratory fluidized bed furnace with a temperature of no more than 1200°C, a complete reduction of chromium and iron to carbides was observed for 7 hours. This result formed the basis for the creation in 2010 of a pilot line with a capacity of 1 ton/hour, including a rotating kiln, a fluidized bed reactor to maintain the material temperature at 1150°C and a leaning electric furnace [9]. The further development of such a process and its commercialization have so far not been reported. Studies of the metallization of fines of chromic ore in a fluidized bed continue to this day, but, unfortunately, the authors do not bring them closer to the cherished goal of obtaining an attractive market metallized material from undocumented raw materials.

An example illustrating the above is the well-known construction project of shop No. 4 at Aktobe Ferroalloy Plant (Aktobe, Republic of Kazakhstan). The construction of the workshop was started in 2010. The new production consists of four DC furnaces of new generation with a total capacity of 440 thousand tons of high-carbon ferrochrome per year. In such furnaces, a different mechanism for the recovery of ore materials is implemented than in furnaces of alternating current carbon, namely mainly liquid-phase. The total cost of the project is about US$ 843 million. During the operation of the furnaces, the following indicators were achieved—the consumption of fine chromium ore is 3850 kg/ton of chromium, the consumption of reducing agent (coal) is 950 kg/1 ton of chromium, which practically corresponds to the performance of the process on alternating current. 7552 kWh/1 ton of chromium against alternating current furnaces versus 6640 kWh/1 ton of chromium, due to the open arc burning on the bath surface and, consequently, large heat losses from the walls and roof of the kiln. Unfortunately, so far, the furnaces have not reached their design capacity, which indicates the absence of convincing arguments in favor of working on non-agglomerated raw materials.

An attempt to avoid agglomeration as part of the reduction smelting process was undertaken in connection with the development of the well-known Romelt process [10], developed in 1979 by employees of the Moscow Institute of Steel and Alloys (Romenets V.A. and others) and implemented in 1985 as large-scale pilot plant at the Novolipetsk Steel Company (NLMK). This technology, unfortunately, did not receive wide distribution as well.

Thus it appears clear that the level of development of new metal smelting processes using pre-metallization of raw materials without agglomeration achieved so far does not allow them to be considered as a serious alternative to blast furnaces and ore-smelting furnaces. The performance of the majority of well-known installations that implement such processes does not reach the declared design levels that justified their commercialization.

Thus, the agglomeration of dispersed iron-containing raw materials, natural and man-made, for its subsequent use as a charge component is a necessary step in the preparation of the charge for blast-furnace and ferroalloy furnaces, which in the near foreseeable future will remain the main metallurgical units for smelting iron and ferroalloys. But is it possible to do without roasting during agglomeration; are there alternatives to the methods described in the earlier chapters? Such methods do exist, and, moreover, historically they preceded sintering and pellet production. This is the so-called cold or unburned briquetting. It is necessary to clearly understand that, like any modern technology, briquetting combines "hardware" and "software". The first, obviously, is the equipment itself, which allows one to create a solid-state structure from fine material, and the second is a set of methods for preparing and processing charge materials (with or without a binder), which ensure that the briquette properties comply with the requirements of one or another metallurgical process. Such properties have been called "metallurgical" in literature. To give the reader an idea of the technology in question from a scientific point of view, we will focus on the mechanism of gaining strength by a briquette, both at the stage of its creation in one way or another, and in the metallurgical unit itself. Therefore, the physical nature of the main industrial briquetting technologies, and focus on the above-mentioned metallurgical properties will be briefly described.

References

[1] Vegman, E.F. 1981. Concise Handbook for Blast-Furnace Operators. Metallurgiya, Moscow (in Russian).

[2] Pavlov. V.V. 2013. The Incongruities of Metallurgy: Scientific Monograph. Ural state mining University. Third ed. Revised and Enlarged. Ekaterinburg: Publishing House of the USU, 212 p. ISBN 978-5-8019-0272-2 (in Russian).

[3] Fedulov, Yu. V. 2002. Alternative Directions for the Development of the Blast Furnace Process in the 21st Century. Steel 10: 14–19 (in Russian).

[4] Patent USSR No. 129213 (C21B 13/14) Device for direct production of iron. Zudin, V.M., Morev, I.I., Voronov, F.D. et al. // Priority 11/24/1959 (in Russian).

[5] Hakue, R., Ray, H.S. and Mukherjee, A. 1991. Fluidized bed reduction of iron ore by coal fines. ISIJ International 31(11): 1279–1285.

[6] Zeller, S., Reidetschlager, J., Ofner, H.P. and Peer, G. 2003. Update on FINMET Technology, Product Information, Siemens VAI.

[7] Nuber Dirk, Eichberger Heinz and Rollingen Bernt. 2006. Circored fine ore direct reduction. Millennium Steel pp. 37–40.

[8] Elmquist, S.A., Weber, P. and Eichberger, H. 2002. Operational results of the Circored fine ore reduction plant in Trinidad. Stahl u. Eisen 122(2): 59–64.

[9] Harman, C.N. 2010. Reduction of chromite fines in solid state using a mixture of gases containing natural gas, hydrogen and nitrogen. Harman C.N. // INFACON 12. –Helsinki, Finland pp. 283–388.

[10] Romenets, V.A. (eds.) et al. 2005. The Romelt Process, Ruda i Metally, Moscow (in Russian).

CHAPTER 2
History of the Industrial Briquetting in Ferrous Metallurgy

Acquaintance with briquetting could begin with reference to the most extensive modern information resource—to Wikipedia. But the curious reader would be disappointed. In the English account, the use of briquette in metallurgy is not paid attention at all, and in the Russian one there are only 15 lines on briquetting in ferrous metallurgy. On the Cambridge Dictionary website (https://dictionary.cambridge.org), the word briquette is explained as—a small block of coal dust or peat, used as fuel. There is no mention of metallurgical briquetting applications. Here reverberations of the Second World War are heard, when in England briquettes from a mixture of coal dust and cement, molded into blocks of $15 \times 15 \times 5$ cm, were used for heating homes.

Let us try to fill this gap. So, what is the briquette? The word briquette (from the French "brique" – brick) means a pressed product. Indeed, as one will see later, the history of briquetting is closely intertwined with the production of bricks and some other building materials. And even the first commercially successful project of briquetting iron ore fines for blast furnaces was based on the equipment used at that time for the production of bricks.

If one goes back to the history of the brick business, it is known that the brick business originated in ancient Egypt. During the period of the so-called Middle Kingdom (21th century–early 19th century BC), when economic conditions no longer allowed the construction of giant pyramids, the raw brick became the building material for their construction. The pyramid of Senusret was built (the base length is 107 meters, and its height is 61 meters) using mainly such a brick. The brick, while already burnt, was also used for the facing of the famous Tower of Babel (Etemenanki ziggurat—"House of the foundation of heaven and earth", II millennium BC). This is stated even in the Bible itself (Genesis, Ch. 11-3): "And they said to one another: "Come, let us make bricks and bake them thoroughly". They used brick instead of stone, and tar mortar. "Come", they said, "let us build for ourselves a city with a tower that reaches to the heavens, that we may make a name for ourselves and not be scattered over the face of all the earth…""

Archaeological excavations on the Indian subcontinent allowed in proving that as far back as the 3rd millennium BC dwellings were built of raw brick. The most famous monument of antiquity, during the construction of which used raw and burnt bricks is the Mohenjo-daro Citadel. A part of the Great Wall of China, which has survived to the present day (3rd century BC), was also built of baked bricks.

In ancient Russia, brick appeared only in the X century. The first Russian building of this material was the Tithe Church in Kiev. A striking example of the use of brick construction in Russia of the times of John III was the construction, beginning in 1485, of the wall and church of the Moscow Kremlin, which was led by Italian masters: "... and they made a brick stove outside the Andronikov monastery, in Kalitnikovo, what to burn brick and how to do, our Russian brick is already longer and harder, when it needs to be broken, it is soaked with water. Lime is densely tied up with tillers, it dries in the morning, and it is impossible to break the knife out." The main element from which the walls and towers of the Moscow Kremlin were erected is the so-called two-handed brick with dimensions up to 31 × 15 × 9 cm and weighing up to 8 kg.

In the Middle Ages in Western Europe, urban growth and the risk of fire were one of the reasons for limiting the use of wood in construction. After the Great Fire of 1666, the English Parliament approved the famous "An Act for rebuilding the City of London", according to which it was forbidden to build wooden houses and houses on the basis of a wooden frame. The brick of the familiar modern form appeared in the 16th century in England. The most famous medieval brick buildings in England are Hampton Court near London (beginning of construction in 1514), several college buildings in Cambridge and Herstmonceux Castle in Sussex (mid-XV century). From XII, brick spread to Holland, Denmark, Northern Germany and other parts of Northern Europe, where there was a shortage of natural stone. Examples of brick structures can be: the Church of Mary (Marienkirche) in Lübeck (Germany), the Basilica of San Ambrolio in Milan (Italy), etc. Brick was also popular in Moorish Spain.

Until the XIX century, bricks were made by hand, the process remained primitive and time consuming. Skilled molders filled the raw clay tins (special forms), which were then dried and burned in floor kilns. They dried the bricks exclusively in the summer, and burned them in kilns made of dried brick raw. One of the first patents on the molding machine was granted in 1741 to William Bailey from Taunton (England, UK patent number 575, 1741). Like other early machines, it was a molding machine largely imitating a manual molding procedure, but with a significantly higher productivity. Bailey's invention included a separate mill to give the clay uniformity before molding; brass or iron molds containing five or six bricks that were filled with a homogenized clay, molded into a brick with a stamp or plunger press; the device was returned for cyclic repetition of the process. Half a century later (March 7, 1792), a patent for a molding machine was granted to David Ridgeway from Philadelphia (USA). A year later, with an interval of 5 days, Christopher Colles and Apollo Kinsley from New York received patents for similar inventions [1]. Historical sources preserved only a description of the device and the principle of operation of the machine, presented by Apollo Kinsley. In 1800, a British patent No. 2368 was issued to Isaac Sanford, former Apollo's countryman, who resettled at Covent Garden. Figure 2.1 shows the image of the scheme of the machine, invented by Apollo Kinsley [2].

The mixed clay was fed under great pressure. Such a machine, successfully operating in Washington in 1819, had a capacity of up to 30,000 bricks per day, while the strength of freshly formed bricks made it possible to do without drying them and send them directly to the kiln. It is

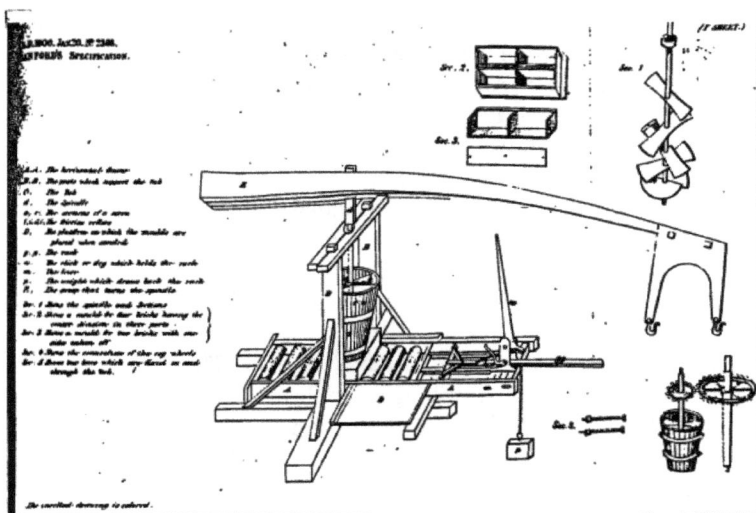

Figure 2.1. Apollo Kinsley brick making machine (from patent No. 2368, 1800, [2]).

Figure 2.2. Degerlein molding machine (England, 1810, [3]). Http://www.binfieldheath.org.uk/the-village/binfield-heath/history/brickmaking.

interesting to note that, despite the efforts of Isaac Sanford, these machines did not gain success in England. Nevertheless, many researchers consider Kinsley's invention to be the progenitor of modern belt-pressing technology.

The concept of a belt press was developed by a molding machine, invented by Joan George Degerlein in England in 1810 (Fig. 2.2, [3]), which made it possible to produce semi-solid clay bricks.

The wet clay mass entering the machine was forced through a special mold in the form of a bar, which was then cut into individual bricks. These first inventions served as the basis for widespread belt pressing, which would play an important role in briquetting.

For the first time, the idea to use a press for the production of briquettes was claimed by the Russian inventor A.V. Veshnyakov in application to the briquetting of charcoal and hard coal [4]. With the date of July 7, 1841, he was granted a "privilege" for a new type of fuel, called "carbolyne". Veshnyakov proposed to subject the charcoal (coal, coke) to grinding and sifting, then add vegetable or animal oil to it, load the resulting mass into bags (made of bast or strong canvas), tightly wrapped with ropes or strong canvas and compress them in the hydraulic press. The obtained pieces were proposed to be further dried. Later, Veshnyakov proposed replacing the bags with cast-iron boxes with holes. A year later, in 1842, Marsais, a Frenchman, invented a coal-fuel briquette with coal tar as a binder [5]. In 1843, a similar patent was granted to Wylam in the UK [6]. The first patent for coal briquetting in the United States was granted in 1848 by William Easby (Fig. 2.3, [7]).

In 1859, the first coal briquette factory was built by SAHUT-CONREUR in Raismes in France, where in 1860 a roll press was used to produce coal briquettes. The concept of such briquetting was formed with the help of two cast-iron rolls, the surface of which was covered with submerged ellipsoidal cells, which, while simultaneously rotating the rolls, should ideally coincide with each other, forming a container for forming the briquetted mass held in it. One of the first patents for the design of a roller press was obtained only in 1884 by J.M. Wilcox in the USA (Fig. 2.4, [8]). The inventor was able to significantly simplify the design of the roller press through the use of dividers, grooves and lifts, which reduced the cost of the production of such presses and contributed to the growth of their use for the production of briquettes.

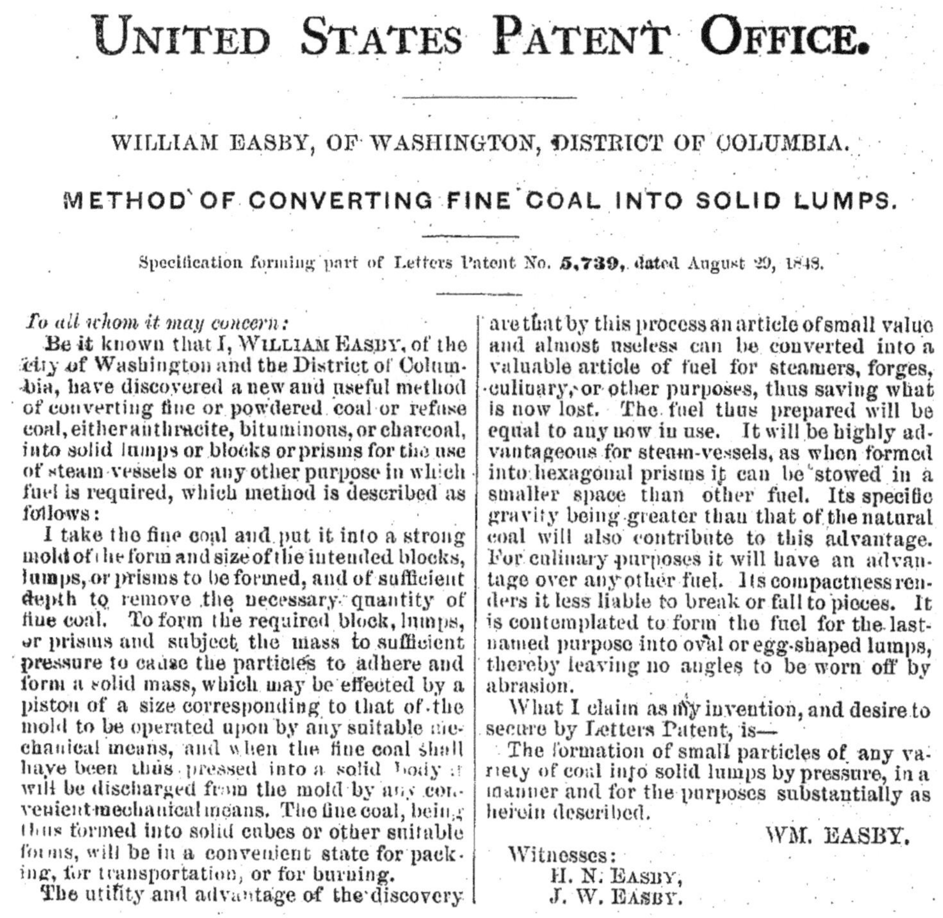

UNITED STATES PATENT OFFICE.

WILLIAM EASBY, OF WASHINGTON, DISTRICT OF COLUMBIA.

METHOD OF CONVERTING FINE COAL INTO SOLID LUMPS.

Specification forming part of Letters Patent No. **5,739**, dated August 29, 1848.

To all whom it may concern:

Be it known that I, WILLIAM EASBY, of the city of Washington and the District of Columbia, have discovered a new and useful method of converting fine or powdered coal or refuse coal, either anthracite, bituminous, or charcoal, into solid lumps or blocks or prisms for the use of steam-vessels or any other purpose in which fuel is required, which method is described as follows:

I take the fine coal and put it into a strong mold of the form and size of the intended blocks, lumps, or prisms to be formed, and of sufficient depth to remove the necessary quantity of fine coal. To form the required block, lumps, or prisms and subject the mass to sufficient pressure to cause the particles to adhere and form a solid mass, which may be effected by a piston of a size corresponding to that of the mold to be operated upon by any suitable mechanical means, and when the fine coal shall have been thus pressed into a solid body it will be discharged from the mold by any convenient mechanical means. The fine coal, being thus formed into solid cubes or other suitable forms, will be in a convenient state for packing, for transportation, or for burning.

The utility and advantage of the discovery are that by this process an article of small value and almost useless can be converted into a valuable article of fuel for steamers, forges, culinary, or other purposes, thus saving what is now lost. The fuel thus prepared will be equal to any now in use. It will be highly advantageous for steam-vessels, as when formed into hexagonal prisms it can be stowed in a smaller space than other fuel. Its specific gravity being greater than that of the natural coal will also contribute to this advantage. For culinary purposes it will have an advantage over any other fuel. Its compactness renders it less liable to break or fall to pieces. It is contemplated to form the fuel for the last-named purpose into oval or egg-shaped lumps, thereby leaving no angles to be worn off by abrasion.

What I claim as my invention, and desire to secure by Letters Patent, is—

The formation of small particles of any variety of coal into solid lumps by pressure, in a manner and for the purposes substantially as herein described.

WM. EASBY.

Witnesses:
H. N. EASBY,
J. W. EASBY.

Figure 2.3. Title page of William Easby patent (1848, [7]). Http://www.briquettepress.com/News/History-Of-Coal-Ball-Briquetting.html.

In literature there is an erroneous idea that the first roller press was invented in 1865, which is illustrated by a drawing from a report on briquetting peat dated to the same year (Fig. 2.5, [8]).

However, it can be seen that the figure shows a piston press, not a roller press. Since 1901, Köppern began to use the roller press for coal briquetting.

The history of industrial briquetting can be divided into the following periods:

- Beginning of the 20th century–the 20s of the 20th century
- 30–50s of the 20th century
- 60s–70s of the 20th century
- 80s–end of 20th century
- These days.

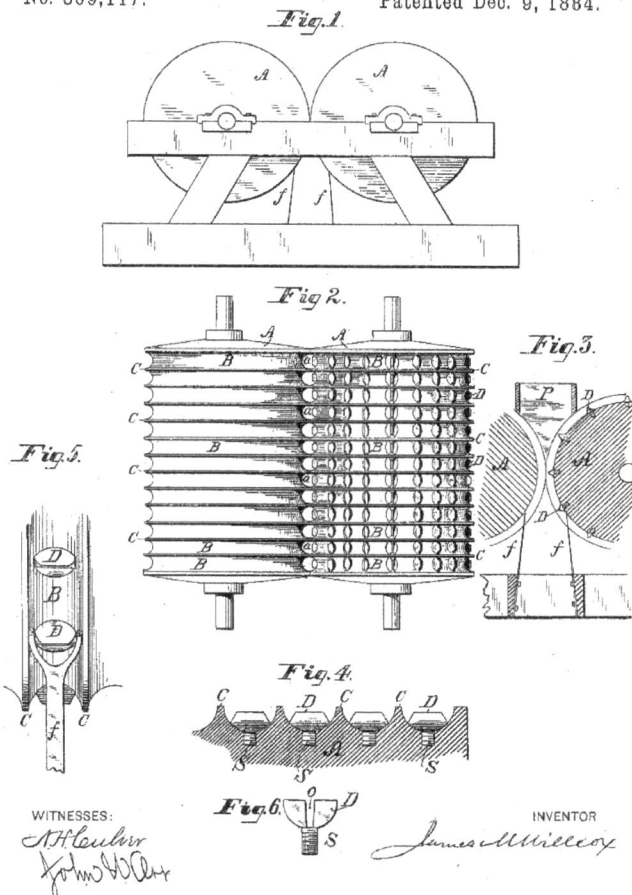

Figure 2.4. Title page of J.M. Wilcox patent (1884, [8]).

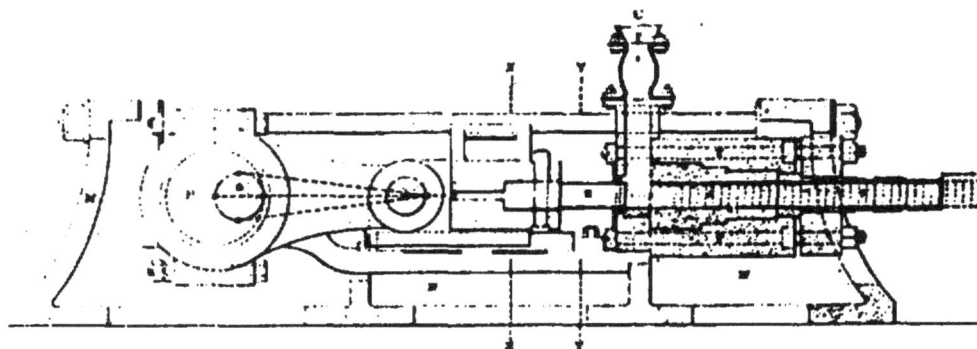

Figure 2.5. Piston press for peat briquetting (1865, [9]).

2.1 Beginning of the 20th Century – the 20s of the 20th Century

At the beginning of the 20th century, metallurgists were faced with the need for the agglomeration of fine iron ore, and it was decided to try to apply the technology used at that time for the production of bricks for these purposes.

It is safe to assume that industrial briquetting originated from the first commercially successful project to produce briquettes from small magnetite iron ore, implemented at the beginning of the last century in Sweden and Finland, mainly due to the talent and efforts of Gustav Gröndal (Fig. 2.6). In Herrang (Sweden) and in Pitkaranta (Finland) briquette factories were built and put into operations as part of the blast furnace units.

Figure 2.6. Johan Gustav Gröndal (February 11, 1859–March 16, 1932). Https://www.geni.com/people/ Gröndahl/6000000041491719847.

Gröndal Johan Gustav was born on February 11, 1859 in Fröslunda (Sweden), died on March 16, 1932 in Stockholm (Sweden). His energy and inventive talent in metallurgy began to emerge with the beginning of his work in 1880 as an engineer at the Pitkäranta Copper-Tin Plant on Lake Ladoga in Finland near the Russian border. Gustav introduced a number of improvements, which soon led to an increase in the company's profits. He invented new methods of enrichment of iron ores, which were previously considered unsuitable for processing. These methods included magnetic separation, briquetting of new kiln designs. Of particular importance was the practice of its ore separators and ball mills, which were distinguished by a high degree of safety in operation. Gröndal became widely known in the 1910s and 20s due to its numerous inventions in the mining industry and fuel technology. Between 1894 and 1932, he issued a total of 65 patents, of which 62 were national. Thanks to his world-patented inventions and new designs, he occupied a special place among metallurgists and industrial entrepreneurs. He implemented a number of projects in Russia, Germany, Austria, England and Manchuria. A number of well-known companies such as Herrängs Gruf AB, Luleå Ironworks AB and Totherhättte electrothermal AB are associated with his name. He also founded The Grondal Kjellin Co. Gustav Gröndal was distinguished by simplicity and personal charm, sense of humor. He always remained an optimist and kept youthful enthusiasm throughout his life.

For the production of briquettes, stamp presses of Sutcliff design (Fig. 2.7, [10]) with a pressure of 30–50 MPa were used. A patent for such a press was received by Edgar Sutcliffe from Lancashire (England) on March 31, 1925. Patent number 1531631 was issued in the United States on May 31, 1925 and was called "Press for the production of bricks, briquettes, blocks, and the like." Sutcliffe himself defined his invention as "presses of the mould and plunger type for the production of bricks,

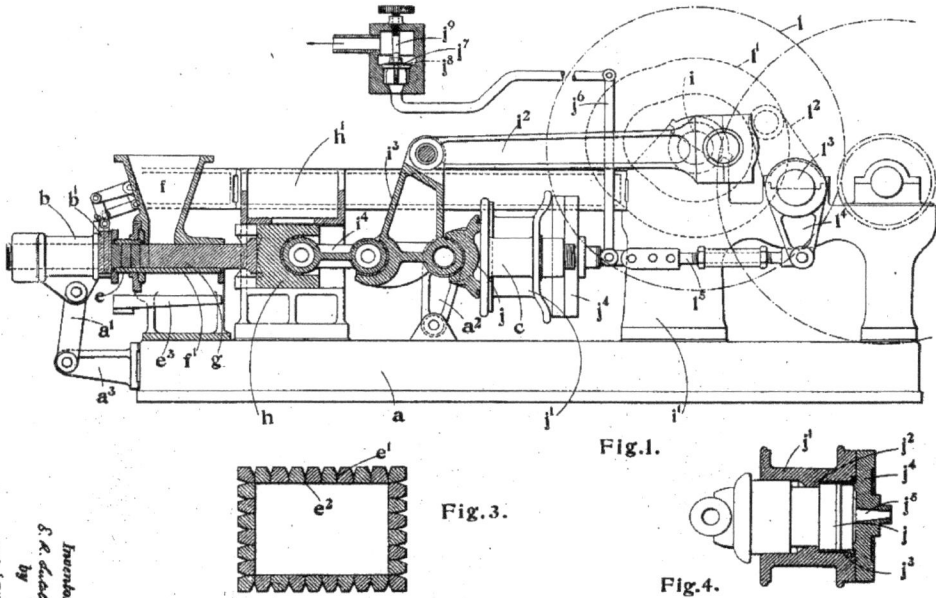

Figure 2.7. Design of Sutcliff stamp press for the production of bricks and briquettes (Patent No. 1531631, England, 1925, [10]; a - base plate; b, c - parts of the crank-slider mechanism, e - mold (mold), f - receiving funnel, g - plunger, h - slider, i - shaft, j - hydraulic cylinder).

briquettes, blocks and like consolidated masses from coal or like carbonaceous substances, concrete, clay or other substances or mixtures capable of being consolidated by pressure". A feature of the Sutcliff press was the possibility of pressing the wetted material. The mixture for briquetting, for example, coal "may consist of one part coal to one and a half to two parts of water or may be a mixture containing say 25% of water". The molded mass was fed into the receiving funnel of the press. The molding of the briquette was carried out with the progressive advance of the plunger that pushed the briquette out of the mold. Electricity consumption for the production of a ton of briquettes was 5 kWh.

The fine concentrate was pressed into briquettes without the use of a binder material, the moisture content in the concentrate was adjusted to obtain briquettes with strength sufficient to remove them from the press and load them on special carts for delivery to Gustav Gröndal's kiln furnaces. On December 28, 1903, he received a patent for the design of a briquette kiln (Patent No. CA85977A). The necessity of burning briquettes to give them the strength required for blast-furnace smelting was due to a number of reasons. Achieving the required mechanical strength without a binder and roasting would require the application of ultrahigh pressures (over 200 MPa), as, for example, in the famous Ronay press, which made it possible to obtain dense and durable briquettes, the porosity of which, however, did not contribute to their good reducibility during heating. Adding organic binders also did not solve the problem. Firstly, most of the known organic binders were very expensive, and secondly, when heated, the organic material burned out or underwent pyrolysis, which negated its astringent effect. The use of inorganic ligaments, in addition to the effect of "dilution"—reducing the iron content in the charge, significantly affects the composition of iron and slag, increasing the yield of the latter.

Gröndal's kiln was in the form of a tunnel, with a track leading down the center and a combustion chamber in the middle. The air required for combustion enters the gas-tight platform at the feed end of the furnace, and, having passed the discharge end, returns above the platforms of the carts loaded with briquettes. The cold air circulating under the platform keeps the wheels and carriage frame cool, heats up as it simultaneously cools the baked briquettes and enters the combustion chamber hot; hot gases, in turn, heat the briquettes and cool themselves to exit the furnace. Such heat recovery increased the efficiency of the furnace. Coal consumption averaged 7% of the briquette mass; the main heat loss was associated with evaporation of water in briquettes. The temperature in the combustion chamber when burning gas reached 1300°C or 1400°C, and with this heating the particles were sintered enough to create a solid, strong briquette. It is important that during briquette roasting their desulfurization was achieved. The time spent on the operation depended on the type of ore and the required degree of desulfurization. When the total iron content in the ore is 38%, it contained 68.3% in the concentrate and 68% in the briquette. For sulfur content, these values were, respectively, 0.066, 0.026 and 0.006%.

The briquette manufacturing process at the factory in Herrang, kilns and briquettes on carts are shown in Figs. 2.8–2.9, courtesy provided, as well as Fig. 2.10, by David Michaud, the owner of the Internet resource 911 Metallurgy Corp [11]. The sizes of briquettes (150 × 150 × 75 mm), exceeding the generally accepted sizes of bricks, attract attention.

Figure 2.8. Briquette factory The Grondal Kjellin Co. in Sweden [10]. Https://www.911metallurgist.com/magnetite-concentration/.

Figure 2.9. Briquettes firing kilns at The Grondal Kjellin Co. in Sweden [10]. Https://www.911metallurgist.com/magnetite-concentration/.

Figure 2.10. Briquette factory The Grondal Kjellin Co. in Guldsmedshyttan (Sweden, [10]). Https://www.911metallurgist.com/magnetite-concentration/.

The carts had a frame structure lined with refractory bricks. Special devices at the front end, a groove at the rear end and flanges on the sides, which were lowered into the oven in a groove filled with sand, allowed the column of such carriages to form, thus, providing an airtight platform.

Briquettes were made using the Gröndal technology and at a lower temperature due to the use of different binding materials. For example, in Pitkäranta (Finland), lime was used as a binder (3–5% of the mass of the briquette). After two weeks of curing, the briquettes were heated to 800°C for subsequent hardening. At a briquette factory in Edison, the USA, resinous substances were used as a binder. The cost of briquette at this factory was about 45 cents per ton. Interestingly, the equipment for iron ore dressing at this plant was designed by Thomas Edison himself.

A factory with a capacity of 60 thousand tons of briquettes per year was soon built in Guldsmedshyttan (Sweden, Fig. 2.10). The factory staff included only 14 people.

Similar factories with a capacity of 40–60 thousand tons of briquettes per year were put into operation in other Swedish cities (Strassa, Bredsjo, Hjulsjo, Lulea and Uttersberg). The largest briquette factory in those days, which was based on the Gröndal technology, was built at Syd Varanger in Norway. The processing plant had a capacity of 1.2 million tons of ore per year and was equipped with 56 ball mills and 200 separators of the Gröndal construction and at the briquette factory the briquettes were fired in 20 furnaces of the above-described construction of the same Gustav Gröndal. The capacity of the briquette factory was 600 thousand tons of briquettes per year. Briquettes were exported to Germany. A factory with a capacity of 100 thousand tons of briquettes, which were also exported to Germany, was built in another Norwegian city - Salangen.

Despite their large size, these briquettes were successfully used in a low shaft blast furnace with a capacity of 50–140 tons/day at the plant in Pitkäranta. In the process of oxidative roasting, magnetite was oxidized to hematite, and sulfur was removed from briquettes (by 98%). Porous briquettes (40% porosity) had high mechanical strength (10 MPa) and high recovery, which led to a reduction in coke consumption and an increase in the productivity of the blast furnace. The success of the project contributed to the rapid spread of such technology. In 1913, there were already 38 such briquetting lines (16 in Sweden, 12 in England, 6 in the USA).

At about the same time, anthropogenic materials from the ferrous metallurgy, such as top dust and mill scale, were also subjected to briquetting. In Germany (Ilseder-Hütte) a pilot plant was created for briquetting these materials in mixture with clay slurry of brown iron ore washing. The mixture was heated and pressed at a pressure of 30 MPa. Briquettes gained strength indoors, reaching a compressive strength of about 10 MPa [11]. Clay iron ore fines, blast dust and cinder were also briquetted without binders, plasticizing the mixture by grinding the components and moistening. Stamp presses were applied for briquetting. Ready briquettes were burned in a shaft furnace (capacity up to 40 tons per day).

The successes in the briquetting of iron ore fines and anthropogenic materials achieved at the very beginning of the twentieth century and the dominance of briquetting as the only industrial sintering technology until the 20s of the last century, allows one to consider this time as a kind of "golden age". Along with the development and modernization of equipment for the production of briquettes, the ways of preparing and processing the charge for briquetting were developed and improved. The operating experience of the first briquette factories showed that the potential for agglomeration without bonding materials is limited due to the insufficiently high strength of raw briquettes and the need for expensive roasting. In 1901, the Russian mining engineer L. Yuzbashev proposed to do without firing and use hydraulic cement as a binder for briquetting ore fines [12].

His theory was soon embodied in a number of Russian factories. As well as in the process of Gröndal briquettes were made on ordinary brick presses. Coal or coke breeze, chopped straw, sawdust and other materials were added to the charge, which, faded when briquettes were heated in a blast furnace, contributed to the growth of their porosity, and consequently, better recoverability. Briquettes were dried at ambient temperatures and, after curing, were sent to the stock yard of the blast furnace. The use of cement made it possible to remove part of the fluxes from the charge, in proportion to the content of CaO and MgO in it. Yuzbashev's method, however, was not widespread due to the relatively high cost of cement at that time. In addition, the cement content in the briquette mass (3–6%) did not ensure the integrity of the briquettes in the zone of high temperatures of the blast furnace. "The first pancake is lumpy". Unfortunately, this bad experience led to the emergence of a "myth" that cement in principle is not capable of providing the so-called hot strength to briquettes due to the fact that Portland cement loses its binding properties at 750–900°C. This appears to be true. Calcium silicate hydrates are destroyed at even lower temperatures (400–500°C). But from the mineral phases formed at such temperatures, under certain conditions, a new liquid high-temperature bond can form. Modern experience in the use of cement as a binder for blast furnace briquettes confirms the validity of this conclusion.

At the same time, at the very beginning of the last century, V. Schumacher (Germany) proposed to use crushed quartz sand (1–5% by weight of the briquette) and quicklime (3–10% by weight of the briquette) as a binder. This powder was thoroughly mixed with ore after wetting, and then pressed at a pressure of 40–70 MPa. Briquettes were strengthened by steam treatment at a pressure of 1 MPa for 2–4 hours in an autoclave at a temperature of 174°C. Schumacher is credited with the authorship of the very principle of autoclaving [13]. Briefly, this is nothing more than the processing of the material at a pressure above atmospheric and at a temperature above 100°C. With this steaming, a new binder of calcium hydrosilicates ($CaO \cdot SiO_2 \cdot nH_2O$) was formed, which was obtained by the reaction $Ca(OH)_2 + SiO_2 + (n-1)H_2O = CaO \cdot SiO_2 \cdot nH_2O$. The compressive strength of such briquettes reached a value of 10–13 MPa. Briquettes maintained integrity when heated to temperatures of blast furnace smelting. The method was introduced at two steel plants in Germany. However, in the practice of briquetting Schumacher's method was not rooted due to its complexity and high cost.

Thus, by the 20s of the last century, the achievement of the strength of briquettes required for smelting iron was mainly provided by firing or other temperature treatment (steaming), and the productivity of briquette factories was limited by the productivity of equipment used at that time to make bricks.

The performance of such equipment did not meet the growing needs of blast-furnace production for agglomerated raw materials. Briquetting was quickly replaced by the sintering technology developed by that time. High performance and manufacturability of the process, the possibility of utilization in it of the inevitable in the smelting of iron and steel dispersed iron-containing materials, led to its rapid growth. Sintering practically supplanted briquetting from ferrous metallurgy. Thus for example, by 1923 almost all the above-mentioned Gröndal briquette factories were replaced by sinter plants.

The use of roller presses for briquetting natural and man-made yellow raw materials that began in the 20s of the last century did not change the situation with briquetting. It is important that roller presses were created specifically for briquetting, that is, they were not "borrowed" from the building materials industry. Roller briquettes at the beginning of the 20th century found sufficient application in low-shaft blast furnaces and in blast furnaces of small volume. Briquettes were made of iron ore fines, limestone and coke dust. In the blast furnaces of the Kushvinskyi plant, the proportion of briquettes in the charge reached 25%. Briquettes in the amount of up to 100 thousand tons per year were used in the charge of the Kerch blast furnace (the capacity of the briquette factory in 1915 was 35 thousand tons of briquettes) and the Taganrog metallurgical plants (the capacity of the briquette factory in 1906 was 30 thousand tons of briquettes) [14].

2.2 30–50s of the 20th Century

In the 1930s, the search for effective and affordable bonding materials continued. It is of interest that such a search had its main purpose... to reduce energy consumption for firing. One of the few attempts to turn briquetting into a real "cold", unburned was associated with the name of the Soviet scientist-metallurgist P.P. Kazakevich, who in 1936 conducted research on the briquetting of flue dust with Portland cement as a binder. But, despite the generally satisfactory results of his research, the method was not widely distributed at that time, since obtaining mechanically strong briquettes required a very long time. Cement myth rooted.

Another noteworthy attempt to carry out briquetting without firing was undertaken in 1936 at the Krivoy Rog briquette factory. Briquettes were made from ore fines of rich hematite ore. To the ore fines 5–10% of crushed iron shavings and 0.5–1.0% NaCl in the form of a solution were added, which accelerated the process of formation of iron oxide hydrates. The briquettes' strengthening was achieved as a result of corrosion and hydration of iron shavings. The hardening of briquettes for some ores was completed within a few hours after their production. For most ores, the curing period was 20–40 hours. Briquettes did not need drying or firing. In literature, this way of briquetting is known as the method of Yarkho, after him [15]. The same method of briquetting was used at the briquette factory built in 1933 in Nizhny Tagil. However, this generally successful way did not become a landmark in cold agglomeration. The obvious disadvantage of the method was the high cost of the additives used and the content of alkaline compounds in the briquettes, which is highly undesirable for the blast furnace.

In 1932, Averkiev and Udovenko published an article entitled "Briquetting of flue dust and fine ores with alkali silicates", in which they suggested using the so-called "liquid glass"—an aqueous alkaline solution of sodium silicates $Na_2O (SiO_2)_n$ [16]. Dissolved liquid glass was added to the briquetted ore, the wetted mass was thoroughly mixed and pressed. The consumption of liquid glass for briquetting, for example, flue dust was not less than 15–18% of the mass of the briquette. Compressed briquettes were dried and fired at 400–500°C, after which they acquired good strength and water resistance. The disadvantages of the method were already revealed at the first full-scale testing. Briquettes could not withstand high temperatures, differing in low porosity. High cost of liquid glass together with its substantial consumption, markedly reducing the iron content in the briquette, made such briquetting economically unjustified and technologically inefficient because it required additional flux consumption and increased the amount of slag in the blast furnace smelting.

A way that allowed to significantly reduce the consumption of liquid glass for briquetting was soon proposed by A.P. Fonyakov [17]. The method was based on the binding properties of silicic acid gels of the general formula $nSiO_2 \cdot mH_2O$, which fall out of the liquid glass solution when treated with a weak solution of calcium chloride. To obtain such a gel, the ore was processed, before pressing, in succession with two solutions—first with a 1–2% solution of liquid glass, then with a 1.5–2% solution of calcium chloride, and finally pressed. Freshly formed briquettes were dried and burned at a temperature of 500–600°C to dehydrate the resulting silica gel. The finished briquettes had high strength (18–22 MPa), porosity up to 21% and met the requirements for blast furnace briquettes. The consumption of liquid glass in terms of the SiO_2 content decreased markedly (to 1.0–2.5%), but this decrease was compensated for by the need for costly roasting. However, this method was not widespread.

In 1957–1958, a series of pilot industrial melts with briquettes in the charge of basic oxygen converters of the Petrovsky plant took place in the Ukraine. Briquettes were made according to the method of Yarkho at the Krivoy Rog briquetting factory. The composition of the mixture—iron ore concentrate, limestone, bauxite and iron shavings fraction 0–3 mm. Raw briquettes gained strength during the week in a warmed warehouse. The supply of briquettes that gained strength was carried out along the path of bulk materials with the first addition of charge materials. Briquettes were also loaded into the converter during purging. In total, over 100 heats were held. In the first 60 heats, the content of calcium oxide in briquettes was 21.5%, and the missing number of fluxes was injected into the converter in the form of lime. The results of the melts performed confirmed the possibility of using ore-limestone briquettes to control the temperature of the metal bath. The process with briquette was more intensive, but emissions were not observed. The slagging process improved. For the next series of heats, the limestone content in briquettes was increased to 35.45% to achieve the required basicity (CaO/SiO_2 = 10.74). Experimental melts showed that briquettes are a useful and effective part of the charge of oxygen converters [18].

The results of using briquettes in open-hearth furnaces will not be dwelled on, as this experience is only of historical interest, since this method of steel production is already in the past. Nevertheless, some of the results of using briquettes in open-hearth heats obtained in the late 50s in the Ukraine continue to be relevant for the briquetting technology as a whole. These heats briquettes were made all at the same Krivoy Rog briquette factory. The results of pilot industrial melts showed that the productivity of furnaces increased by 6.3% compared with the performance in the base period (when melting on the ore), while reducing the duration of melting when using briquettes by 2 hours. The results made it possible to conclude that the use of briquettes in the charge of open-hearth furnaces is preferable to sinter and pellets.

In the 1950s, the first known attempt to produce iron ore briquettes was made using the vacuum extrusion technology well known in the ceramic industry. A wetted mixture of iron ore and bentonite was fed into the vacuum chamber and then into an extruder, from the die of which oblong cylindrical briquettes exited [19]. However, this method did not receive its distribution at that time. Unsuccessful attempts to use extrusion for the agglomeration of raw materials in the iron and steel industry were also made in the USSR in the 60s. At the same time in Germany, the production of a tape-vacuum press was mastered, which included two screws and a vacuum chamber located between them. The first screw mixed and homogenized the mixture, and the second molded the briquettes, forcing the mixture through the holes of the die. The press was used for briquetting of ore and coal charges and concentrates of poor iron ores; however, its low productivity (10 tons of briquettes per hour) limited its use in metallurgy.

In the absence of new, breakthrough developments in briquetting in the 30–50s of the last century, its suppression by sintering continued. In addition, since the 1950s, the industrial production of pellets began to develop rapidly—a technology that effectively pelletized fine concentrate. As of the middle of the last century, only about 0.5% of all iron-containing natural and man-made materials of ferrous metallurgy were briquetted [17]. Nevertheless, interest in briquetting has not waned, which was reflected in 1953 of the creation of International Briquetting Association (International Briquetting Association)—a public organization uniting metallurgists and manufacturers of briquetting equipment [20], later renamed the Institute for Briquetting and Agglomeration. The progress of briquetting in the world was facilitated by the IBA science and technology conferences held in the USA every two years. For many years, IBA was headed by Richard Komarek (President and Founder of K.R. Komarek—the largest manufacturer of roller presses in the world).

2.3 60–70s of the 20th Century

Interest in briquetting resumed in the 60–70s of the last century due to the complicated environmental situation in the areas of metallurgical production, associated mainly with harmful emissions from sinter plants. In 1961, the Donetsk Briquette Factory was built in the city of Donetsk (Ukraine). In the 1960s–1970s, briquetting factories at the Alchevsk and Magnitogorsk plants and the KMAruda plant were also commissioned. At Magnitogorsk Metallurgical Plant, mill scale was briquetted along with lime using a roller press, and the resulting briquettes were dried in a tunnel kiln. At the KMAruda combine, magnetite concentrate (Fe - 59.68%, SiO_2 - 5.5–14.0%) was briquetted together with quicklime. The mixture was mixed and maintained in silos, in which the lime was quenched with moisture concentrate. The lever press was used for briquetting. Raw briquettes were autoclaved at a pressure of 8 atmosphere and a temperature of 174°C.

In Germany blast furnace flue dust was briquetted with sulfite liquor as a binder (5–7% of the mass of the briquette). The briquetting factory, equipped with Humboldt roller presses (with a capacity of 10–15 tons per hour), produced up to 200 tons of briquettes each day. To achieve the required mechanical strength, the briquettes were fired at a temperature of 600–900°C. The resulting briquettes were melted in blast furnaces. The share of briquettes in the ore part of the charge reached 18%.

In the second half of the twentieth century, several experimental and industrial briquetting units for ore-fuel charges according to the Weber method [17] were also working in the Germany. Ore fines mixed with carbonaceous reducing agent (high-volatile coal and tar) and binders (Sulfite-Alcohol Bards (SAB) concentrate - 5%, hydrogenated coal tar or acetylene sludge - more than 20%) are mixed and briquetted in a roll press (20–70 MPa). Raw briquettes were dried at 250°C, and then subjected to semi-coking for one hour in retorts of the Humboldt company, where sand with a temperature of 700–800°C was used as a heat carrier. Briquettes were used to smelt iron in low-pit furnaces with a capacity of up to 120 tons each day.

The production of ore and coal briquettes became widespread in the GDR. At the Max Hütte enterprise, briquettes were produced from fines of poor iron ore (53% of the mass of the briquette, fraction 0–2 mm, total iron content 23–33%) with the addition of coke breeze (5% of the mass of briquette, 0–2 mm). Acetylene silt (42% of the mass of the briquette) was used as a binder, which also served as a fluxing additive. Briquettes were made with a roller press with a pressure of 50–70 MPa. Raw briquettes were carbonized by blowing off the exhaust gases from the blast furnace Cowper at a temperature of 120–150°C for 30–120 minutes. The compressive strength of briquettes increased from 6–9 MPa to 15–17 MPa after such processing.

The essence of the carbonization process as applied to briquetting is that calcium hydroxide (Ca (OH)$_2$), the basic mineral of slaked lime, loses crystalline moisture when heated to 530–580°C, as a result of which the binding properties are lost and the briquettes lose strength. Carbonization proceeds according to the scheme:

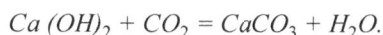

$$Ca\ (OH)_2 + CO_2 = CaCO_3 + H_2O.$$

leads to the formation of secondary limestone, which is characterized by a dissociation temperature in the range of 860–920°C, which determines the higher strength of the briquettes. Air carbonization is an extremely slow process and may require from a week to a month or more to achieve the required strengths of briquettes. To accelerate the carbonization process, the Weiss method is used, which consists in a two-stage treatment of briquettes with carbon dioxide under pressure. At the first stage, briquettes are treated with cold gas, and at the second stage, heated to a temperature of 90–100°.

Tests have shown that with an increase in the content of coke breeze, the strength of briquettes decreased, with an increase in the iron content in the ore and pressing pressure, the sludge consumption decreased. Achieving the values of compressive strength of 20 MPa required the grinding of ore to the class – 1 mm. With an ore fineness of 0–5 mm, the strength of briquettes averaged 17 MPa. When heated, the briquettes did not crack; till 400°C their strength increased, at a higher temperature - decreased. Full-scale testing conducted using such briquettes at two blast furnaces of the Max Hütte plant showed that the melting progress was improved both when replacing part of the traditional batch with briquettes and when melting 100% of the briquettes. The productivity of the furnaces increased by 20%, coke saving was 14.7%.

At another metallurgical plant of the GDR in West Kalb, built after the WW2, briquettes containing 75% of iron ore fines, 20% limestone and 5% of coke breeze were smelted in 10 low-shaft furnaces of 78 m³. For hardening briquettes, carbonization was used. The share of briquettes in the charge of such furnaces reached 30–40%. The productivity of the furnaces, at the same time, increased by 4.1%, the consumption of coke decreased by 11%. The furnace course was smooth, the pressure of the top gas increased without increasing the pressure of the blast, which indicated a high gas permeability of the briquette layer of the charge.

One of the most important features of this stage in the development of the technology of briquetting was the emergence and wide distribution of the technology of the so-called hot briquetting. This meant the molding of a softened, partially metallized mixture heated to high temperatures. The possibility of producing briquettes from metallized hot material was the result of the development and improvement of the design of roller presses. The interest in briquetting partially metallized material was due to hopes for increasing the efficiency of blast furnaces and steelmaking furnaces. The briquetted charge was distinguished by a higher iron content, uniformity of chemical and granulometric composition, resistance to secondary oxidation, etc.

One of the first successful projects of hot briquetting, which was important for the development of briquetting technology in general, was implemented in the UK in the mid-60s. In a blast furnace with a volume of 1798 m³, briquettes made of iron ore concentrate by hot briquetting on a semi-industrial plant were melted. The hot briquetting unit was located at the Steel of Wales Company (now Tata Steel) in Port Talbot [21–22]. It consisted of a rotary metallization furnace with a length of 25.6 meters and a roller press (Fig. 2.11).

The share of briquettes in the blast furnace charge was 25%. The total number of the melted briquettes is 12 thousand tons. In fact, it was one of the first experiments using blast furnace briquettes. The operation of a large blast furnace did not deteriorate with the introduction of briquettes into the charge. However, coke consumption in the experimental period (25% of the briquettes in the charge) increased and amounted to 567 kg per ton of pig iron versus 547 kg when operating without briquettes. Iron smelting also decreased from 2281 tons per day when operating without briquettes to 2234 tons with 25% of briquettes in the charge.

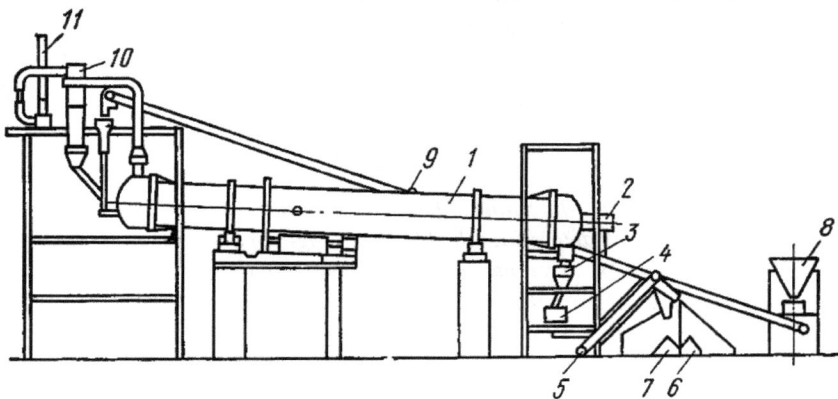

Figure 2.11. Hot briquetting plant in Port Talbot (UK, [21]). 1 - rotary kiln, 2 - burner, 3 - intermediate bunker, 4 - roll press, 5 - elevator, 6 - briquettes, 7 - screening of briquettes, 8 - iron ore bunker, 9 - secondary air supply, 10 - cyclone, 11 - exhaust pipe).

In the same years, the factory of hot briquetting was commissioned in the United States by United States Steele. Iron ore fines (particle size less than 0.635 mm) were heated in a rotary kiln, and then fed to a roller press using a screw feeder. After sieving on the vibrating conveyor, the briquettes were supplied by a bucket elevator to a rotating refrigerator.

Hot briquetting was also applied by another American company - Dofasco. But in this plant, a mixture of iron ore fines, mill scale and flue dust were heated to 1000°C in a fluidized bed reactor. The heated mixture entered the roll press of Komarek with a capacity of 10 tons per hour. One of the important results was that the temperature dependence of the wear of the roller press tires was revealed. It turned out that at temperatures of the briquetted mass below 535°C, sleeves worked without replacement for a long time.

In the 1960s, several more companies in the United States commissioned hot briquetting lines— Dominion Foundries and Steel, Gray Iron Foundry, and others [23–25].

No less important achievement of the metallurgy of the 60–70s is the development and implementation of another method of briquetting associated with heat treatment. When briquetting ore-coal mixtures, the coal component of the charge, heated to a plastic state, played the role of a binder. In modern literature, this method is also called hot briquetting, which often leads to confusion and mistakes. More precisely, the essence of such a method is defined by the term "thermal briquetting" proposed at the Moscow Mining Institute (MGI), in which in the 60s of the last century, the features of this method of briquetting and the properties of briquettes were fully studied [17, 26–27].

The basis of the method known from the practice of continuous coking is the formation of a solid from coal of plastic mass upon reaching a certain temperature level. For example, hard coal is plasticized at 350–460°C, peat - at 300–400°C.

In experiments on thermal briquetting at MGI, different iron ore concentrates were used with an iron content of 65–68% and mixed with peat or black coal (Fig. 2.12). The prepared mixtures, with or without the addition of fluxes (from 5 to 15% hydrated and quicklime), were heated in electric molds to 300–430°C. One of the main results of the research was that the quality of thermo-briquettes was mainly influenced by their fuel component. The chemical composition of iron ore concentrates practically did not affect the quality of thermal briquettes. The mechanical strength of such briquettes decreased with a decrease in the content of the fuel component in them. An increase in the pressing pressure from 20 to 120 MPa led to a noticeable increase in compressive strength and a less pronounced increase in abrasion strength.

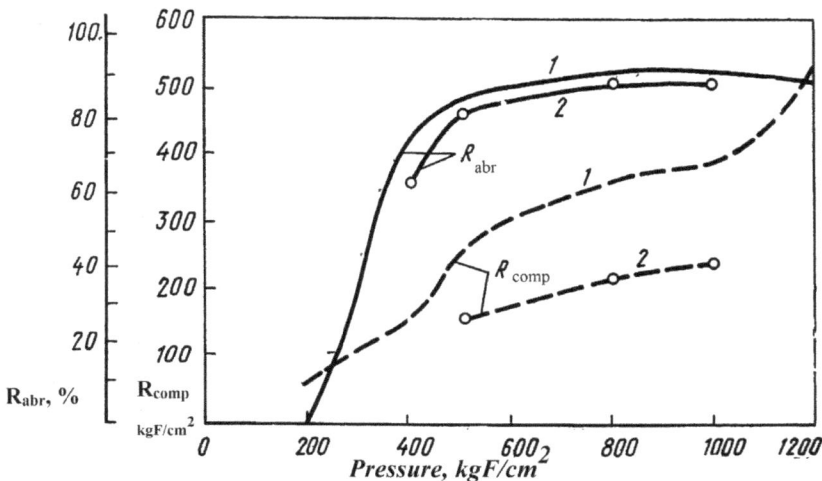

Figure 2.12. Influence of pressure on the mechanical strength of thermal briquettes (1 - briquettes with peat, 2 - briquettes with coal, R_{abr} - abrasion resistance, R_{comp} - compressive strength) [28].

The addition of fluxes to the charge contributed to the growth of the mechanical strength of the briquettes due to the manifestation of the binding properties of lime. At temperatures of thermal briquetting, the formation of calcium hydrosilicates takes place, as in the Schumacher process described above. The content of calcium oxide in thermal briquettes was 1–3%, with 10–15% in the initial charge.

It was also found that the porosity of thermal briquettes depends on the particle size distribution of the charge components and the applied briquetting pressure. Porosity reduced with decreasing particle size and with increasing pressure. The absolute values of the porosity of thermal briquettes ranged from 19 to 39%.

The behavior of thermal briquettes during heating was studied in a resistance furnace with a graphite heater in a nitrogen atmosphere. The reduction of briquettes of two different compositions was studied—75% of ore and 25% of peat and 70% of ore and 30% of peat. For both compositions, the degree of reduction on heating to 1500°C reached values of 94–98% within 4.5 minutes. The high thermal stability of thermo-briquettes was explained by the formation of a framework of metallized phases.

Significant interest shown in those days to thermal briquetting in the UK, Germany, Holland and Japan led to the emergence of a number of industrial plants in which the mixture heated to temperatures for plasticizing coal was briquetted with roller presses. Interest in such briquetting remains today.

Regarding the term "hot briquetting", it should be noted that this concept should be referred to the processes of production of so-called Hot Briquetted Iron (HBI), which is a product of direct reduction iron production processes, such as, developed in the same time period (in 1966) the famous Midrex process [29]. Direct reduction iron briquetting is the only passivation method that complies with the standards of the International Maritime Organization (IMO) governing the rules of maritime transport. The fact is that, due to its high porosity, spongy iron exhibits a pyrophoric property, to combat which, its pressing in HBI is applied. The most popular in the world for the production of HBI received the Köppern roll press [30]. Sponge iron after reduction in the Midrex reactor is fed at a temperature of about 700°C to pressing with a specific force of up to 180 kN/cm. The continuous strip of briquettes coming out of the roller press is further divided into separate briquettes.

The historical picture of briquetting technology in 60–70 years would be incomplete without mentioning another important milestone—the beginning of the use of briquetting for smelting ferroalloys. An interesting experience was gained in the ferroalloy industry, showing the promise of using briquettes in the charge of ore-smelting furnaces. The first industrial experience of using briquettes based on manganese ore concentrates in the charge of the ore-smelting furnace was obtained in 1961, when 160 tons of roll briquettes made of manganese concentrate (30.9% Mn; 4% Fe; 0577% P; 26, 95% SiO_2; P/Mn = 0.0018; 0.33% S; 1.9% CaO; 4.05% BaO) were used in the charge of a 2500-kVA furnace of the Zestafoni Ferroalloy Plant (RFP) [31]. The results of the heats showed that this type of charge component is quite effective. In 1966, pilot-industrial studies on the smelting of ferrosilicon manganese in industrial ore-smelting furnaces were carried out at the same plant using ore briquettes of manganese concentrate from the Chiatura deposit [32]. For industrial experiments, briquettes were obtained on a roller press with a capacity of 5 tons per hour from ore with a particle size of 5–0 mm on a binder of sulfite-alcohol bards (SAB) with a density of 1.2 g/cm^3. With an ore moisture content of 4% and an addition of 8% binder, the mixture was stirred for 15 minutes. And in this case heat treatment was required. The fact that the binding properties of SAB are caused by polymerization processes occurring in it, led to the formation of long chains of molecules in the body of the briquette. The polymerization reaction in the presence of manganese is more realized at temperatures of 160–180°C. For iron ore briquettes, the polymerization temperature reaches 200°C. The achievement of the required strength values of briquettes required their drying at temperatures of 50–300°C.

On the charge with ore briquettes, silicomanganese was smelted in a three-phase open ferroalloy furnace with a capacity of 16.5 MVA. The furnace worked normally and stably, the gas permeability of the charge was good, the flame was distributed evenly throughout the furnace. After 112 hours of experimental melting, it was concluded that sufficiently strong briquettes can be obtained from manganese ore of this size suitable for use in the charge of ferroalloy furnaces. When working on ore briquettes, the furnace productivity increases, the power consumption is reduced, the reducing agent consumption is also lowered. In 1970, at the Dnepropetrovsk Metallurgical Institute, parameters of the smelting of marketable silicomanganese on a briquetted charge of a mixture of manganese oxide concentrates I and II of the Nikopol grade at 1:1 and on sinter made from the same mixture were investigated [33]. For briquetting, the mixture of concentrates of a fraction of 10–0 mm was ground to a size of 3–0 mm. Briquettes were made on a semi-industrial roller press at a pressure of 500 kg/cm^2, the binder was a mixture of bitumen, fuel oil and SAB in an amount of 10% by weight of the charge. The mixture was mixed in tanks with steam heated (to activate the polymerization of the SAB). Briquettes of two compositions were prepared and melted: with an excess of the reducing agent (coal) in the amount of 50%, introduced to create a skeleton in the briquette and increase its strength (mixture of concentrates—54.5%, river sand—9.1%, coal—27.3%, a mixture of bitumen and mazut—3.6%, SAB—5.5%), with a stoichiometric amount of reducing agent necessary to reduce silicon and manganese (a mixture of concentrates—60.6%, river sand—10.1%, coal—20.2%, a mixture of bitumen and fuel oil—3.6%, SAB—5.5%). Laboratory tests showed that the physical properties of raw briquettes were superior to those for fired briquettes, as they refused to fire the briquettes.

The values of mechanical strength and heat resistance were higher for raw briquettes. Commodity silicomanganese was decided to be smelted on raw briquettes. Semi-industrial smelting on briquettes of the above composition and on sinter was carried out in a three-phase open ore smelting furnace with a capacity of 1.2 MVA at a voltage of 81.60–85.89 V and a current of 6000–6500 A with a continuous process with a closed top chamber. Briquettes with an excess of reducing agent had a high electrical conductivity and when penetrating they gave a significant increase in the current load, which led to the rise of the electrodes and the opening of the top. To eliminate this phenomenon, 10% of manganese concentrate was added to the briquette sample. In this way, 14 tons of briquettes with an excess of reducing agent, 1.2 tons of concentrate and 200 kg of dolomite were melted. The smelting of silicomanganese on briquettes containing a stoichiometric amount of reducing agent, occurred without any complications. The load was steady; In addition to briquettes, nothing was additionally introduced into the furnace. Comparative data of heats with different charges (briquettes, sinter) showed that smelting of silico-manganese on briquettes should be preferred to smelting on sinter.

The results of the pilot heats formed the basis for the construction and commissioning in 1976 of a briquette factory at the site of the Zestafoni Ferroalloy Plant. Manganese ore briquettes, in ways with the addition of gas cleaning dust and without it, were produced by roller presses Sahut Conreur (France). However, this first industrial experience was unsuccessful. Due to the increased wear of the bandages of roller presses, their high cost and the their independent production not being possible, this briquette factory was closed.

Briquetted charge was used also for smelting ferrosilicon. At the Zaporizhia plant of ferroalloys, briquettes were made from sand, coking coal and iron ore concentrate. The mixture was heated and stirred in a steam mixer and then pressed with a roller press. Raw briquettes were burned at a temperature of 600–800°C for 10–13 hours to achieve a reduction degree of 70–80%. The smelting of ferrosilicon on the briquetted charge was carried out in a 3.5 MVA furnace. Briquettes were also used for smelting carbon black ferrochrome, ferrosilicochromium and ferromanganese. When working on briquettes, the furnace productivity increased, the power and reducing agent consumptions decreased [33]. The smelting of ferroalloys from briquetted charge became widespread in the United States, France, Japan and Germany [34].

Summarizing the contribution of the 60–70s to the development of briquetting technology, it can be said that heat treatment of the charge or raw briquettes (in one form or another) remained in most of the implemented projects of briquette factories as a mandatory component of briquetting. The use of roller briquettes in blast furnace production did not lead to an increase in the scale of the industrial use of this technology in the iron and steel industry. The main reason for the low competitiveness of briquetting in those years was the low productivity of roller presses compared to the performance of sintering tape (1500–10000 tons per day) and firing machines (2500–9000 tons per day). In the ferroalloy industry, where the requirements for productivity and for the strength of briquettes are much lower than in the blast furnace, growth in briquette production was observed. As a result, by the mid-70s, the production of briquettes was already about 2% of the total volume of agglomeration in ferrous metallurgy [35].

2.4 The 80s–the End of the 20th Century

In the last two decades of the last century, one of the main reasons for developing briquetting technology, was still the difficult environmental condition in the regions where metallurgical plants are located, primarily due to harmful emissions from sinter plants (more than 50% of all harmful emissions). In the structure of emissions of pollutants of the integrated metallurgical enterprise for all the most significant components of emissions (dust, carbon monoxide, sulfur dioxide, nitrogen oxides), sintering is dominated by 31.1% dust, 77.8% CO, 61% SO_2, 26% NOx [36].

In addition, it turned out that not all types of fine anthropogenic materials of ferrous metallurgy can be considered as a full-fledged charge material for sinter production. First, this applies to blast furnace and converter sludge. The presence of zinc and lead in such materials prevents their use in sinter production. The use of dusts and sludge from electric furnaces for the production of sinter also involves serious difficulties due to fluctuations in the chemical composition of such materials, high dispersion, low iron content, the presence of non-ferrous metals, etc. The "capricious" charge for sintering is also the flue dust due to its poor crumpling ability due to the presence of carbon. Much of the flue dust is carried away with sinter gases and repasses to the sludge. Another problem in the sinter production was also the preparation of small fractions of the so-called return (sinter fines) to the sinter production.

A similar situation turned out to be relevant for another technology of agglomeration—the production of pellets. This was most acute in Sweden, where by the end of the last century blast furnaces worked exclusively on pellets. The last sinter plant in Sweden ceased to exist in 1995 [37].

The limited or inability to use the above materials for sinter production, the accumulation of reserves of iron-rich pellet fines together with the increased cost of waste disposal (up to US$ 10–15 per ton of waste) in the early 90s justified efforts to find effective methods of briquetting such materials. Briquetting was destined to become a recycling technology—the "assistant" of sintering and pellet production. The search led to the emergence in the arsenal of briquetting a new method of forming briquettes—vibropressing.

Thus, in Germany, during 1982, in the framework of a project for the effective recycling of dust and sludge with a high zinc content, known as the "Oxy-Cup", a study was launched aimed at choosing the most efficient briquetting technology. The result of the research, which continued in 1995 in Aachen, was the choice in favor of the technology of vibropressing for the production of carbon-containing self-reducing briquettes (C-Brick, [29]). The possibility of using vibrations for agglomeration was well known by then. First, as early as 1902, vibration was used in the United States in the production of concrete, and in 1927 in France a patent was issued for a method of vibrating concrete compaction. In the same year, a method for producing agglomerated masses using vibration was patented in the USA (Fig. 2.13, [38]).

In 1988, the first vibropress briquetting factory with a capacity of 110 thousand tons of briquettes for ferrochrome smelting was commissioned at Vargön Alloys in Sweden [39]. Since 1991, this

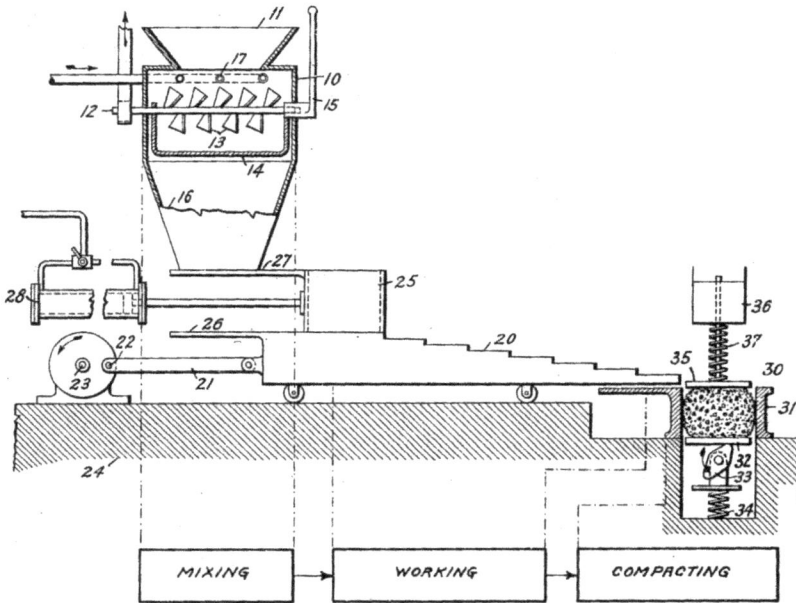

Figure 2.13. Illustration for the US patent No. 1647075 on the method of production of agglomerated mass using vibration [38].

method of briquetting has been used in France by Eurometa SA for the agglomeration of ferroalloys fines [40]. Briquettes from ferroalloys fines were used in the foundry business.

According to this method, briquettes were produced at low pressures (0.02–0.1 MPa) and simultaneous exposure of the compression mixture to vibration (frequency 30–70 Hz, amplitude 0.2–0.6 mm). For molding, multi-die molds with sizes from $20 \times 20 \times 20$ mm to $500 \times 1500 \times 1500$ mm were used. The manufacturing cycle lasts no more than 30 seconds. Cement was used as a binder.

The components of the briquette were dust and sludge from the blast-furnace production, converter and electric-smelting production, oily mill scale, magnetic slag fractions after desulfurization and other iron-containing materials. Carbon-containing components—coal, coke breeze, anthracite, petroleum coke. Briquettes together with large-sized (up to 1 m in diameter) metallurgical slag scrap were then melted in a special cupola operating on an oxygen-rich blast. The share of briquettes in the cupola charge was up to 70%.

In 1999, a pilot cupola with a capacity of 210 thousand tons per year was built in Duisburg (Germany) at a loaded charge consisting of 75% of briquettes and 25% of buckets skulls. The satisfactory metallurgical properties of the briquettes produced by this new method contributed significantly to the commercial success of the Oxy-Cup process, which today is rightly regarded as providing a solution to the problem of recycling zinc-containing dusts and sludge. In the period from 2005 to 2011, 6 Oxy-Cup process furnaces with briquette factories were built in the world (three each in Japan and in China). However, the high cost of the main equipment and engineering of Oxy-Cup and the insufficiently high performance of the vibropress (up to 20 tons per hour) restrain its further wide distribution.

The choice of vibropressing as a method of briquetting pellet fines also came from specialists of Merox Company (Sweden, [41]). One of the first industrial briquette lines of vibropressing was put into operation at the SSAB plant in Oxelösund (Sweden) in 1995 [42]. Briquettes 60×60 mm in size, hexagonal in cross section are used as a component of the blast furnace charge (60–100 kg of briquettes per ton of iron). The composition of the briquette is 25% top dust, 50% mixture of scrap metal, desulfurization scrap, converter sludge, pellets fines and 5–8% aspiration dust. Binder -

Portland cement (10–12%), moisture content of briquettes 5–8%. The particle size of the briquetted mixture is 0–5 mm. The limitation for the proportion of briquettes in the blast furnace blast furnaces of SSAB is the zinc content in the sludge. Studies conducted at the Technical University of Lulea showed, briquettes from pellets fines had abnormal swelling (volume increase by 20%), similar to swelling of the pellets themselves, when reduced in atmosphere of the CO and H_2 mixture [43–45].

The vibropressing briquetting technology debuted in this way turned out to be commercially successful, but still could not be considered 100% cold agglomeration technology, since the briquette's curing required their thermal treatment at temperatures of 80–120°C. An important result of the above-described vibropressing experience was also the restored "reputation" of cement as a binder.

But perhaps the most important event of the last two decades of the last century in briquetting was the entry into the market of companies specialized in providing briquetting services for ferrous metallurgy materials on a so-called "toll" basis, when briquettes are made by the manufacturer at their own expense from the raw materials provided by the metallurgical enterprise.

The pioneer of such a business in briquetting was the company NRS (National Recovery System, USA), which in the early 70s began to search for a solution to the problems of utilization of oxide materials of ferrous metallurgy by briquetting. Several compositions of briquetted fluxing additives and different compositions of briquettes from materials with high iron content were worked out. These briquettes began to be widely used in the US recycling practice in the 90s of the last century. Since the end of the 80s, the NRS has been looking for efficient and inexpensive binders for briquetting in blast furnaces. The combined binder of molasses and non-gypsum cement, for which the patent was obtained, was used in the production of 15000 tons of experimental blast-furnace briquettes for smelting on blast furnace No. 8 by US Steel at Gary Works [46]. To achieve the required hot strength in blast furnace smelting, from 14 to 19.5% (mass) of such a binder was added, as indicated in the patent itself. For the composition consisting of mill scale (41%), steelmaking slag (19%) and flue dust (40%), 11.5% of cement and 8% molasses were required, which considerably diluted the mixture. In addition, the briquetted charge had to be dried to a moisture content of 3%, since it is impossible to form a wet mass using a roll press. Such low moisture content turned into a lack of moisture for cement hydration. Cold strength of briquettes, measured according to ISO 3271–1985 E tumble-testing was not less than 70% (the proportion of pieces of briquettes with a size of more than 6.3 mm after the 200 turn of the drum at a speed of 25 revolutions per minute). The share of briquettes in the blast furnace charge, which did not violate its operation, reached 10%. This level was enough for recycling anthropogenic materials. The success of the pilot campaign allowed the NRS to build four briquetting factories in the USA and one in the UK at the Port Talbot plant.

In early 1993, US Steel began working at the Edgar Thomson plant, the Bethlehem Steel company in Pennsylvania, and the Inland Steel Company. The fourth factory in the United States began operations in 1996 at the National Steel Corporation's plant. In 1997, the factory started operating in the UK.

There was no sintering plant at Edgar Thomson's US Steel plant, and all the generated iron containing wastes were buried or stored in dumps. The above-mentioned increase in the cost of waste disposal has provided an economic opportunity for their on-site processing by roller briquetting. A briquetting factory was designed to produce 200 thousand tons of briquettes per year. Productivity of the pressing equipment is 40 tons of briquettes per hour. A patented composition of molasses and cement was used as a binder. And it was with this mixture that the first problem arose for blast furnace specialists. Starting from about the sixth month of the operation, the smells emanating from the blast furnace gas cleaning system became a serious problem. The source of the smell was identified. It turned out to be a protein that got into the system with molasses. It was decided to switch to a purely inorganic binder, as a result of which odors ceased. By 1998, up to 10% of briquettes were used in the blast furnace charge. For the converter briquettes, a combined binder from a mixture of lime and molasses was also used, but no problems with odor were noted.

The briquetting factory at the Inland Steel plant had to solve the problem of recycling 100% of all iron-containing anthropogenic materials—current and accumulated earlier. The need to eliminate dumps at the site of the enterprise and the rising cost of disposal outside of its borders stimulated the search for a suitable recycling method, which resulted in a choice in favor of roller briquetting. The factory capacity was 250,000 tons of briquettes. Briquettes were used in the charge of blast furnaces and oxygen converters. When using blast furnace briquettes at this enterprise there was also a problem with odor. In addition, there were elevated levels of ammonia in the wastewater. As in the case with Edgar Thomson's US Steel plant it was also decided to abandon the combined binder.

Another briquetting factory built by NRS was located at the site of the National Steel plant. The sinter plant was closed shortly before that and all the anthropogenic materials formed during the operation of the three blast furnaces and two converters were accumulated or buried. The roller briquetting factory was designed for production of 300 thousand tons of briquettes per year, of which 80% of briquettes were intended for blast furnace charge. Briquettes were made from blast furnace and converter sludge, top dust, mill scale, coke breeze and other materials. The share of briquettes in the blast furnace charge reached 6.5%.

Roller briquetting was also used at the NRS briquette factory built in Port Talbot (UK). The rising costs for the disposal of waste was again the main cause for their processing. Unlike the aforementioned briquetting factories, zinc-containing converter sludge was briquetted here. Other anthropogenic materials were added to the sintering mixture. Another feature of briquettes was the high content of coal (up to 30% by weight). The capacity of the factory is 180 thousand tons per year of briquettes per year.

In 1993, the NRS also built a roller-press briquetting factory at the Lake Erie Works plant in Canada [47].

In the summer of the same year, the Bethlehem Steel plant was built and put into operation a briquetting line based on Stiff Vacuum Extrusion (SVE), a technology widely used in the United States for the production of bricks. The factory was intended for the agglomeration of converter dust that has been stored in the company's dumps for many years, and on reaching a critical level of its accumulation, the owners of the enterprise were forced to recycle this material. Briquettes were used as a component of the blast furnace charge. The capacity of the line was 100 thousand tons of briquettes per year [48]. The performance of the extruder supplied by J.C. Steele and Sons reached 20 tons per hour. The main advantage of extrusion, which led to the choice of this technology for briquetting, was the very low cost of basic equipment. Cement was used as a binder, which added US$ 8 per ton to the cost of briquette. The cost of manufacturing tons of briquette ranged from US$ 11 to 14. In total, the line produced about 70 thousand tons of briquettes and was dismantled in 1996 after the liquidation of Bethlehem Steel. Disputes about the expediency of liquidation of the company have not yet subsided, and the blast furnace, where the extrusion line was located, is a gloomy monument to the once successful enterprise. The equipment of this line still functions as part of a Texas Instruments bauxites briquetting plant [49].

Due to the efforts of NRS, the scale of briquetting in the world has increased significantly. In essence, the company created a market for briquetting services, to which other service companies soon entered. Another important achievement of this legendary company was that the briquetting was finally... cold. Preparation of briquettes for metallurgical processing did not require any heat treatment, if one does not take into account the drying of the charge for a roller press.

It was also very important that the NRS Company become the "godfather" of a fundamentally new technology of briquetting—stiff vacuum extrusion. But, as often happens, this industrial extrusion briquetting debut went unnoticed and unappreciated by the metallurgical community, and the NRS Company, despite the undoubted success of the project at Bethlehem Steel, no longer used stiff extrusion in its projects.

However, in 1996, one of the largest briquetting factories for the production of extrusion briquettes made from aspiration dusts of ferronickel production and lateritic nickel ore fines with an annual production of 700 thousand tons was put into operation at a ferronickel plant in Colombia.

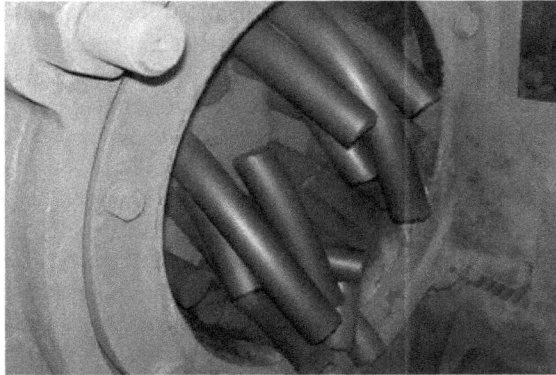

Figure 2.14. Extrusion briquette manufacturing in Colombia.

(Fig. 2.14). Initially this enterprise, which has been operating since 1992, used pelletizing and roll-press briquetting to agglomerate the fines of lateritic nickel ore and gas-cleaning dust of electric furnaces. In 1996, after carrying out a series of comparative tests of various methods of agglomeration and processing of these materials (pelletizing, agglomeration, plasma furnaces and SVE), the final choice was made in favor of the SVE technology, which made it possible to obtain an agglomerated product of the required size and shape with cold and hot strength sufficient preserving the integrity of briquettes when they are processed in rotary kilns and when melted in electric furnaces. This technology allowed to achieve high performance with a minimum cost of the resulting product. The company installed three Steele-90 extruders to produce 700 thousand tons of briquettes per year. A characteristic feature of the briquetting of the lateritic nickel ore using the SVE technology is the possibility of obtaining strong briquettes without the use of a binder. The main equipment of extrusion lines has been operating for 20 years without replacement. Nickel production in ferronickel in Cerro Matoso is 50 thousand tons per year, of which 10 thousand tons are brought in by briquettes. Nickel content in commercial products is 35%.

Important results that contributed to the development of briquetting in the ferroalloy industry were obtained in the process of developing and commercializing technologies for producing dust briquettes and smelting manganese ferroalloys from them at the Zestafoni Ferroalloy Plant [50]. The need to dispose of dust and sludge from furnace gas cleaning was due to the large volume of its formation (50 thousand tons annually). Taking into account the metallurgical evaluation of waste, their dispersed state and physicochemical properties, it was proposed to make briquettes from the mix of the dry dust, earlier from dehydrated sludge and third-grade manganese concentrate. It has been established that to obtain mechanically strong briquettes (8–12 MPa), the optimum briquetting parameters are: moisture content of the charge 4–6%, binder content (SAB) 6–8%, amount of the fine component (dust, sludge) 30% and minimum compacting pressure of 19.6 MPa. The comparative kinetics of the recovery of sludge briquettes and manganese sinter was investigated. It was found that briquettes gave the greatest degrees of reduction at different temperatures. The use of manganese and silicon from waste was increased due to the presence of bound carbon in them and the low intensity of silicate formation, which is inevitable when using sinter. The results of high-carbon ferromanganese heats showed that the capacity of the furnace in the case of using briquettes increased from 73.33 to 75.67 tons/day, and the specific consumption of electricity decreased by 90 kWh/ton. Coke consumption decreased by 34 kg/ton.

Thus, by the end of the last century, the main competing industrial technologies of briquetting were defined: roller-press briquetting, vibropressing and stiff extrusion. The market of manufacturers of equipment and services of briquetting according to the "toll" scheme was formed. Briquetting has become firmly established in all branches of the steel industry. It is also important that, by this time, the success of technological developments was largely ensured by specialized scientific institutions (universities and laboratories). The role and importance of the Institute for Briquetting and Agglomeration (IBA) has increased, which contributed to bringing together the efforts of scientists, metallurgists, equipment manufacturers and the growing popularity of briquetting. In 1999, the 26th IBA Conference took place, which was opened by Wolfgang Pietsch, author of the encyclopedic monographs on agglomeration in industry published in the 1990s [51].

2.5 21st Century

These trends were also manifested at the beginning of this century.

NRS continued to dominate the briquetting market. In 2004, another factory was built in the United States at the site of AK Steel, and since December 29, 2005, the NRS became a structural unit of Harsco Metals [47]. By 2011, Harsco Metals already owned a dozen briquetting factories around the world. Without exception, these briquetting factories were used for the production of briquettes roller pressing. Seven briquetting factories, of which five with the capacity from 150 to 250 thousand tons of briquettes a year, were working in North America at the sites of the company's customers (US Steel Edgar Thomson Works, US Steel Lake Erie Works, ArcelorMittal - IH East, US Steel Great Lakes, AK Steel - Ashland, US Steel Mon Valley) and two more with a capacity of 100 thousand tons of briquettes per year were located at the sites of Harsco Minerals. Several briquetting factories operated in Europe (two in the UK - at Port Talbot and Scunthorpe; three in France, one in Slovakia) and in Asia. Briquettes were used in the charge of blast furnaces and oxygen converters. Blast furnace briquettes were produced in five factories—US Steel Lake Erie Works, US Steel Great Lakes, AK Steel - Ashland, US Steel Mon Valley and ArcelorMittal - IH East. Briquette charge included mill scale, blast furnace and converter sludge, bag filter dust. ArcelorMittal - IH East used bauxite and flue dust briquettes. The share of briquettes in the blast furnace charge averaged from 3 to 6% (the maximum value was 12%). Fresh oily mill scale briquettes were added to the charge of converters, for example, at the Harsco Metals factory in Fort-sur-Mer (France). The share of briquettes in the charge was 1%. The briquetting factory in Ugine (France) produced 10 thousand tons of mill-scale and aspiration dust briquettes for steel-smelting Electric Arc Furnaces (EAF).

The work of Harsco Metals was promoted by the changed prices for charge materials. When comparing the prices of coke and iron ore fines in the late 80s and by 2011, it was clear that briquette began to be a valuable component of the charge of metallurgical aggregates, and not just a means of recycling anthropogenic materials. In fact, the cost of coke breeze increased over this period from US$ 45–60 to 350–400 per ton (almost eight times), the cost of iron ore fines—from US$ 20–25 to 170–180 per ton (more than seven times), while the cost of waste disposal increased from US$ 10–15 to 30–35 per ton (twice). The total volume of briquettes produced by Harsco Metals exceeded 1.5 million tons per year in 2011.

In 2010, the company made an unsuccessful attempt to enter the market of briquetting services in Russia. It was planned to carry out a project for the production of almost 800 thousand tons of briquettes for blast furnaces from a mixture containing flue dust, aspiration dust, converter and blast furnace sludge and sinter fines. A mixture containing additives developed by Harsco Metals was recommended as a binder (the composition was not disclosed). The proportion of binder in the briquette mass exceeded 17%, which would lead to a substantial "dilution" of the blast furnace charge. This project was never implemented.

A similar picture began to take shape at other briquetting factories of the company, which led to the fact that Harsco Metals completely ceased production of blast furnace briquettes, and the production of briquettes for converters was preserved only in the factories in Fort-sur-Mer (France) and in both factories in Great Britain. The Harsco Metals briquetting factory in Kosice (Slovakia) changed owners in 2015 and is currently producing briquettes for converters that are manufactured by the Phoenix Services [52] service company, founded in 2003.

Thus, roller compaction was almost completely superseded from the blast furnace segment of the briquetting market. However, roller-press briquettes are still successfully used in steelmaking processes (in converters and in EAF).

Roller presses continued to be used in the ferroalloy industry. In particular, roller presses are used for agglomeration of metallized, and, therefore, abrasive fines of crushing of ferroalloys [53]. In recent years, the efforts of many manufacturers of ferroalloys have been directed towards the development of efficient technology for the briquetting of crushing ferrosilicon fines [54–55].

Nowadays, one more application of roller briquetting is the agglomeration of dispersed oxide and metallized materials of Direct Reduction Iron (DRI) production processes.

In 2002, POSCO launched a rolling briquette factory to produce 500 thousand tons of briquettes per year from coal fines as a replacement for expensive coking coal in the FINEX process [56].

In 2011, a roller briquette factory was built in Brazil as part of an industrial demonstration plant operating according to the Tecnored process [57]. Briquettes are used in the charge of low shaft furnaces of the process, as an alternative to cold bonded pellets. In 2015, the plant successfully reached a design capacity of 75 thousand tons of pig iron per year.

In 2016, Qatar Steel put into operation a roller-press factory equipped with a South Korean-made press (Jeil) to agglomerate metallized dusts and sludge formed during the direct production of iron in the Midrex reactor. By 2018, the factory produced 200 thousand tons of briquettes, which were successfully smelted in the company's electric furnaces.

Primetals completed a series of studies that confirmed the viability of using roller-press briquettes as components of the Midrex process charge and in 2015 announced plans to build a briquetting factory in Corpus Christi (Texas, USA) as part of a DRI reactor construction project (Midrex). The factory, commissioned in 2016, produces 160 thousand tons of briquettes from dust, sludge and pellet fines per year.

Emirates Steel has announced the construction of a mill for the production of briquettes for the agglomeration of metallized and oxide materials generated in the Midrex process. Briquetting services will be provided by Phoenix Services [52].

Another interesting application of roller-press briquetting, unfortunately not crowned with success, is associated with the PIZO (Pig Iron Zinc Oxide) process developed by Heritage Technology Group, in which briquettes from a mixture of EAF dust and carbon were melted in induction furnaces. A pilot plant was built at the site of the Nucor plant in Blytheville (USA) [58]. The briquettes are fed into the molten iron bath, in which the reduction of iron oxides by the carbon of the briquettes takes place. Volatile metal oxides (zinc, lead, cadmium, etc.), evaporate at the operating temperature of the furnace (1300–1500°C) and are collected by bag filters. The zinc-rich material obtained in this way can then be used as a raw material in zinc plants. The zinc content in this material reaches 67%. Commodity products of the PIZO process are, in addition, cast iron and slag. Since the construction of the pilot plant in 2006, neither the technology developer company nor the customer has announced a transition to the full-scale project stage.

In general, for well-known reasons, the use of briquettes made of iron oxides in the charge of induction furnaces is impractical. In the body of such a briquette, eddy currents will not be induced, leading to the heating of a traditional metal mixture of induction furnaces in an alternating electromagnetic field.

A method for efficiently melting such briquettes in induction furnaces was proposed in 2012 [59]. It was proposed to equip an induction furnace with a graphite electrode and a plasma burner. Melting of the electrically conductive mixture at the bottom of the crucible is carried out by arc energy, followed by switching on the inductor and adding briquettes in portions with the formation of a protective charge layer between the arc and the crucible wall with a decrease in discharge power as the briquette is metallized with hard carbon and hot slag is formed. The proportion of briquettes can reach 90% of the mass of the charge.

The melting in induction furnaces of metallic wastes from the foundry and engineering (grinding cuttings, iron and steel chips, etc.) was successfully implemented in Western Europe as part of the BRICETS project in 2004. For smelting iron and steel using such waste, an induction crucible furnace is recommended. Two types of lumpy charge materials were tested—briquettes and pellets. Balance smelting on briquettes made from metallic materials implemented at a number of Scandinavian enterprises showed their acceptable efficiency. At the same time, research in the framework of the BRICETS project [60], conducted in the UK with pellets, showed the futility of their use as an additive in an induction furnace charge. Currently, the results of the BRICETS project are implemented in commercialized projects in the SWEREA Group and SKF (Mekan and Gothenburg, Sweden). Reducing (or completely eliminating) the cost of disposal of waste (grinding sludge) contributes to the success of the process. The SKF plant in Mekan used cast iron briquettes with the addition of grinding sludge (5%) and aspiration dust (2%). The quality of cast iron does not lead to any complaints.

As for briquetting in iron metallurgy, until 2011 the lion's share of blast-furnace briquettes was produced by the method of vibropressing.

In Russia, the first vibropressing factory was commissioned at the enterprise JSC Tulachermet in 2003 [61].

A series of campaigns on the use of vibropress briquettes of different component composition on a cement binder in a 1000 m^3 blast furnace charge was carried out in 2003 at NLMK [62–64].

In 2004, Progres Ecotec (Czech Republic) began to produce vibropress blast furnace briquettes for ArcelorMittal.

In 2010, the Kosaya Gora Iron works put into operation a vibropress factory with an annual capacity of 120,000 tons for the production of briquettes from a mixture of concentrate, ore fines, flue dust and a binder - Portland cement (not less than 10% of the mass of the briquette) [65].

In March 2012, a vibropressing line for the production of briquettes for blast furnaces was operating at the enterprise of the SSAB Company in Finland (Raahe Works) [66].

The results of full-scale testing and industrial use of vibropress briquetting technology show the fundamental possibility of achieving the required level of metallurgical properties of agglomerated products. However, this method of briquetting creates some significant technological limitations, overcoming which is either difficult, or leads to a large increase in the cost of briquette. As of the beginning of 2016, briquettes in this way in industrial volumes are not produced in Russia.

The ferroalloy industry has also maintained an interest in vibropressing, aided by the success of the factories mentioned earlier, Vargön Alloys and Eurometa. Since 2001, Aaltvedt Betong AS's vibropress factory has been operating in Norway, producing briquettes for Eramet [67]. Manganese ore fines and gas cleaning dust are briquetted. Binder - Portland cement (10–12% of the mass of the briquette).

An example of the successful application of vibropress briquetting technology for the agglomeration of manganese-containing anthropogenic materials, including ferroalloy fines, is of the South African company Assmang at the plant in Cato Ridge [68]. Vibropress briquettes made of silicomanganese on a "toll" basis were also produced in the USA for the Felman Company [69].

In 2006, the technology of stiff vacuum extrusion briquetting again entered the "arena". In Brazil, a briquette factory for the production of extruded briquettes from one lateritic nickel ore fines and aspiration dusts of ferronickel production was commissioned—a complete analog of the factory previously built in Colombia with a capacity of 700 thousand tons of briquettes per year. The factory was built on the site of the company Vale. Thus, the total annual production of briquettes using stiff vacuum extrusion was 1.4 million tons, compared with the total production of briquettes at Harsco Metals factories.

Despite such obvious commercial success, until 2009, stiff extrusion technology remained terra unrecognized for the steel industry. The choice in its favor could be based more on risk than on the results of a systematic study of the metallurgical properties of extrusion products. This gap was eliminated as a result of research carried out by Bizhanov A.M. and others [49, 70–79]. In the period 2009–2013 stiff extrusion has been systematically studied and adapted to the conditions of most industrial processes for smelting iron, steel, ferroalloys and direct iron production. The choice

Table 2.1. Comparison of briquetting technology.

Process characteristics and properties of briquettes	Briquetting units and their characteristics		
	Vibropress	Roller-press	Stiff extrusion
Maximum performance	30 t/h	50 t/h	100 t/h
Service life (cost of parts to be replaced, US $/ton)	1 year (n/d)	1 year (1.5)	1.5 year (1.0)
Compaction pressure	0.02–0.10 MPa	40–150 MPa	3.5–4.5 MPa
Cement share in briquette, %	8–10	15–16	4–6
Heat treatment of raw briquettes	80°C (10–12 hours)	Drying of charge	Not required
Returns	absent	30% of production	absent
Briquettes shape	Prism, cylinder	pillow	Any shape
Briquette size, mm	80 × 80	30 × 40 × 50	5–50
Charge moisture content, %	less than 5%	less than 10%	12–18%
The ability to store raw briquettes in a pile	absent	possible	possible
Utilities: Electricity Natural gas Heat Compressed air	42.6 kWh/t 47 m³/t 0.3 GCal/t 90 m³/t	45.0 kWh/t 0 0 0	33 kWh/t 0 0 0 0

in favor of this technology was made cognizant and scientifically grounded. For the designation of extrusion products, the term BREX (extrusion briquette) was proposed, for which a trademark registration certificate was issued [80].

Some parameters of roller and vibropress briquetting with briquetting using a stiff extrusion method can be compared (Table 2.1).

The high performance of stiff extruders, the possibility of agglomeration without pre-drying the charge, and the lower consumption of binder caused a growing interest in extrusion briquetting technology.

In April 2011, an industrial briquetting line was commissioned at the Suraj PL plant in Rourkela (West Bengal, India), producing brex for their use in a blast furnace. The management of the company decided to implement a project for the production of brex from a mixture of converter sludge, blast furnace dust, Portland cement and bentonite. The main objective of the project was to increase the economic efficiency of a small blast furnace by replacing the purchased ore with cheap brex.

In the period from 2011 to 2018, five more briquette factories were commissioned, producing brex. Among them two factories for the production of blast furnaces, two for the smelting of silico-manganese and ferromanganese and one factory for the production of brex for the smelting of ferrochrome. A distinctive feature of the projects carried out in the interests of ferroalloy companies was their pronounced environmental focus. Extrusion agglomeration made it possible to effectively utilize gas cleaning systems dust accumulated over decades, significantly improving the ecological situation in the area of ferroalloy production.

The world's largest factory for the production of blast furnace brex was built in 2018, Lipetsk (Russia) at the site of NLMK. Its capacity is 700 thousand tons of brex. With its commissioning, the total annual production of brex will exceed 3 million tons.

The first experience of the operation of SVE lines in comparison with the experience of roller-press and vibropress briquetting factories made it possible to designate some specialization of the known methods of briquetting. Namely, for the briquetting of dry materials with organic binders, when the achievement of high hot strength of briquettes is not required, roller-press briquetting allows to obtain mechanically strong briquettes due to the application of high pressure (up to 150 MPa). Vibropressing allows agglomeration of a mixture containing a substantial part of large fractions (up to 10 mm and more), but requires special measures to preserve the strength of raw briquettes

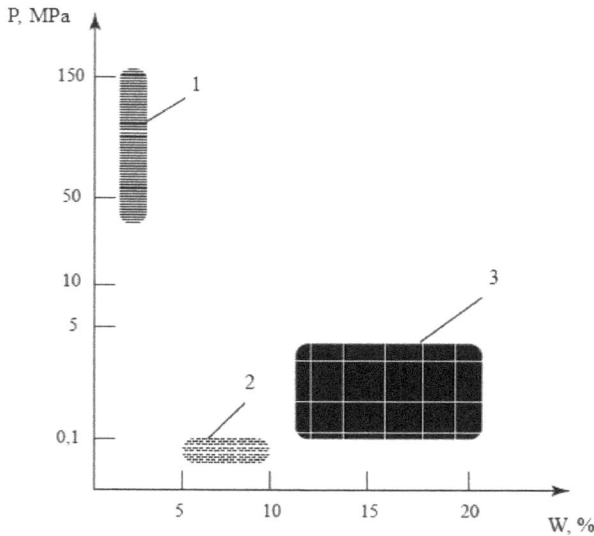

Figure 2.15. Comparison of technologies of briquetting according to the parameters of briquetting (applied pressure and moisture content of the charge). 1 - roller pressing, 2 - vibropressing, 3 - stiff vacuum extrusion.

(moving on pallets, heat and moisture treatment) and does not allow moisture of the briquetted mixture to exceed 12–15%. Stiff extrusion is irreplaceable when agglomerating thin materials and is able to agglomerate mixtures with a moisture content of up to 20% at compacting pressures an order of magnitude lower than in a roller press (3.5 MPa). In particular, this technology allowed the agglomeration of EAF dust and gas cleaning dust of ferroalloy production, which practically could not be briquetted with either roller presses or vibropresses. The possibility of agglomeration of wet materials allows either to completely abandon the drying of raw materials, or to significantly reduce the cost of such drying. A comparison of these technologies in parameters such as the applied pressure and moisture contents of the agglomerated mixture is given in Fig. 2.15.

The success of briquetting as an industrial technology was largely promoted by the results of scientific research aimed at improving and extending the service life of equipment, as well as studying the metallurgical properties of briquettes and the search for new methods of sintering. Specialized studies of the characteristics of briquetting and the behavior of different briquettes in metallurgical aggregates were carried out at the University of Lulea (Sweden), at the Moscow Institute of Steel and Alloys (MISiS), in scientific laboratories of the largest metallurgical companies. Research by Swedish scientists substantiated the expediency of using vibropressing to produce blast briquettes with high metallurgical properties. The results of research carried out at MISiS made it possible to determine SVE as the best available technology in the production of agglomerated metallurgical raw materials and as the only briquetting technology alternative to sinter production.

Another major trend in the modern stage of development of the technology of briquetting is the emergence of a global market for product briquettes. Mining enterprises began to receive significant commercial benefits from the sale of briquettes, including for export.

It should also be mentioned that the entry into the market of briquetting chemical companies offering polymer binders have been successfully tested for mechanical and hot strength. BASF announced the commercialization of several of its products and supplies binders for a number of briquette factories [81]. A similar product is also being offered by Legacy Hill [82]. Despite the very high cost of such binder materials necessary to achieve the required strength, the proportion of such a binder is so small (0.2–0.3% by weight of the briquette) that its use can be economically justified. The importance of increasing the content in the briquettes of the main element (iron, manganese, chromium, etc.) due to a decrease in the proportion of binder should be taken into account. In particular, it is very important for the market of commodity manganese briquettes.

References

[1] List of Patents. Inventions and Designs, issued at the United States from 1790 to 1847. Washington, 1847. Printed by J&G.S. Gideon.

[2] Watt, Kathleen Ann Nineteenth Century Brickmaking Innovations in Britain: Building and Technological Change. PhD Thesis, University of York, 1990.

[3] Electronic resource. http://www.binfieldheath.org.uk/the-village/binfield-heath/history/brickmaking.

[4] Man and Izv, G. 1841. 5: 180–184 (In Russian).

[5] Berthelot, Ch. 1938. Epuration sechage, agglometionet broyage du charbon, Paris: chez Dunod. 22ᵉ partie: Appareillage modern pour la fabrication des boukets et briquettes: Les Ayelliers d'agglometaion des mines de Mariemont-Bascoup (Belgique), p. 321.

[6] Hatheway, A.W. 2012. Remediation of Former Manufactured Gas Plants and other Coal-Tar Sites. Taylor & Francis Group, CRC Press.

[7] Electronic resource. http://www.briquettepress.com/News/History-Of-Coal-Ball-Briquetting.html.

[8] Wilcox, J.M. 1884. Machinery for compressing and molding pasty substances, No. 309,117. Patented Dec. 9, 1884.

[9] Eriksson, S. and Prior, M. 1990. The briquetting of agricultural wastes for fuel, FAO Environment and Energy Paper 11, FAO of the UN, Rome.

[10] Edgar Rouse Sutcliffe. 1925. US Patent No. 1,531,631. Press for the Production of Bricks, Briquettes, Blocks and the like. Patented Mar. 31, 1925.

[11] Electronic resource. https://www.911metallurgist.com/magnetite-concentration/.

[12] Yuzbashev, L. 1901. A method of obtaining various ores and fossil combustible artificial lump ore and artificial lump fuel from fines // Gornyi zhurnal 2(6): 257–288 (In Russian).

[13] Lotosh, V.E. and Okunev, A.I. 1980. Unburnt agglomeration of ores and concentrates. Section SRNTI: Production of ferrous metals and alloys. Moscow. Science, 215 p. (In Russian).

[14] Astakhov, A.G., Machkovsky, A.I., Nikitin, A.I. and others. 1964. Handbook of Sinter. Kiev: Technique. 448 p. (In Russian).

[15] Ravich, B.M. 1975. Briquetting in Ferrous and Non-Ferrous Metallurgy, Metallurgy, Moscow 232 (in Russian).

[16] Averkiev, N.D. and Udovenko, N.V. 1932. Briquetting of top dust and dust ores by alkali silicates. ONTI.

[17] Vegman, E.F. 1981. Concise Handbook for Blast-Furnace Operators. Metallurgiya, Moscow (in Russian).

[18] Godinsky, N.A., Kushnarev, N.N., Yakhshuk, D.S. and others. 2003. Experience of using iron-containing briquettes in electric steel-making production // Metallurg 1: 43–45 (In Russian).

[19] Tupkary, R.H. and Tupkary, V.R. 2013. An Introduction to Modern Iron Making, Khanna Publishers, Delhi.

[20] Electronic resource. http://agglomeration.org.

[21] Sharp, K.C. and Davies, O.G. 1966. Journal of Metals 1: 79–86.

[22] Metal Bulletin, 1965, #5014, pp. 171–173.

[23] Skilling Mining Review, 1968, 57(24): 16.

[24] Eng. and Mining Journal, 1972, 173(8): 78–81.

[25] Harris, Z.S. 1972. Scrap Age 29(9): 131–154.

[26] Ravych and Yarkho, N.A. 1964. Steel 2: 118–119 (In Russian).

[27] Yarkho, N.A. and Ravich, B.M. 1962. Bulletin "TSIINCHM" 7(435): 41 (In Russian).

[28] Butt, Yu.M. and Krzheminsky, S.A. 1953. Sat. Proceedings ROSNIIMS, number 2, Moscow p. 75 (I Russian).

[29] Yaroshenko, Yu.G., Gordon, Y.M. and Khodorovskaya, I.Yu. 2012. Edited by Yu.G. Yaroshenko. Ekaterinburg: LLC "UIPTS". 670 p. (In Russian).

[30] Electronic resource. http://www.koeppern-international.com.

[31] Khazanova, T.P. 1961. Production of manganese alloys from poor oxide and carbonate ores / Khazanov, T.P., Shearer, G.B. and Lyakishev, N.P. // Development of the USSR Ferroalloy Industry. - Kiev. P. 122 (In Russian).

[32] Khvichia, A.P. 1970. Smelting of silico-manganese from ore briquettes in a furnace with a capacity of 16.5 MVA / Khvichiya, A.P. and Mazmishvili, S.M. // Steel. 2: 138 (In Russian).

[33] Sukhorukov, A.I., Sosedko, P.M. and Khitrik, S.I. 1970. // Steel. 2: 135 (In Russian).

[34] Kozhevnikov, I.Yu. and Ravich, B.M. 1991. Agglomeration and the foundations of metallurgy. - M.: Metallurgy, 296 p. (In Russian).

[35] Ravich, B.M. 1982. The Briquetting of Ores, Nedra, Moscow (in Russian).

[36] Frolov, Yu.A. and Sintering. M. 2016. Metallurgizdat, 672 p. (in Russian).

[37] Electronic resource. https://www.jernkontoret.se/en/the-steel-industry/production-utilisation-recycling/raw-materials/.

[38] Bowen, R. 1924. Process of fabricating agglomerated mass. US Patent No. 1,647,076. Filed Jan. 5, 1924.

[39] Electronic resource. http://www.vargonalloys.se/index_eng.html.

[40] Electronic resource. http://www.eurometa.fr/en/.

[41] Electronic resource. https://www.ssab.com/company/merox/about-merox/history.

[42] De Bruin, T. and Sundqvist, L. 1998. ICSTI/Ironmaking Conference Proc., pp. 1263–1273.

[43] Sharma, T. et al. 1991. Effect of porosity on the swelling behavior of iron ore pellets and briquettes. ISIJ Int. 31: 312.

[44] Sharma, T. et al. 1992. Effect of reduction rate on the swelling behavior of iron ore pellets. ISIJ lnt. 32: 812.

[45] Sharma, T. et al. 1994. Swelling of iron ore pellets under non-isothermal conditions. ISIJ lnt. 32: 960.

[46] US Patent No. 5100464, 1992.

[47] Jim Torok. 2013. Briquetting for the Steel Industry: Then and Now. 32nd Biennale Conference of the Institute for Briquetting and Agglomeration, 25–28 September 2011, Curran Associates 20.

[48] Steele, R.B. 1993. Agglomeration of Steel Mill By-products via Auger Extrusion. Proc. 23d Biennial Conference. IBA, Seattle, WA, USA, p. 205–217.

[49] Kurunov, I. and Bizhanov, A. 2017. Stiff Extrusion Briquetting in Metallurgy // Springer. P. 169.

[50] Mazmishvili, S.M. 1992. Development and industrial development of technologies for producing dust briquettes and smelting manganese ferroalloys from them / Mazmishvili, S.M. and Simongulashvili, Z.A. // Izvestiya VUZov. Ferrous Metallurgy 12: 43 (In Russian).

[51] Pietsch, W. 2005. Agglomeration in Industry. Occurrence and Applications Wiley. 375 p.

[52] Electronic resource. http://www.phxslag.com.

[53] Noskov, V.A., Bolshakov, V.I., Maimur, B.N. et al. 2004. Pilot-industrial production of briquettes from screenings of ferroalloys at JSC "NFZ" // Metallurgical and Mining Industry 3: 124–126 (In Russian).

[54] Hycnar, J.J., Borowski, G. and Józefiak, T. 2014. Conditions for the preparation of stable ferrosilicon dust briquettes. Inżynieria Mineralna – Journal of the Polish Mineral Engineering Society 33(1): 155–162.

[55] Electronic resource. http://www.silingen.eu/products/.

[56] Electronic resource. http://poscoenc.com/file_download/download/project_list_steel_plants.pdf.

[57] Noldin, Jr J.H. 2012. An overview of the new and emergent ironmaking technologies // Millennium Steel. pp. 19–25.

[58] US Patent No. 2007/0157761 A1, James Bratina, Jul. 12, 2007.

[59] Patent of the Russian Federation No. 2518672, 2012.

[60] Electronic resource. https://cordis.europa.eu/project/rcn/55374/factsheet/en.

[61] Titov, V.V., Murat, S.G. and Kiselev, N.I. 2007. Ecology and Industry, 1: 16–21 (In Russian).

[62] Kurunov, I.F. and Kanaeva, O.G. 2005. Briquetting—a new stage in the development of the technology of agglomeration of raw materials for blast furnaces // Bulletin of scientific, technical and economic information "Ferrous metallurgy" 5: 27–32 (In Russian).

[63] Kurunov, I.F., Shcheglov, E.M., Kononov, A.I., Bolshakova, O.G. etc. 2007. Study of the metallurgical properties of briquettes from man-made and natural raw materials and evaluation of the effectiveness of their use in blast-furnace smelting. Part 1. // Bulletin of scientific, technical and economic information "Ferrous metallurgy" 12: 39–48 (In Russian).

[64] Kurunov, I.F., Shcheglov, E.M., Kononov, A.I., Bolshakova, O.G. etc. 2008. Study of the metallurgical properties of briquettes from man-made and natural raw materials and evaluation of the effectiveness of their use in blast-furnace smelting. Part 1. /// Bulletin of scientific, technical and economic information "Ferrous metallurgy" 1: 8–16 (In Russian).

[65] Electronic resource. http://www.kmz-tula.ru/articles-20100728.html.

[66] Timo Paananen and Erkki Pisilä. 2015. Improved raw material efficiency in hot metal production; Düsseldorf, 15–19 June 2015, METEC-ESTAD, pp. 1–8.

[67] Leif Hunsbedt and Peter Cowx. 2016. NyKoSi Agglomeration Seminar, November 2016 22–23 November, 2016.

[68] Davey, K.P. 2004. The development of an agglomerate through the use of FeMn waste. Proceedings of Tenth International Ferroalloys Congress, INFACON-X: Trans-Formation through Technology, Cape Town, South Africa, pp. 272–280.

[69] Electronic resource. http://www.fpiwv.com.

[70] Bizhanov, A.M. and Kurunov, I.F. 2017. Extrusion Briquettes (Brex), a new stage in the agglomeration of raw materials for ferrous metallurgy, Moscow, Metallurgizdat, 234 p. (In Russian).

[71] Patent of the Russian Federation No. 2495092, 2013.

[72] Patent of the Russian Federation No. 2506327, 2014.

[73] Patent of the Russian Federation No. 2499061, 2013.

[74] Patent of the Russian Federation No. 2506326, 2014.

[75] Patent of the Russian Federation No. 2506325, 2014.

[76] Patent of the Russian Federation No. 2502812, 2013.

[77] Patent of the Russian Federation No. 2501845, 2013.

[78] Patent of the Russian Federation No. 2504588, 2013.

[79] Patent of the Russian Federation No. 2579706, 2016.

[80] BREX. Certificate of trademark (service mark) No. 498006, application No. 2012706053 of 02.03.2012. Right holder A.M. Bizhanov (in Russian).

[81] Villanueva, A., Hoff, S., Michailovski, A. 2018. Novel Binder Technology from BASF: Study of Bentonite Modification. SME Annual Meeting Feb. 25–28, 2018, Minneapolis, MN.

[82] Electronic resource. https://ambershaw.ca/our-technology.

CHAPTER 3
Basic Materials for Briquetting

The most important metallurgical properties of the briquettes are: cold strength, porosity, reducibility and softening, hot strength.

Cold strength allows the briquettes to maintain integrity and not to collapse with the formation of fines during transportation, handling and loading into the furnace. The porosity of briquettes affects the reaction surface of the interaction of the oxides of its components with a gaseous reducing agent. Reducibility—the ability of the components of the briquette to give oxygen to the gaseous reducing agent. Good reducibility of briquettes allows to reduce the degree of development of direct reduction and the specific consumption of coke for iron smelting. Softening of briquettes can affect the gas permeability of the charge column in the furnace, therefore, the higher the temperature at the beginning of the softening of the briquette, the less the contribution of this part of the charge to the increase in resistance to gas flow. The metallurgical value of briquettes increases with a decrease in the temperature difference between the beginning and the end of their softening. Hot strength—the ability of briquettes to preserve integrity as much as possible without destruction in the process of heating and phase transitions. Before considering the examples of the manifestation of these properties of briquettes in various metallurgical furnaces and reactors, the analysis of the basic materials used for briquetting in ferrous metallurgy will be focused on.

Dispersed materials of natural and anthropogenic origin, which can be used as components of the briquette mixture, are formed at all stages of the metallurgical process from the extraction and enrichment of ores, in the production of sinter and pellets, in the preparation and introduction of the charge in metallurgical furnaces and reactors, during gas cleaning of exhaust gases and shipment of the finished product to the consumer.

3.1 Mining and Enrichment of Ores

For the majority of furnaces and reactors of ferrous metallurgy used today, there is a lower limit for the size of the charge pieces. For example, for blast furnaces, the allowable size of the charge pieces is at least 10–12 mm. With the extraction and further enrichment of brown iron ore and hematite iron ore, the yield of fractions larger than 10 mm does not exceed 30% and the formation of a fine fraction is large, with particle sizes less than 0.05 mm, which makes the direct use of such materials in the charge (without preliminary agglomeration) economically meaningless due to the gas flow from the working space of furnaces and ore particles reactors.

Iron ore fines and concentrates are suitable charge components for briquetting. Depending on the iron content, there are rich iron ores (Fe content 60–65%), ores with average iron content (45–50%) and poor ores (less than 45%). With increasing iron content in the briquette, its metallurgical value also grows. It is known that with an increase in the iron content in the blast furnace charge by 1%, the productivity of the blast furnace increases by 2–2.5%, and the specific coke consumption decreases by 1–1.5%.

The properties of briquettes are also influenced by the features of the crystal structure of iron ore minerals, the content of gangue and impurities. The difference in the mineralogical structure manifests itself predominantly at the stage of heating briquettes due to the restructuring of the crystal structure of iron oxides (as, for example, when hematite is reduced to magnetite). On the briquetting of iron ore fines and concentrates, such features, as a rule, are not affected.

The composition of gangue of iron ore fines and concentrates may include iron-free quartz minerals, aluminosilicates, garnets, pyroxenes, calcite, etc. The main components of the gangue are silica (SiO_2), alumina (Al_2O_3), lime (CaO) and magnesium oxide (MgO). When melted in a furnace, gangue forms slag, whose properties are largely determined by the so-called basicity - (CaO + MgO)/(SiO_2 + Al_2O_3). For example, the so-called "self-melting" ore with a basicity of about one is considered the most valuable for blast-furnace smelting.

Impurities, which are usually found in iron ores, can significantly affect the properties of briquettes. It should be noted that, in contrast to sintering and the production of pellets, which imply mandatory high-temperature processing of iron-containing materials, which allow to remove harmful impurities, in cold briquetting the chemical composition of the charge remains unchanged. Harmful impurities include—S, P, As, Zn and Pb. Mn, Cr, Ni, V, W, Mo and other elements are considered useful.

The characteristics of other types of ore fines and concentrates (manganese, chromium and nickel) will be considered when studying the properties of briquettes based on them in the following sections.

3.2 Sinter and Pellet Production

Disperse materials of interest for briquetting are formed in the processes of sinter and pellet manufacturing, crushing, handling and transportation. These include dust and sludge from sinter production and induration machines and screenings of sinter and pellets.

The specific yield of dusts from sintering machines can be from 10 to 24 kg per ton of sinter. At the preparatory sites 9 kg of dust per ton of sinter is formed. The output of dust in the production of pellets reaches 7–9 kg per ton of pellets, in the preparatory areas—5–16 kg per ton of pellets.

Dust consists of particles of sinter (pellets) and particles of raw materials for such industries (ore, coke, coal, limestone). The typical chemical composition of dust of sintering machine: iron content—45–55%; SiO_2—5–10%; sulfur—6–10%; CaO—10–20%; Al_2O_3—1–2%. Particle shape is spherical with a granular surface (magnetite) or oblong (hematite). More than 46% of such dust particles have a size in excess of 40–50 microns. The size of the dust particles of roasting machines is smaller. About 40% of particles are larger than 40–50 microns.

The small fraction of the sinter (the fraction of the finished sinter with size smaller than 5 mm) is formed when it comes off from the sintering belt, after sieving the return from the cooled sinter and when it is loaded into the skip of the blast furnace. This material has the chemical composition and mineralogical composition of the finished sinter.

The indurated pellets fines are formed during classification, transportation, unloading, storage and subsequent loading of pellets into the blast furnaces and metallization reactors. The typical granulometric composition of pellets fines—8% of particles are larger than 6.30 mm, 60% have sizes in the range of 0.15–6.30 mm and 32% of particles are smaller than 0.15 mm. The iron content in the pellets fines is 65% on average. Some features of the mineralogical composition of iron-containing phases of pellet screening can lead to an abnormal swelling of briquettes on a cement binder when they are heated in a reducing atmosphere.

3.3 Coke Production

In the process of coking coals coke breeze, coke dust and other dispersed materials are formed. The output of dust in the production of coke is 2–4 kg per ton of coke. When furnaces are loaded, about 0.4 kg of dust per ton of coke is produced, and in the production of finished coke—0.7–0.8 kg of dust per ton of coke. The chemical composition of dust and fines corresponds to the composition of coke. Coke breeze and dust are among the most popular reducing agents used in briquetting.

3.4　Blast Furnace Production

The most promising dispersed materials for briquetting generated in the blast furnace production include flue dust, wet gas cleaning sludge, sludge and dust generated in the sub-bunker rooms of blast furnace. Flue gas dust is dust carried through the blast furnace top by the blast furnace gas. It is formed as a result of the abrasion of pieces of the blast furnace charge during its movement in the shaft of the furnace. Dust in the sub-bunker rooms is formed during transportation and handling of the charge components. Sludge—as a result of hydraulic cleaning and wet cleaning of emissions of aspiration systems.

The total formation of flue dust and sludge is about 64 kg per ton of iron. The output of sludge of wet gas cleaning of blast furnace gas—26 kg/t of hot metal. The output of top dust is on average 38 kg/ton of hot metal.

About 30% of flue dust particles have a size in the range of 150–250 microns. The iron content varies widely (40–50%) and depends on the type and properties of the main components of the blast furnace charge. An important distinguishing feature of flue dust is the presence in its composition of noticeable amounts of carbon introduced by particles of unreacted coke, which makes this material a promising component of self-reducing briquettes. The carbon content in flue dust can exceed 30% (mass). Hematite, magnetite and calcium ferrite are the main iron-containing phases of flue dust. Iron ore grains have an irregular shape with sizes ranging from 0.5 to 150 microns. Fragments of coke are relatively large in size (from 350 microns to 1 mm).

The iron content in the sludge of the wet gas cleaning of blast furnace gas varies widely (20–55%), there may be a significant presence of undesirable impurities such as zinc and lead.

Iron content in sludge of sub-bunker rooms of blast furnace is 33–35% and its composition is close to the composition of the sinter charge. The granulometric composition of such sludge is close to the composition of the dust of sinter plants.

3.5　Steelmaking

During steelmaking processes in oxygen converters and electric furnaces, dispersed materials are formed, which can serve as valuable charge components of briquettes. These include converter sludge, dust and sludge from the Electric Arc Furnace (EAF). In some cases, steelmaking slags can also be added to the briquette mixture, however, due to their low iron content, their use in briquetting is limited.

In the exhaust converter gases, the dust content is very high (on average, up to 30 g/m^3, and in some cases up to 60 g/m^3). The main minerals of such dust are maghemite, magnetite and wustite. The material itself is a black powder with spherical particles. The average particle size is 0.08 microns; 10–15% of the particles are larger than 10 microns. Large fractions of dust are particles of the charge materials, droplets of slag and the shell of carbon monoxide bubbles carried by the gas stream. Sludge is formed during wet gas cleaning of converter gases. In such materials, in the process of drying, sludge granules are formed, on the surface of which there is a tricalcium silicate ($3CaO \cdot SiO_2$), the granular structure of which prevents the strong adhesion of particles with cement gel during briquetting, which leads to low mechanical strength of briquettes based on converter sludge. Giving such a briquette the required strength may require increased consumption of the cement binder. The strength properties of briquette based on converter sludge are improved when using a combined binder consisting of equal parts of hydrated lime and molasses (3% by weight of the briquette). Converter sludge is preferably briquetted in a mixture with a blast-furnace sludge. The strength of the briquette consisting of a mixture of equal fractions of the converter and blast-furnace sludge (47% of the mass of the briquette each), with such a binder, when tested for five-fold dropping from a height of 2 meters onto a steel base, exceeded 95% (the fraction of particles with dimensions more than 4.75 mm), while the strength of the briquette on the basis of only converter sludge (85% by mass) with a cement binder (15% by mass) after similar tests did not exceed 89%.

The specific yield of the EAF dust is on average 8–10 kg per ton of steel and can reach 20 kg/t with oxygen blowing. Dust is formed in the zone of arcs and is a vapor of metal condensing in the cold parts of the tract, condensed vapors of slag and particles of the charge removed from the bath by a stream of exhaust gases. The content of Fe_2O_3 in such dust is 10–50%. The dust particles are predominantly spherical in shape. The bulk of the dust particles in the EAF dust have a size in the range of 0.2 to 20 microns.

A distinctive feature of EAF dust is its high content of Zinc Oxide (ZnO), which can exceed 20% of the mass of the material, which is unacceptable for blast furnace briquettes. The sale of dusts to the industrial enterprises of zinc producing industry is possible with a content of at least 15% zinc (18.7% ZnO).

The structural features of the particles and the particle size distribution of the EAF dust create some difficulties in its briquetting with the method of vibropressing with cement as a binder associated with significant values of the specific surface of the particles (16000 cm^2/g), which leads to a noticeable manifestation of the capillary effect and dewatering the briquette. Due to the lack of moisture, the hardening of the cement stone is impeded and the briquette does not gain strength. Such a feature of the EAF dust is not an obstacle for its briquetting with a roller press and the method of stiff vacuum extrusion.

3.6 Rolling Production

During the hot rolling process, scale is formed on the outer surface of the plates, sheets and profiles. The output of mill scale of rolling production is about 2% by weight of the finished product. The shape of the particles is scaly. The size of the main mass of particles does not exceed 0.2 mm. The chemical composition of scale is close to pure magnetite (65–72% Fe). The main phases in the scale are wustite and magnetite, and a thin layer of hematite on the surface of particles that are in contact with an oxidizing atmosphere. The ratio between the amounts of these oxides in the scale can be different depending on the conditions under which the metal is oxidized.

The scale may be contaminated with lubricating oils and cutting fluids (coolant) as a result of their leakage from the lubrication system of rolling mill equipment. The presence of oil determines the high hydrophobicity of the particles of scale surface. The amount of oily scale in developed industrial countries is about 0.5% of the mass of rolled steel, in Russia it reaches 2.0% or about 600 thousand per year.

The hydrophobicity of the surface of particles of oily scale reduces the effectiveness of the use of cement as a binder during its briquetting. The de-oiled scale can be used to produce washing blast furnace briquettes.

3.7 Ferroalloys Production

In the production of ferroalloys, different dispersed materials are also formed, which are suitable components of briquette charges (dust and sludge, fines of pellets and sinter, fines of crushing of ferroalloys). In the ferroalloy industry, dry (open and semi-closed ore smelting furnaces) and wet systems (closed and sealed furnaces) are used. Open and semi-closed ore-reducing furnaces are equipped with a dry gas cleaning, closed and sealed furnaces are equipped with a wet gas cleaning.

The specific dust emissions of ferroalloy furnaces are 8–30 kg per ton of alloy. The total formation of dust in the areas of preparation of raw materials, warehouses of charge materials and finished products reach a total of 1.1 kg/t of alloy. Particles of large fractions (about 10%) have an irregular shape, since they were formed outside the arc zone in the absence of phase transitions. Such dust contains particles of charge materials (ore raw materials, fluxes and coke). The size of almost 90% of the dust particles of gas cleaning of ferroalloy furnaces does not exceed 10 microns. The particle shape is mainly spherical.

The content of manganese oxides in dusty gas cleaning systems in the production of silico-manganese is 21–34% (mass) and in the production of ferromanganese is 20–25% (mass). The content of chromium oxides in the dust during the smelting of ferrochrome can reach 22–45% (mass). The content of microsilica in the dust of gas cleaning production of ferrosilicon reaches 98%. As well as flue dust, gas cleaning dust of ore smelting furnaces may contain a large amount of carbon (10–20% and higher). The use of such dusts as components of the briquette mixture can help reduce coke consumption in the smelting of ferroalloys. Dust containing microsilica can exhibit binding properties, resulting in lower consumption of binder materials.

Before shipment to consumers, ferroalloys are subjected to crushing and sieving on mechanized sieves to obtain the size of pieces required by the technology. When crushing is used, for example, a jaw crusher with a slit width of 100 mm, up to 10–15% of the fines of ferroalloys with particle sizes less than 3 mm can be formed, direct recycling in the form of a charge component of the ore smelting furnace is impossible. The fines of crushing of ferroalloys and dusts, captured by systems of aspiration of sites of fractionation of ferroalloys, are promising components of the briquette mixture due to the high content of the metal phase.

3.8 Direct Reduced Iron (DRI) Production

Sources of sludge formation in direct iron production processes are flue gas treatment systems (Venturi tubes); briquette cooling system; hydraulic washing of belt conveyors of the finished product transportation system; de-dusting scrubbers of the briquetting system.

The sludge is mixed in a common collector and the mixture is then transported to a sludge collector. Moisture content of sludge reaches values of 30–40%. The total iron content in sludge can exceed 55% (total), metallic iron content—22%.

Metallized fines are formed in the process of briquetting at the breaking of the briquette tape, at the destruction of the briquettes of Hot Briquetted Iron (HBI). In part, it is represented by pellets that have not fallen into the gap between the sleeves of the roller press. The proportion of particles of such a metallized material with a particle size of less than 4 mm exceeds 92%. The total iron content in HBI fines exceeds 87%.

The number of materials suitable for briquetting is not limited to the list above, which contains a brief description of the materials widely used in practice as components of industrial briquettes in ferrous metallurgy.

One of the most essential components in briquetting is a binder. The practice of industrial briquetting has shown that, of the variety of binding materials, cement, slaked lime (sometimes in combination with molasses) and some types of organic ligaments (lignosulfonate, starches) are most common in ferrous metallurgy. The use of such materials allows reaching a compromise between the cost of their acquisition and the level of briquette strength achieved with their help.

3.9 Binders

Some of the issues related to the use of binders and briquetting without binders (hot and thermal briquetting) were discussed above. At present, the processes of briquetting of iron ore concentrates, some ores (concentrates) of non-ferrous metals and ore-fuel charges are carried out with and without the introduction of binders.

Without introducing binders, ores and industrial wastes are briquetted, which include components with a binding capacity, or ores subjected to very fine grinding followed by pressing at high pressures. While briquetting ores and ore-fuel charges, inorganic (mineral) binders are most widely used.

Methods for briquetting natural and technogenic finely dispersed materials with the introduction of organic binders, as well as briquetting methods at relatively low pressures have been developed and successfully applied.

When using some types of binders for briquetting, the use of different fillers, plasticizers, surfactants, etc. can significantly accelerate the course of the cementation process and improve the quality of the resulting briquettes.

Binders for briquetting in metallurgy must meet the following basic requirements;

1. Possess high cementitious ability with fast setting;
2. Not be ballasted in briquettes;
3. Be as cheap and affordable as possible; not difficult to use;
4. To provide sufficient mechanical strength, water resistance and heat resistance of briquettes;
5. Do not create harmful working conditions for service personnel, that is, comply with sanitary and hygienic requirements;
6. Do not introduce harmful impurities (ash, sulfur, phosphorus, etc.) in such quantities that will affect the quality of the metal being smelted;
7. Do not significantly reduce the metal content in briquettes;
8. Do not deteriorate the quality of briquettes under the influence of high temperatures (up to 1600°C) in both reducing and oxidizing environments;
9. Do not worsen the reducibility of the material in briquettes;
10. Do not deteriorate the conditions for melting briquettes, in particular, not to increase the consumption of fluxes.

The most widespread in briquetting as applied to blast furnace, steel-making and ferroalloy industries are cements and lime (including in combination with molasses).

Different types of cements are used as binders: Portland cement, hydraulic, bauxite, ore, slag Portland cement, etc.

Cement is a product obtained by fine grinding clinker with a small amount of gypsum. The clinker is formed as a result of firing (at t = 1450°C) a mixture consisting of lime and clay or other materials of similar composition and sufficient activity. A typical clinker has the following composition: 63–66% CaO, 21–24% SiO_2, 4–8% Al_2O_3, 2–4% Fe_2O_3, 0.5–5% MgO, 0.1–0.3% P_2O_5, 0.3–1% SO_3.

Portland cement is a widely used hydraulic binder based on calcium silicates (up to 70–80%). The density of Portland cement is 3.1–3.2 g/cm³. An increase in the CaO content in its composition contributes to an increase in the rate of cementation and final strength with a slight decrease in water resistance.

Slag Portland cement consists of finely ground Portland cement clinker, slag and gypsum stone. In terms of its physical and mechanical properties, slag cement is similar to Portland cement, but differs from the latter in higher heat resistance, slightly lower density (2.8–3 g/cm³) and low cost (by 15–20%).

Cement consumption is 3.5–6.0% for extrusion briquetting and can exceed 10–12% for vibrocompression, for which cement is the only acceptable binder.

Alumina cement, which is obtained by firing a raw mixture consisting of bauxite and limestone, has also proved to be good. Depending on the alumina content, ordinary (up to 55% Al_2O_3) and high-alumina cement (up to 70% Al_2O_3) are isolated. However, this grade of cement is not widely used due to its high cost.

The main disadvantages of using cements as binders are a relatively long period of setting and hardening of green briquettes, as well as a rather large content of silica in their composition.

Lime is the most common type of binder and is used both in quicklime and in dry form (slacked lime). The main advantages of lime are that it does not introduce harmful impurities into the composition of the charge, it is a flux during melting and has a low cost, since the raw material for its production (limestone) is widespread in nature.

Lime is a product obtained by firing limestone containing no more than 8% impurities at a temperature of 900–1200°C. This process takes place with significant heat consumption by the reaction:

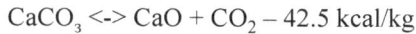

$$CaCO_3 <-> CaO + CO_2 - 42.5 \text{ kcal/kg}$$

The slaking of burnt lime is carried out by adding water in an amount of at least 30%. The reaction of the formation of calcium hydroxide is exothermic and proceeds with a significant release of heat:

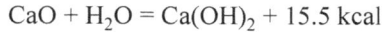

$$CaO + H_2O = Ca(OH)_2 + 15.5 \text{ kcal}$$

When slaking lime, there should not be more than 10% of non-slacked lumps. Unslacked pieces can be formed as a result of underburning or overburning of limestone, which is equally harmful. Unburned pieces are not slacked in water, and entering into briquettes can serve as centers of their destruction, burnt pieces are slacked very slowly, and getting into briquettes, during aging or melting, can cause their cracking and destruction.

During open storage in a warehouse, the surface layers of lime are hydrated under the influence of atmospheric moisture, and the resulting layer of slaked lime begins to carbonize under the influence of carbon dioxide contained in the air, which entails a partial loss of the binding capacity of lime. Therefore, lime should be stored only in closed storage rooms.

The main reaction that turns lime into a cementitious material in briquettes is the reaction of transferring calcium hydroxide when interacting with carbon dioxide into calcium carbonate:

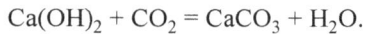

$$Ca(OH)_2 + CO_2 = CaCO_3 + H_2O.$$

This reaction occurs when the briquettes are aged in air (natural carbonization) or during special treatment of the briquettes with gases containing an increased amount of carbon dioxide (artificial carbonization). An intensive carbonization process occurs only with a certain amount of moisture in the charge. In this regard, fast or, on the contrary, slow drying of briquettes leads to a slowdown in the rate of their carburization and a decrease in strength.

During hydrothermal (autoclave) treatment of briquettes, cementation of their structure is carried out due to calcium hydrosilicate formed by the interaction of calcium hydroxide with silica gel contained in the ore.

The consumption of quicklime in roll briquetting is 10% at a pressing pressure of 35–40 MPa and decreases to 5% with an increase in the pressing pressure to 85–100 MPa. The optimal time for heat treatment of briquettes at 250°C is 2–3 hours, at 500°C–0.5 hours. The mechanical strength of briquettes increases when used for briquetting ores with a high silica content and more finely ground lime. With extrusion briquetting, lime consumption does not exceed 5–6%.

When briquetting ores, the consumption of fresh slaked lime is approximately 5–10%, the amount of added water is 8–12%.

Further, when considering different aspects of briquetting in metallurgy, examples of the use of other types of binding materials will be given.

In addition to cement and lime, in a number of cases, polymers are used to solve the problems of agglomeration of natural and technogenic raw materials. Various organic materials (coal, tar, pitch, petroleum bitumen, starches, etc.) are also used as a binders. In particular, a binder based on starch has proved good for agglomeration of ferrosilicon fines.

CHAPTER 4
Methods for Preparing Materials for Briquetting

An important role in achieving the required level of metallurgical properties of briquettes (mechanical and hot strength) belongs to the methods of preparing the briquetted charge.

Among the goals achieved at this stage of briquette production:

- correspondence of moisture content and granulometric composition of the briquetted mixture to the technological features of the used briquetting method.
- uniform distribution of the density of the briquetted mixture.

The moisture content of the briquetted mixture, depending on the type of briquetting equipment, was described earlier and is illustrated in Fig. 2.15 (section 2.5). A distinctive feature of stiff vacuum extrusion is the possibility of briquetting moistened materials, which allow eliminating the need for preliminary drying, for example, of sludge. In roller briquetting, there are restrictions on the moisture content of the briquetted mixture (less than 5%), which limits the use of hydrated binders in this method. There may simply not be enough moisture for curing. Restrictions on moisture content, not very pronounced, are also typical for vibrocompression.

The influence of the granulometric composition of the briquetted mixture is also manifested in different ways. In roll briquetting, a significant pressing force is used (from 10 MPa), in contrast to vibrocompression, where the magnitude of the pressing force is negligible. But both of these technologies allow agglomeration of a mixture with particle sizes reaching (and exceeding) 10 mm. On the contrary, in extrusion briquetting, the very possibility of effective briquetting depends on the plastic properties of the mixture, which are significantly influenced by the granulometry of the mixture components.

4.1 Mixture Granulometric Control

One basic criterion for determining the suitability of extrusion material is its plasticity—a trait that assures that it can be effectively pushed through the holes in die. A necessary condition for the status of plasticity is complying with the granulometric composition and moisture content requirements for SVE. In some cases, additional pulverization may be required in order to achieve the desired plasticity of material.

Unlike a roller-press and vibropress briquetting, shear stress plays an important role in SVE agglomeration. Shear stress occurs when the mixture is processed in the screw feeder, in pug mills and then in the extruder. In [1], based on a comparison of coal briquette porosity values in different pressing options (compression and its combination with torsion), it was found that more dense briquettes (less porous) are formed in the combined pressing option (under identical values of applied pressure). With full compression, a large proportion of energy is expended on the elastic deformation of the particles themselves, while, in the presence of shear stress, the convergence of particles on the surface forces activation distance is more effective. In full compression of close-packed particles, each particle only comes into contact with its immediate neighbors and is subjected to compression load. Under shear stress, the particles of the adjacent layers are subjected to abrasion due to contact with irregular surfaces, which can lead to crushing, the opening up of new surfaces, and hence, to an increase in the number of contacts between particles of the mixture.

In order to identify the possible impact of shear stress on the change of particle size distribution of briquetted material in conditions of SVE, we have compared the results of pulverization of coke fines in three different ways: in a hammer mill, in a roller crusher, and by double shearing through shearing plate in the extruder (Fig. 4.1). Samples of original coke breeze had the following properties: 8.97% moisture content; 13.28% ash content; volatile content 3.45%; 0.61% sulfur content. Since coke breeze particle were larger in sizes than what is usually required for extrusion, they were pulverized.

Figure 4.1. Equipment and tools for materials pulverization (top left-hammer mill; top right-roller crusher; bottom left-shearing through the shearing plate of extruder; bottom right-shearing plate).

Wet screens were used to determine particle size distributions of raw materials. Moisture content was measured using a moisture balance. A calibrated electronic scale with a density measuring attachment was used to determine pellet density. The results of the granulometric analysis are presented in Fig. 4.2.

As evidenced above, the grinding of coke breeze is maximized after double extrusion through a shearing plate in an extruder. In this case, the effect of deep grinding is achieved through the application of high shear stresses. The use of a hammer mill for such a material was found to be ineffective, and the granulometric composition of the ground material differed weakly from that of the initial coke breeze.

Based on these results, it was decided to study the effects of the grinding method on the strength of the brex produced from coke breeze pulverized in different manners. Three portions of brex of the same composition—94% of coke breeze, 5% of Portland cement and 1% bentonite—were manufactured. The only difference between the brex was the manner in which the coke breeze was

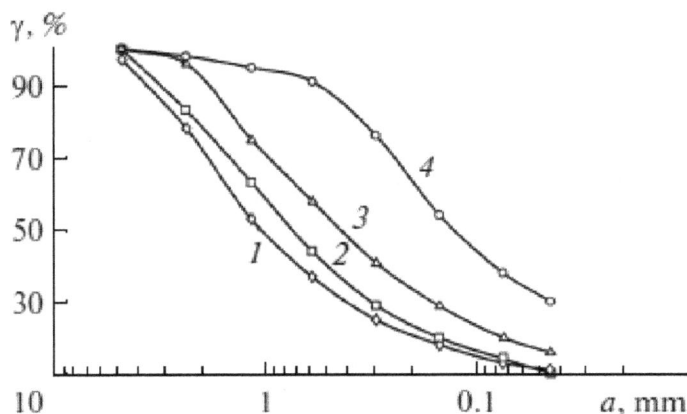

Figure 4.2. Granulometric composition of coke breeze in the following states: (1) initial and (2–4) after additional grinding in a hammer mill, in a roll crusher, and double extrusion in an extruder, respectively. γ is the yield of the oversize mass, and a is the mesh size.

Table 4.1. Extrusion parameters and physical properties of coke breeze brex.

Sample of brex (grinding method)	Moisture content, %	Temperature °C	Vacuum, mm Hg.	Density, g/cm³	Compressive strength, kgF/cm²
No. 1 (roller crusher)	16.5	30.56	15.24	1.63	37.76
No. 2 (double shearing)	16.7	33.33	17.78	1.67	34.32
No. 3 (hammer mill)	16.6	55.56	81.28	1.63	20.25

ground. The brex was classified assigning the following numbers in accordance with the method of grinding: No. 1 was coke breeze ground by roller crusher; No. 2 was coke breeze sheared twice through the shearing plate with a plurality of holes; No. 3 was coke breeze ground by hammer mill. Extrusion parameters and physical properties of brex are shown in Table 4.1.

It follows from these data that the extrusion of the first two mixtures was carried out in accordance with similar processing parameters and that the extrusion of the coarser particles in mixture No. 3 (hammer mill) is accompanied by an increase in the temperature of the material. As a result, the brex made of the largest particles turned out to have the minimum strength in axial compression tests. As for energy consumption, the extrusion of mixture No. 1 was most efficient. Its excellent extrusion ability can be attributed to the material particle shape after grinding in a roll crusher, which favors the plane-parallel orientation of particles.

The difference between the tensile strengths of samples No. 1 and No. 2 is insignificant and only indicates an earlier beginning of cracking in brex No. 2. It also follows from this data that the density of brex after double extrusion exceeds that of brex samples made of ground and milled coke breeze by 2.5%. Obviously, the dense packing of brex No. 2 particles is the result of a high degree of material grinding. In contrast to brex samples No. 1 and No. 3, no mixture dewatering was detected during extrusion in this case. Differences in compressive strengths of brex samples prepared from differently-treated coke breeze of the same batch—can result from a number of factors related to different particle sizes, shapes and surface relief. The shape of a particle after grinding depends on material characteristics and the grinding method, including time [2–4].

For example, it is generally accepted that, after grinding in roll crushers, the material mainly consists of angular particles, whereas the material particles usually maintain the same size and a rounded shape after ball and hammer milling. In roll crushers, grinding occurs under the compressive, shear, and rubbing forces. As a result, one has rough particles with sharp projections, many edges and corners, and (correspondingly) a large contact surface form. When the material is ground by hammer mill, the particle surfaces are polished by way of impact and the particles acquire a rounded shape.

For example, researchers [5] studied the effect of grinding device on oil coke particle shape of (1) low-porosity coke with thick cell walls without visible cracks and (2) fissured coke whose porosity and cell wall thickness of which were distributed over a wide range. In the first instance, grinding method did not affect the shape of particles 200–600 μm in size. In the second case, it was revealed that the particle shape is dependent on particle size in different ways when ground in a hammer mill and in a roll crusher (Fig. 4.3) [5].

These curves reflect the following dependence of the shape factor: $\phi = 1.1 V^{1/3} N^{1/6} A^{-1/2}$, where V is the specific particle volume (cm^3/g), N is the number of particles in 1 g substance, and A is the specific surface area (cm^2/g). The material particles ground in a hammer mill are characterized by stable high shape factors, which indicate the closeness of most of the particles' forms to a rounded shape in the indicated size range. For the particles of the material ground in a roll crusher, the shape factors in the 140–600 μm particle size range are lower and minimal at a size of ~ 350 μm, which supports the conclusion of a more non-uniform particle shape after grinding in a roll crusher.

As in [5], a porous and fissured material was used. Figure 4.4 shows an image of coke breeze particles taken with a JEOL JSM_6490 LV Scanning Electron Microscope (SEM). In Fig. 4.5, the particle shapes of the initial coke breeze and the coke breeze after additional grinding were compared. The initial coke breeze particles were characterized by hillocks (rounded projections), which are typical of coke; they belong to internal (not opened) pores located under the upper layers of the carbon material. It is seen that the particles after a roll crusher and double extrusion through an

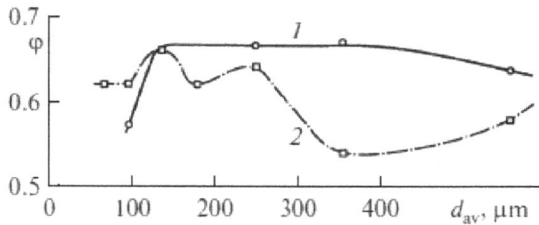

Figure 4.3. Particle shape factor ϕ vs. the average particle size upon grinding in (1) hammer mill and (2) roll crusher [5].

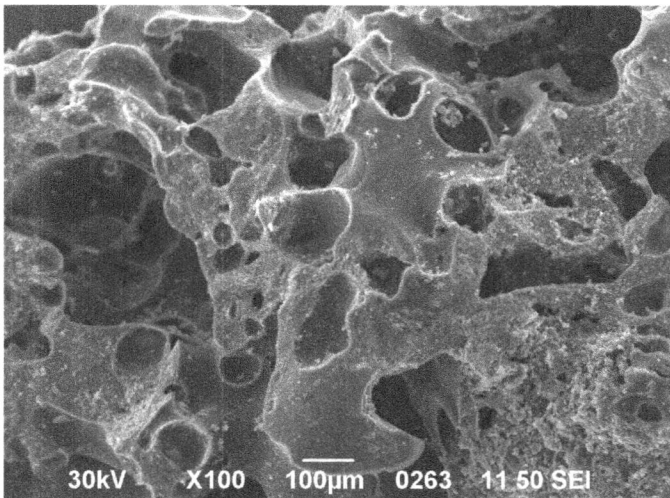

Figure 4.4. Structure of coke breeze particle (scanning electron microscopy).

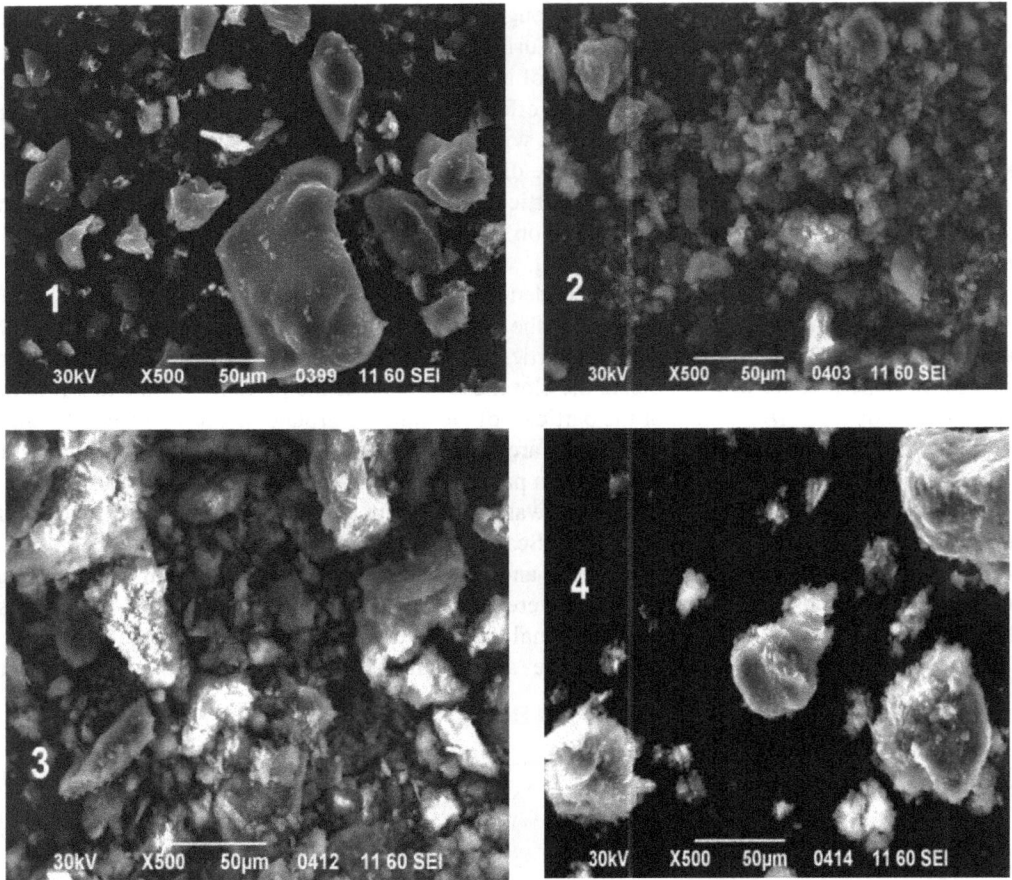

Figure 4.5. Micrographs of (1) initial coke breeze particles and after additional grinding (2) in roll crusher, (3) by double extrusion through an extruder, and (4) in a hammer mill.

extruder have a non-uniform angular shape, whereas the coke breeze particles ground in a hammer mill have a rounded shape.

The brex samples were subjected to tensile splitting tests on a bench type one column electromechanical Instron 3345 tensile testing machine at a load of 5 kN. When studying the statistics of brex orientation distribution in a charge, it was found [6, 7] that this type of external load is most probably possible for a cylindrical brex. Figure 4.6 shows the results of testing specially prepared cylindrical specimens of brex 1–3 25 mm in diameter and 20 mm in height. It is seen that, at approximately the same carrying ability, the brex specimens' reactions to an applied load are different.

The difference in the maximum loads can be related to defects in the specimens. However, the difference in the characters of behavior can have radically different causes. Brex No. 2 demonstrates ductile fracture, which is indicated by the existence of a yield plateau, i.e., the horizontal component of the brex No. 2 curve. This phenomenon is thought to be explained by a "relay-race" grain to grain gliding transfer in accordance with the Hall–Petch equation stipulating an inverse relationship between grain size and yield strength [8]. In this case, a grain boundary is a barrier to a dislocation motion, which causes dislocation nucleation and development in a neighboring grain. In other words, the larger the number of barriers to be overcome, the lower the dislocation motions dynamics and the higher the crack development resistance.

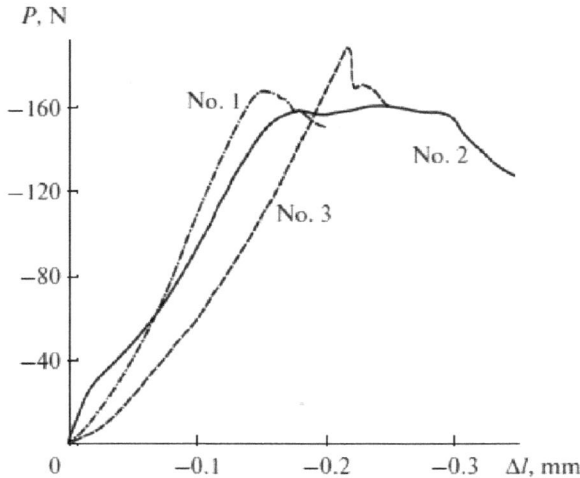

Figure 4.6. Load P–displacement Δl curves for cleavage tensile tests of brex.

Apparently, there exists a threshold particle size, below which cracks cannot propagate in brex. Obviously, the granulometric composition of the brex No. 2 mixture favors this scenario due to the highest content of thin particles among the other mixtures (see Table 4.1). The integrity of brex No. 2 is retained even after tests; therefore, high impact strength can be expected. During this type of loading, local fracture zones will not cause full fracture and debris formation in a specimen. The load–displacement curves of brex No. 1 and No. 3 decrease significantly after crack initiation, which points to more brittle fractures of these specimens. The brittle fracture in brex No. 1 develops more slowly than it does in brex No. 3 because of a smaller average particle size. Note that these circumstances are very important for briquetting. Several stages of charging–discharging related to throwing down briquettes can be required to supply the briquettes to a furnace even in one incident. When the impact toughness of an agglomerated product is increased, the logistics of supplying it to the site, including those that are far away, can be largely simplified and, hence, be made cheaper. It is important for stiff extrusion that a change in the behavior of brex under external mechanical action was achieved using same agglomeration equipment.

In addition to the physical mechanisms that weaken dislocation propagation, leading to a high strength of brex made of small particles, the distribution of a binder in the brex body also substantially contributes to its strength. It is clear that the low strength of brex No. 3 during compression can be explained by the small amounts of contacts between particles. Figure 4.7 shows micrographs of the particle surfaces in brex No. 3 (grinding in a hammer mill) and brex No. 2 (double extrusion through a shearing plate). It is clearly visible that, because of the lower particle surface roughness of brex No. 3 as compared to brex No. 2, the number of hillocks with a binder (cement and bentonite) on their surface (bright aggregates) is significantly lower. (Since minerals contain heavy chemical elements, they manifest themselves as bright precipitates in micrographs.) Therefore, the surface relief of the brex No. 2 particles also favors good adhesion of particles due to the large binder volume that covers the sites of particle–particle contacts.

When studying the particle surfaces of brex No. 2, bentonite fibers were detected, which were described for the first time in [9]. Figure 4.8 shows the surface of brex No. 2 particle and the surfaces of glass microspheres subjected to soft rolling (without crashing) in a roll crusher which imitated shear stresses. This treatment promoted the development of bentonite fibers covering the particle surface. After this preliminary treatment of a magnetite concentrate and bentonite mixture, the fraction of bentonite required for the given strength decreased twofold (from 0.66 to 0.33% concentrate mass). The appearance of this structure in brex No. 2 is likely to result from the shear stress applied to the material during the auger extrusion. The effect of the appearance of bentonite

Figure 4.7. SEM micrographs of the particle surfaces in brex No. 3 (a) and No. 2 (b).

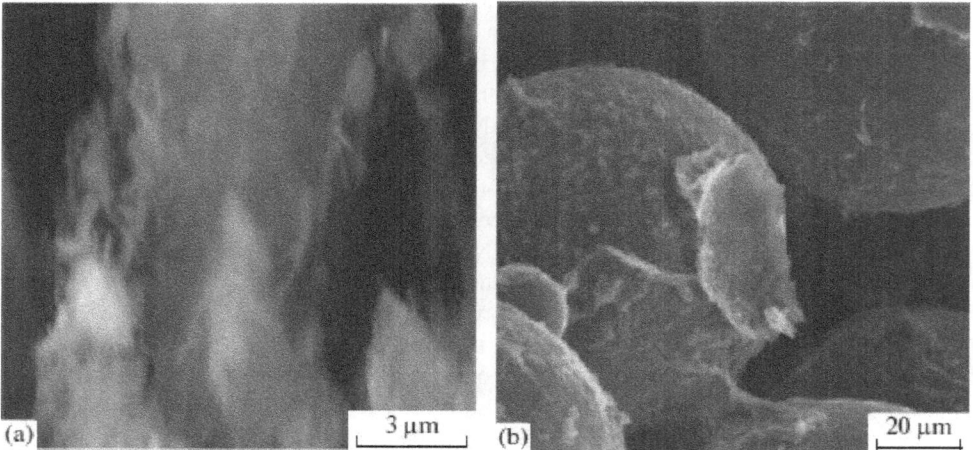

Figure 4.8. Bentonite fibers on (a) the particle surface in brex No. 2 and (b) on the surface of glass microspheres [9].

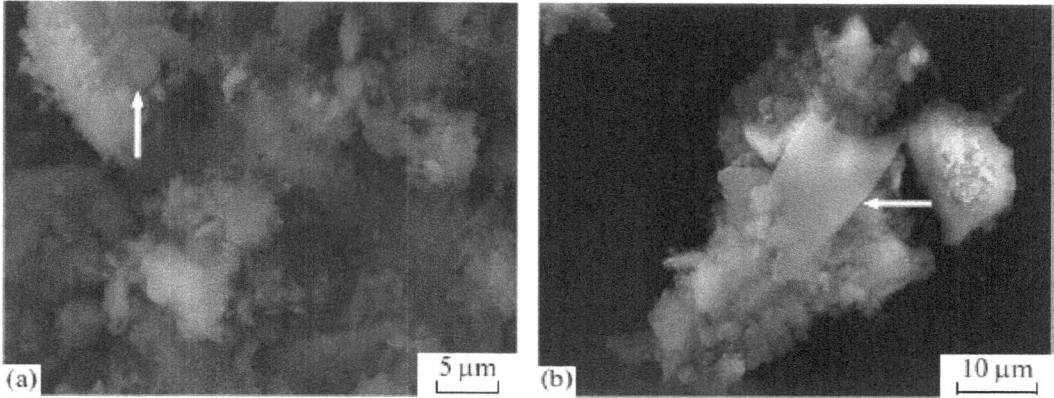

Figure 4.9. Bentonite particles in brex No. 1 (a) and No. 3 (b).

fibers is less pronounced in brex No. 1 and No. 3, which is likely to be associated with a lower fraction of thin particles. Brex No. 1 and No. 3 have predominantly lamellar and flake-like bentonite particles (Fig. 4.9).

4.2 Homogenization of the Mixture

The importance of homogenization of the briquetted mixture using the example of extrusion briquetting will be shown.

The change in the properties of brex induced by the addition of bentonite to the composition of the mixture to be briquetted also manifests itself after the preliminary homogenization of the mixture during its "souring" (storage of the moisturized mixture with added bentonite for a certain time). This method allows achieving a high degree of homogeneity of mixture properties for subsequent briquetting. In some cases, the mechanical strength of brex made of homogenized charge can significantly increase. The effect of souring on the strength of brex of various compositions (Table 4.2) were studied. Table 4.3 gives the results of tensile splitting tests of brex one week after its manufacture. These results demonstrate that the use of bentonite in combination with Portland cement improves the strength properties of brex. The strength of brex increases substantially in some cases. Furthermore, the manner of their fracture also changes and signs of viscoplastic fracture appear. In this case, brex can better withstand the impact loads that appear during their transportation to the sites of their application.

Table 4.2. Brex compositions for testing of the souring efficiency.

Brex composition/No.	1	2	3	4	5	6
BF sludge	42.8	41.8	41.8	41.2	47.8	28.3
BOF sludge	39.8	38.8	38.8	38.2	43.7	25.4
Iron-ore concentrate						29.3
Mill scale	13.0	13.0	13.0	12.1		10.7
Portland cement	4.0	4.0	6.0	8.0	8.0	5.8
Bentonite	0.4	0.4	0.4	0.5	0.5	0.5
Microsilica		2.0				

Table 4.3. Values of tensile splitting strength of brex.

Tensile splitting strength of brex, MPa	1	2	3	4	5	6
Without souring	0.86	1.93	2.08	1.00	1.01	0.77
After souring	2.45	3.83	5.76	1.88	1.29	1.26
The ratio of strength values	2.85	1.98	2.76	1.88	1.28	1.64

The results of the study on the effects of shearing stress on the properties of extruded mixture allow using shearing through the shearing plate for the homogenization of the mixture prior to its agglomeration with a binder.

To demonstrate the effect of such operations on the homogenization of the mixture, the values of compressive strength on brex composed of manganese ore concentrate with added baghouse dust resulting from silicomanganese production with and without homogenizing the earlier-sheared mixture were compared.

The granulometric composition of brex components is listed in Table 4.4. The composition of brex is given in Table 4.5.

Table 4.4. Granulometric composition of brex components.

Material	Fraction, mm/Yield, mass. %					
	−20...+10	−10...+5	−5...+2	−2...+1	−1...+0,5	−0,5
Manganese ore concentrate, %	3.42	25.42	32.58	18.95	9.70	9.93
Bag house dust of SiMn production	fine dust, size less than 0.063 mm					

Table 4.5. Composition of brex.

Composition of brex	No. 1	No. 2	No. 3
Manganese ore concentrate	80	66	56
Bag house dust	14	28	38
Portland cement	5	5	5
Bentonite	1	1	1

Table 4.6. Compressive strength of brex No. 1–3.

Strength/ brex No.	No.	No. 1 – soured	No. 2	No. 2 – soured	No. 3	No. 3 – soured
Day 1 (MPa)	2.9647	2.7579	2.6890	5.7571	4.0955	6.1708
Day 3 (MPa)	4.8608	4.8608	6.0674	7.8255	6.1019	10.2042
Day7 (MPa)	6.6741	7.6187	13.1345	14.3411	10.1698	15.7200

A laboratory extruder has been used for brex sample production. The sheared mix was subjected to the souring over the course of 4 hours. Table 4.6 shows the results of the measurement of the compressive strengths of brex No. 1–3 produced from this mix with and without souring. The increase in compressive strength a week after manufacturing was 14.2% for brex No. 1; 7.62% for brex No. 2, and 54.5% for brex No. 3.

The study of the brex structure by means of scanning electronic microscopy revealed differences in the structure and distribution of pores. Figure 4.10 presents the structure of the brex No. 2.

One can see that pore size is significantly smaller in the brex made from a sheared and soured mix. Our results confirm the efficiency of using a shearing extruder in preparing the briquette charge for souring.

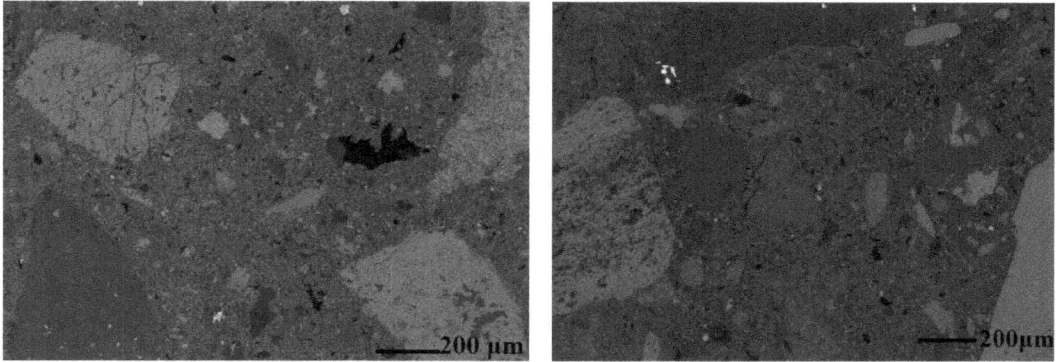

Figure 4.10. Scanning electron microscopy of the brex No. 2 structure. Left - without souring; right - with souring during 4 hours after shearing.

References

[1] Gregory, R. 1960. Briquetting coal without a binder. Colliery Guardian 201, No. 5191.

[2] Kaya, E. 2002. Particle shape modification comminution. Kaya, E., Glogg, R. and Kumar, S.R. Kona 20: 185–195.

[3] Ulusoy, U. 2003. Role of shape properties of calcite and barite particles on apparent hydrophobicity. Ulusoy, U., Hicyilmaz, C. and Yekeler, M. Chem. Eng. Process 43: 1047–1053.

[4] Ulusoy, U. 2003. Determination of the shape, morphological and wettability properties of quartz and their correlations. Ulusoy, U., Yekeler, M. and Hicyilmaz, C. Mineral Eng. 16: 951–964.

[5] Beirne, T. 1954. The shape of ground petroleum coke particles Brit. Beirne, T. and Hutcheon, J.M. J. Appl. Phys. 13: 576.

[6] Bizhanov, A.M., Kurunov, I.F., Durov, N.M. et al. 2012. Mechanical strength of BREX: Part I. Metallurg 7: 32–35.

[7] Bizhanov, A.M., Kurunov, I.F., Durov, N.M. et al. 2012. Mechanical strength of BREX: Part II. Metallurg 10: 36–40.

[8] Malygin, G.A. 2007. Solid-State Physics 49(6): 961–982.

[9] Kawatra, S.K. 2002. Effects of bentonite fiber formation in iron ore pelletization. Intern. J. Miner. Process 65: 141–149.

CHAPTER 5
Requirements for the Metallurgical Properties of Briquettes

5.1 Mechanical Strength

The briquette from the moment of its creation and until it is introduced as a component into the charge of a metallurgical furnace is subjected to a number of mechanical influences: shock-abrasive, dynamic and crushing loads during conveyor transportation for storage, stacking, as well as when loading into wagons and feeding into the receiving hoppers of the furnaces. In a metallurgical furnace or reactor, the briquette must withstand the pressure of the overlying charge layers and maintain its integrity until the required metallization level is reached. In accordance with this, the set of methods for testing briquettes for mechanical strength includes methods for the quantitative determination of impact strength, abrasion and compression strength.

The impact strength of briquettes is determined by repeatedly dropping a briquette from a certain height onto a metal or concrete surface. In accordance with GOST 25471-82 [1], to determine the drop strength, as well as for iron ores, sinter and pellets, a triple drop from a height of 2 meters is used. According to the UK standards, 20 kg of briquette samples are dropped from a height of 2 meters four times and the content of the fraction smaller than 5 mm is determined. In practice, a modification of this method is used, which assumes at least 4–5 drops, and the strength of briquettes for blast furnace production is considered acceptable if as a result of such a test no more than 5–10% of fines (pieces with dimensions less than 5 mm) are formed. This number of drops, obviously, does not fully reflect the frequency of possible shock effects on the briquette in the conditions of a metallurgical enterprise. Depending on the configuration of the briquette factories and their location relative to the blast-furnace and steel-making shops, the number of dumps may exceed 15–20. In addition, a significant effect on the destruction of a briquette during impact depends on the shape of the briquette, the angle of its incidence on the surface, etc. GOST 2787-75 [2] defines a method for determining the crumbling of briquettes for steelmaking, made from secondary metals, by dropping three times from a height of 1.5 m onto a metal or concrete slab, however they should not crumble by more than 10%. Of the five briquettes dropped, at least four must pass the test. If the test results are unsatisfactory, out of 10 re-dumped briquettes, eight should pass the test.

It should be kept in mind that the mass (or volume) of the briquette has a significant effect on the nature of the destruction of the briquette during such a test. When falling from a height, briquettes of larger volume and mass are characterized by higher values of kinetic energy. It is clear that the least dense part of the briquette undergoes destruction, and the degree of destruction is proportional to the number of such inhomogeneities in the structure, which obviously increases with an increase in the volume of the test sample. The probability of briquette destruction increases exponentially with the growth of its volume. The experience of testing the impact strength when dropping briquettes shows that the destruction of suitable samples occurs mainly with the formation of one large piece (half a briquette in size), 2–3 pieces of much smaller size (grain size 5–15 mm) and fines (less than 5.0 mm).

The impact strength of the briquette is also determined simultaneously with its abrasion in the so-called "drum" sample in accordance with GOST 15137-77 "Iron and manganese ores, sinter and pellets. Method for Determining Strength in a Rotating Drum". The method is based on mechanical processing of briquettes in a rotating steel drum (diameter 1000 mm, length, 500 mm, 200 revolutions) and subsequent determination by sieve analysis of changes in the grain size distribution of the sample (with dimensions less than 0.5 mm, from 0.5 mm to 6.3 mm and over 6.3 mm), which characterizes the ability of briquettes to resist impact and abrasion during transportation and overloading. The relative proportion of the fraction with particle sizes less than 0.5 mm characterizes the resistance of the briquette to abrasion, and the proportion of the coarse fraction (+6.3 mm) characterizes its impact strength. A similar test procedure for iron ores, sinter and pellets is provided by ISO 3271: 2015 and the Indian standard IS 6495: 2003. The quality of the sinter is considered satisfactory if its impact strength according to this method is not lower than 70%. For briquettes, the permissible impact strength values start from 60%, and the abrasion resistance should be no more than 15%. For pellets, impact strength in a drum test exceeding 90% is considered optimal, and abrasion resistance is not higher than 5%. Such differences in the rejection limits of strength of various agglomerated products are associated with the difference in their physical and mechanical properties and structure. Pellets (size 5–25 mm) and sinter (size 5–40 mm) are rather inhomogeneous in particle size distribution and have high porosity and fracturing. As products of granulation and firing, they contain components of different physical and mechanical properties. The small volume (mass) of the pellet determines its high resistance to loads during drum tests. Therefore, for pellets, high critical values of impact strength and abrasion are accepted. Sinter, on the other hand, is a less durable porous material with a spongy structure, which leads to lower values of the rejection limit. Unlike pellets and sinter, briquettes are homogeneous in size and shape agglomerated product with a higher density, which significantly changes the nature of briquette destruction when tested in a drum according to the method adopted for pellets or sinter. Briquettes are heavier than pellets or sinter and will therefore be more destructive in a rotating drum. Since a standard drum test assumes testing 15 kg of agglomerated product, it is clear that both the number of briquettes loaded into the drum and the volume occupied by them are significantly less than when testing the pellets, which leads to more intensive destruction of the briquettes.

The presence of such large differences in the nature of the destructive loads during drum tests of pellets, sinter and briquettes requires an adjustment of the test procedure to bring it in accordance with the conditions of real production. Methods for showing the strength in a rotating drum, developed to determine the resistance to shock-abrasion loads for pellets and sinter, do not allow adequate and objective assessment of the strength properties of briquettes.

The compressive strength of the briquette determines the ability of the briquette to maintain its integrity under load during storage in a stack and its reaction to the pressure of the overlying layers of the charge in furnaces before softening as a result of reduction processes. With regard to pellets, such a test is carried out in accordance with GOST 24765-81 "Iron ore pellets. Compressive Strength Determination Method". The method assumes the presence of a special machine, including a device for creating a compressive load (with registration of its magnitude) and satisfying the following requirements: the working parts of the sample holder plates, between which the pellets are placed during testing, must be flat, made of hardened steel and installed in mutually parallel planes, the speed of movement of the compression support should be at least 5 and not more than 75 mm/min, the maximum compressive load on the pellets should be 500 kg. This GOST corresponds to the ISO 4700: 2015 standard.

For blast furnace processing to be considered permissible, the compressive strength of the pellet should be in excess of 150 kgF per pellet [3]. The choice of this dimension of the strength of the pellet is due to the sphericity of its shape, which presupposes the point-like application of the crushing force. The compressive strength of the briquette is measured in kgF/cm^2 (or in MPa), since the applied crushing force is distributed over a part of the briquette surface and can vary depending on the shape of the briquette and the way the load is applied. For cylindrical briquettes, it is necessary to separately consider the options for applying a crushing load in a testing machine to a load standing vertically and on a briquette lying on at the side surface. In the first case, the axial compression strength of the briquette is measured, in the second—the tensile strength during splitting [4–5]. Compressive strength is determined from the ratio $R = F/A$, tensile splitting strength is described by the formula $R = 2F/\pi A$, where F is the compressive force, A is the area of the working section of the sample.

The acceptable compressive strength values for briquettes depend on the type of metallurgical furnaces and reactors. The internal standards of some large iron producers adopted a compressive strength of 6.0 MPa, as, for example, defined in TU 0320-007-55978394-03 "Metallurgical briquettes for blast-furnace processing" of Tulachermet Company [6]. For blast furnace briquettes produced in Sweden, the lower allowable compressive strength is 5.8–6.0 MPa. This strength is approximately sufficient for the briquette to withstand the pressure of the overlying layers of the charge with a height of 30 meters or more.

A number of publications indicate overestimated criteria for the compressive strength for blast-furnace briquettes of 150 kgF/cm² and higher, which is a consequence of an incorrect analogy with the abovementioned minimum pellet compressive strength (150 kgF per pellet). The inaccurracy is associated not only with the different dimensions of the compressive strength of the pellet and briquette, but also with the difference in the structure of these agglomerated materials. The experience of operating blast furnaces with briquettes shows that for some compositions of briquettes, the compressive strength did not exceed 35 kgF/cm² [7–8] and their use in the charge did not lead to an increase in dust generation.

When choosing the value of its compressive strength as a criterion for the suitability of a briquette, it should be kept in mind that there is no direct correlation between the property of a briquette to resist compressive force and its impact strength. A briquette with high values of compressive strength may not withstand the rejection limit for impact strength during dropping, and, conversely, briquettes that are easily crushed can form insignificant amounts of fine fractions when dropping a large height.

It is clear that there is no single and universal criterion for the mechanical strength of a briquette. The various options for the manifestation of its strength properties considered above show that the determination of the suitability of the briquette should be carried out exclusively taking into account the features of the logistics of its delivery to the loading devices at specific metallurgical enterprises.

Due to the fact that the destruction of briquettes is mainly due to the impact of shock (dynamic) and crushing (static) loads, the determination of the impact strength and compressive strength makes it possible to reliably assess the suitability of the briquettes. Partially destroyed briquettes are already subjected to abrasion.

The growing spread of briquette technology in the world of ferrous metallurgy requires the creation and unification of methods for testing briquettes, taking into account the characters of production (absence of high-temperature processing), structure (presence of a binder) and properties of such agglomerated materials. Until now, the determination, in particular, of the strength properties of blast furnace, steelmaking and ferroalloy briquettes is carried out using the methods adopted for sinter, pellets and coke, often without any correction of test conditions and rejection limits. Empirical data are used based on experience accumulated mainly in the 20th century and for conditions required at present. As a result, the conclusion about the acceptability of certain types of briquettes as components of charges for metallurgical furnaces and reactors is deprived of scientific validity and objectivity.

The mechanical transfer of test methods for sinter, pellets and coke to determine the strength of briquettes leads to the spread and rooting of unfounded ideas about the need to follow the criteria for the mechanical strength of hardened pellets. As noted earlier, the pellets must have a compressive strength of at least 150 kgF per pellet. Despite the indicated differences in the physical nature of the manifestation of the mechanical strength of the pellet and the briquette in compression, in practice, when forming technical specifications for the design of briquette factories as part of blast furnace shops, a value of 150 kgF/cm² and higher is often indicated as a rejection limit of compressive strength.

For the first time, the question of testing briquettes for mechanical strength from the standpoint of the validity of using the techniques adopted for sinter and pellets was discussed in detail and comprehensively in [9]. On the basis of a comparative assessment of the methods of testing for strength in compression, for dropping and a drum test, the expediency of a significant adjustment of the methods of conducting such tests on briquettes, the need to adjust the rejection limits depending on the shape and size of briquettes was substantiated. Generally agreeing with the conclusions of the authors, we note that most of the conclusions are based on speculative conclusions and refer mainly to roll briquettes, which in no way diminishes the importance of the conclusions made in the article.

In [10], on the basis of numerical modeling, a comparison was made of the behavior of sinter, pellets and briquettes during the above kinds of testing of agglomerated materials—compressive strength, drum test and drop strength.

To assess the stress-strain state and the level of contact stresses on the surface of pellets and briquettes (using the example of brex) in the backfill, the following model should be considered. The pellets are presented in the form of spheres with a diameter of 11 mm, briquettes have cylindrical surfaces with a diameter of 30 mm and a length of 60 mm. The mechanical characteristics of the material of the pellets and brex are presented in Table 5.1. For the pellets, Young's modulus was 64.1 GPa (Young's modulus of hematite 359 MPa, porosity 0.35, morphological factor 4), for brex 55.4 GPa (Young's modulus of magnetite 175 MPa, porosity 0.25, morphological factor 4). The calculation was carried out on based on the results of work [11–12]. The value of the porosity of the pellets is tabular, the porosity of the briquettes was measured using the Stiman method [13]. Poisson's ratio—0.2 for hematite and magnetite is selected on the basis of the results of work [14].

From the total mass of the backfill, a representative volume is extracted, including a repetitive sequence of particles. Corresponding kinematic boundary conditions of symmetry are determined along the boundaries of the cut of the representative volume of the backfill. An external pressure is applied to the upper surface, which is equivalent to the pressure created by the 20 meter high particle backfill. Taking into account the accepted value of the bulk density of 2200 kg/m³, the value of the external distributed pressure on the representative volume of particles was 440 kPa for 20 m, respectively.

The computational model is implemented on the basis of the finite element method using an implicit integration scheme for the equations of motion in time. Two-way contact interactions are determined between the particles. The normal behavior of the contacting surfaces is described by a penalty model with the formation of contact stiffness based on the stiffness of the contacting finite elements. The tangential component of contact interactions is represented by the Coulomb friction model with a friction coefficient of 0.3. The adaptation of the finite element mesh for the zones of contact interactions has been carried out (Fig. 5.1).

It is advisable to assess the stress-strain state for a central pellet or briquette surrounded by neighboring particles and located far from the boundary conditions (Fig. 5.2).

The distributions of von Mises stresses, contact pressures and nodal contact forces in the pellets at a filling height of 20 m (Fig. 5.3) are given below.

The von Mises stress distributions, contact pressures and nodal contact forces in briquettes at a filling height of 20 m are shown in Fig. 5.4.

Table 5.1. Mechanical characteristics of materials.

Material	Young's modulus, MPa	Poisson's ratio
Pellet	64 100	0.2
Briquette	55 400	0.2

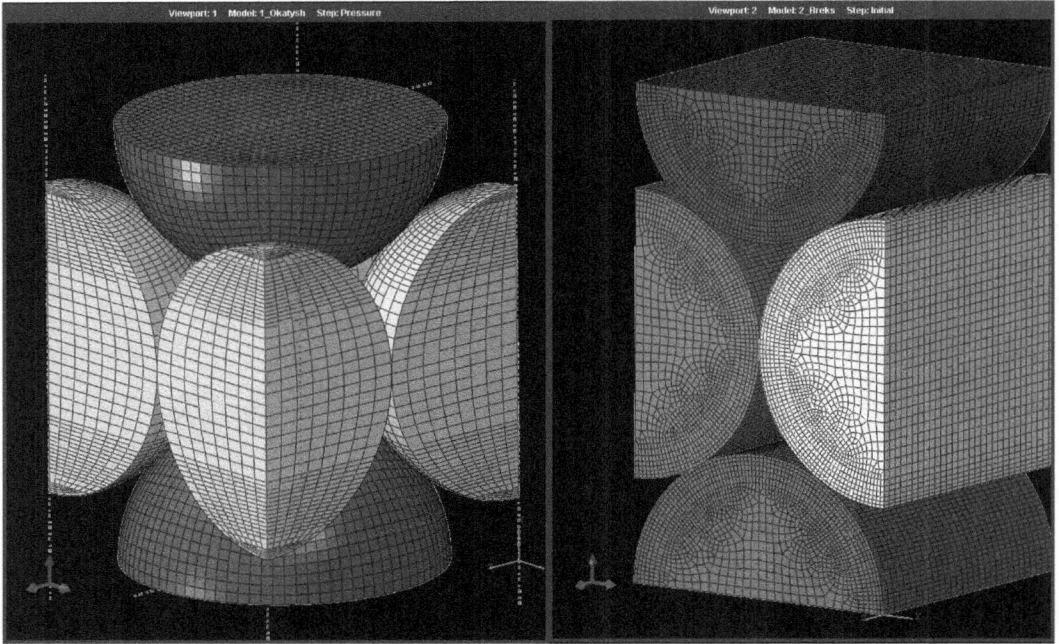

Figure 5.1. Finite element models of pellets (left) and briquettes (right).

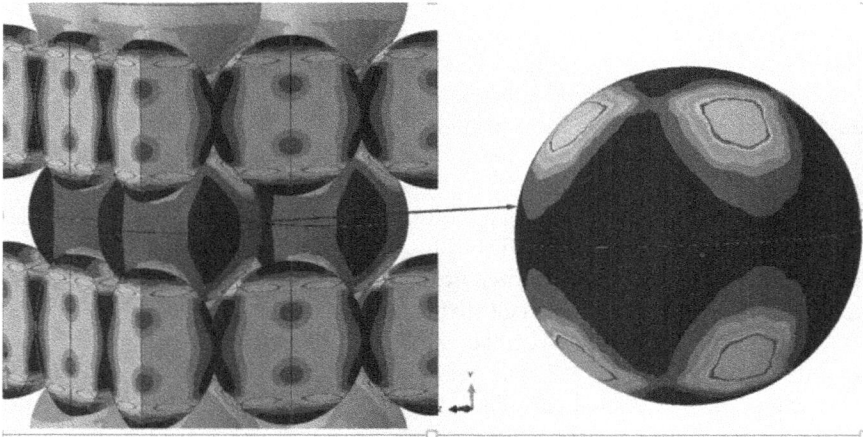

Figure 5.2. Distribution of stresses in the backfill.

The simulation results are shown in Table 5.2. The level of the maximum effective stresses in the backfill for the pellets does not exceed 11 MPa at a backfill height of 10 m and 20.2 MPa at 20 m, for briquettes 3.6 MPa at 10 m and 7.2 MPa at 20 m, which is significantly lower than the ultimate strength of the material. The relatively high values of stresses in the pellets are explained by the pronounced local nature of the contact effects from the overlying layers. Each of the pellets has eight local point contact zones. Briquettes in a stable position have a wider contact zone along the generatrix of the cylinder, which leads to a decrease in stress concentration.

It can also be seen that the general level of mechanical stresses far from the contact areas both in the pellets and in the briquettes has a fairly similar value, which is in the range of 1–3 MPa.

Thus, the maximum values of stresses arising from the action of the overlying layers of the charge on the briquette significantly differ from the effect experienced by the pellet, and are in the range close to the values of the compressive strength of industrially produced briquettes.

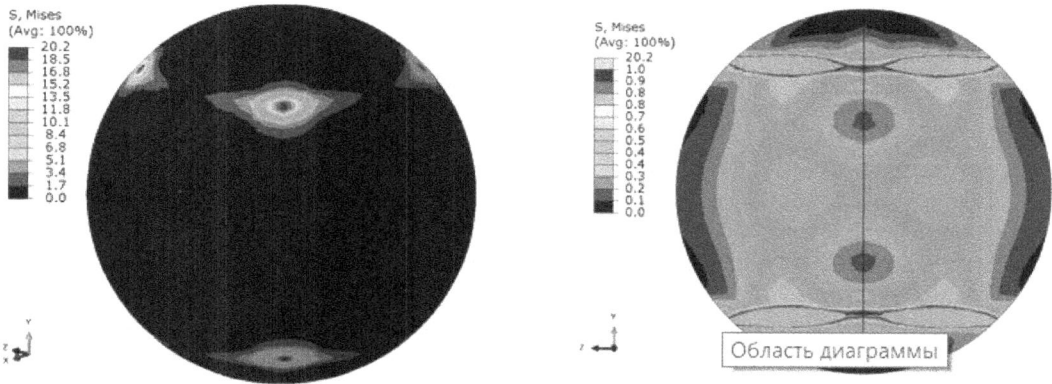

Figure 5.3. Distribution of stresses in the pellet at a filling height of 20 m. Left: stress distribution without palette limits; Right: the distribution of stresses in the section and the application of the limit of 1 MPa.

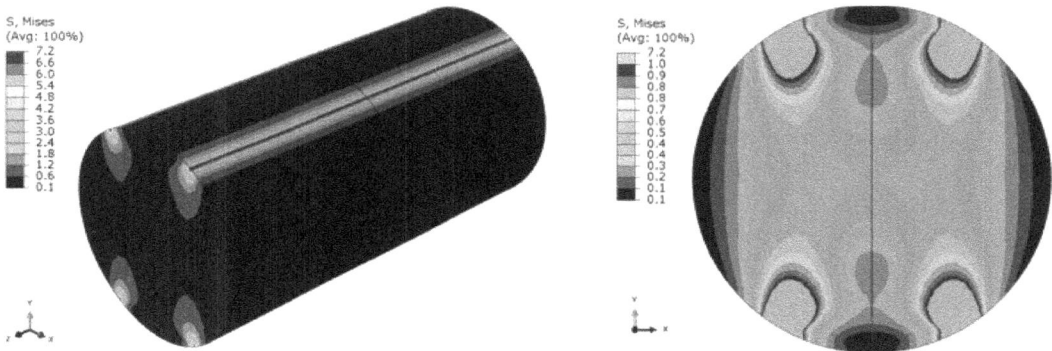

Figure 5.4. Distribution of stresses in the briquette at a filling height of 20 m. Left: stress distribution without palette limits; Right: stress distribution with 1 MPa constraint applied (end view).

Table 5.2. Mechanical characteristics of the material.

	Backfill height, m	Maximum stresses, MPa	Maximum contact pressure, MPa
Pellets	10	11	24.6
	20	20.2	41.3
Briquettes	10	3.6	5.5
	20	7.2	10.8

The rotary steel drum test methods adopted for sinter and pellets are also often used to assess the acceptability of the strength of blast-furnace briquettes. This approach does not take into account the significant differences in the structure of such lumpy products. The small volume (weight) of the pellets determines their resistance to loads during a drum test and high values of the rejection limits (the proportion of classes with particles less than 0.5 mm in size is not more than 4–6%, the proportion of classes of particles larger than 5 mm is not less than 90–95%). Sinter is a more easily destructible porous spongy material. Accordingly, the quality of the sinter is considered acceptable if, after testing, at least 55–65% of the +5 mm class and 6–8% of the –0.5 mm class are formed. Unlike pellets and sinter, briquettes are lumpy products that are homogeneous in properties, size and shape with a density and characteristic dimensions that exceed the density and characteristic dimensions of pellets and sinter. This circumstance, first of all, determines the specificity of the destruction of the briquette when tested in the drum. Briquettes are heavier than pellets and sinter, so they will experience a stronger impact resistance. Moreover, in the accepted test methods for drums, the number of materials is 15 kg. The amount of loaded briquettes and their volume are significantly less than when testing the pellets. Each briquette, when tested in a drum, experiences the same level of breaking loads, which significantly exceeds the effect on pellets or sinter, which have a wider range of sizes and less weight. This combination of conditions leads to more intensive destruction of briquettes and a decrease in the yield of a class of particles larger than 5 mm.

On the basis of the SIMULIA Abaqus software complex, a digital model has been created, which makes it possible to study the movement of particles during the drum test.

The drum sample is a method for assessing the strength of sinter and pellets, which consists in mechanical processing of particles in a rotating steel drum and subsequent determination by sieve analysis of the change in the particle size distribution of the sample, which characterizes the ability of the ore to resist impact and abrasion. A full-scale experiment is regulated by GOST 15137-77. A sample weighing 15 kg is loaded into a drum with a diameter of 1000 mm and a length of 500 mm. The rotation of the drum at a frequency of 25 rpm is determined. On the inner surface of the drum, along the entire generatrix of the cylinder in the longitudinal direction, two equally spaced steel corners $50 \times 50 \times 5$ mm in size are welded.

According to the dimensions presented in GOST 15137-77, a geometric model of the inner surface of the drum was built. Its movement is completely determined by a control point set at the center of mass of the geometric model.

Two standard sizes of agglomerates were investigated: spherical pellets with a diameter of 15 mm and cylindrical brex with a length of 75 mm and a diameter of 25 mm. The cylindrical shape of the briquettes was replaced by a spherical one with a diameter of 41.3 mm with the condition of the equivalence of the sample mass. The number of pellets in a sample of 15 kg was 2360 pieces, briquettes – 90 pieces. The density of pellets and briquettes, taken in the calculations, was 3.6 kg/m^3 and 4.5 kg/m^3, respectively.

The particles were modeled using the discrete element method, which reduces the description of body motion to the motion of a material point located at the center of mass. The contact interaction is determined between the particle spheres and the drum surface using the Hertz model. Particles are placed inside the drum to determine the force of gravity. Next the rotation of the drum in the gravity field with a frequency of 2.618 rad/s is determined, which corresponds to 25 rpm specified in GOST.

Figure 5.5–5.8 shows the change in the position of the particles (pellets on the left and briquettes on the right) during the drum test:

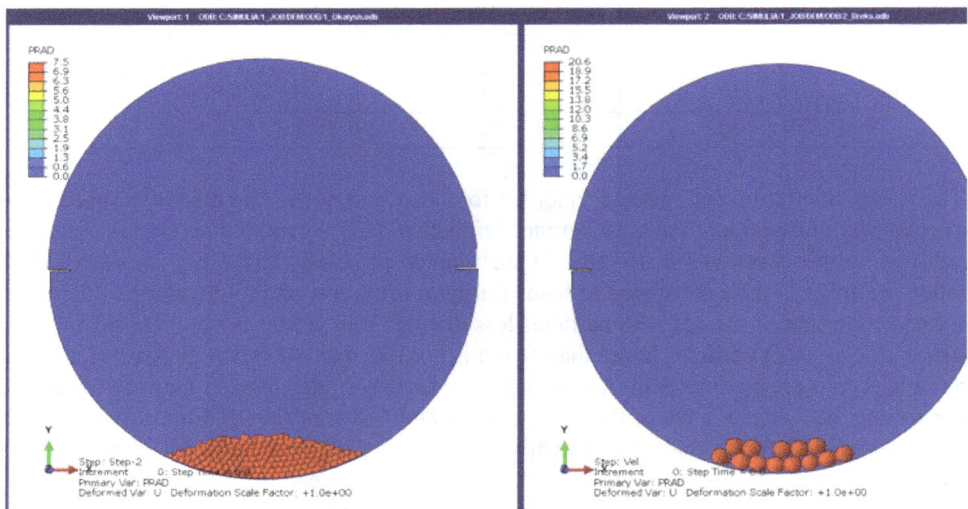

Figure 5.5. The position of the particles under the action of gravity at the initial moment of time.

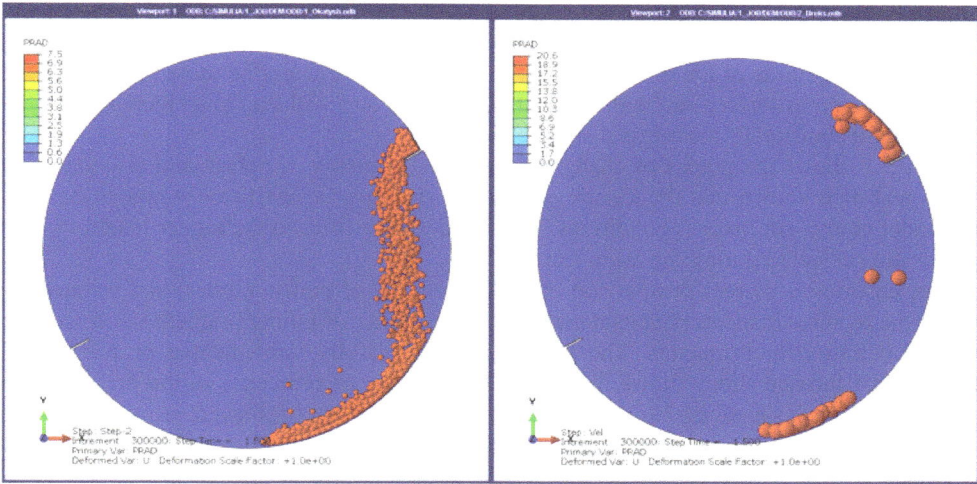

Figure 5.6. Particle position: 1.5 seconds rotation.

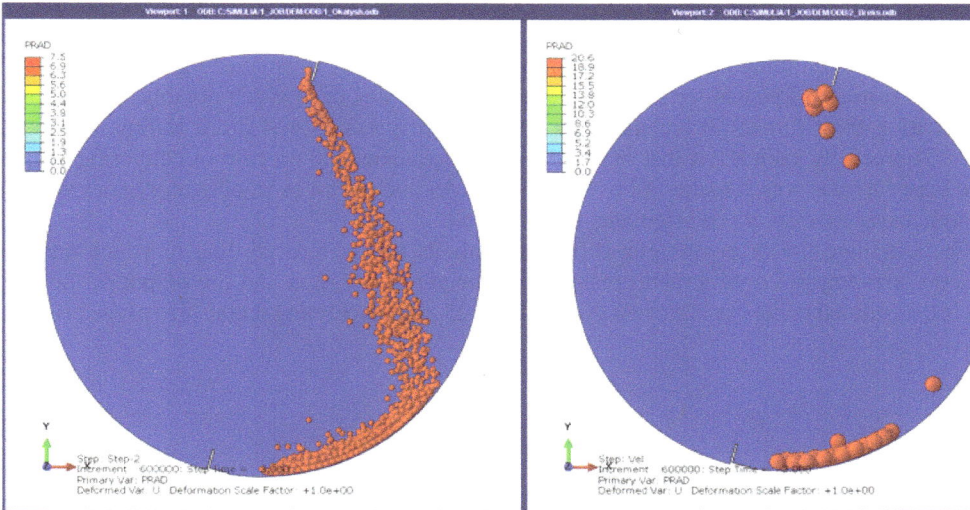

Figure 5.7. Particle position: 3 seconds rotation.

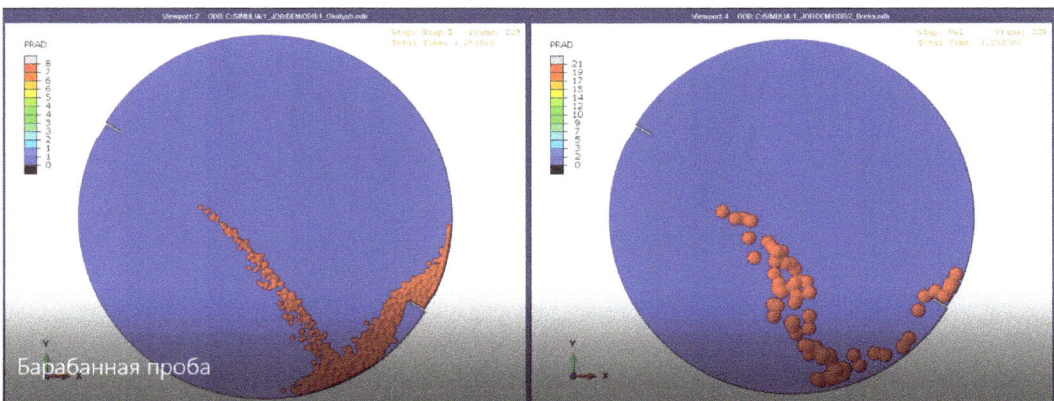

Figure 5.8. Particle position: 9 seconds rotation.

It can be seen that the briquettes fall mainly on the steel surface of the drum, while the pellets fall on the "cushion" of the earlier fallen pellets. It is convenient to evaluate the level of impact effects by the magnitude of kinetic energy. The higher this parameter, the more force the particles contact with the drum surface and with each other, which in turn will lead to an increase in the level of grinding and abrasion of the original sample.

Figure 5.9 shows the graphs of changes in the kinetic energy of pellets and briquettes during the drum test. It is shown that the kinetic energy of briquettes is significantly higher than that of pellets under equivalent process conditions. This condition testifies to the higher load to which the briquettes are subjected during the drum test.

Thus, ensuring a correct comparison of the test results of pellets (sinter) and briquettes in a drum sample requires mandatory consideration of the specifics of falling briquettes and "softening" the rejection limits for briquettes. The practice of successfully implemented briquette projects as part of blast furnace shops shows that ensuring the share of briquettes in the ore part of blast furnace charges at the level of 70–100 kg per ton of pig iron was achieved using briquettes that showed strength in the drum sample at the level of 70–75% (the share of particles with a size above 5 (6.3) mm).

In the process of transporting briquettes to the blast furnace, they can be repeatedly dropped from a great height, which makes it important to impart impact strength to them to a degree significantly exceeding the impact level in a rotating drum, and to carry out corresponding tests for strength during dropping. Changing the logistics of delivering briquettes to the blast furnace to prevent multiple dumping can be impossible or extremely costly in an existing plant. Recently, the method of testing briquettes for strength during dropping has been given notably less importance in evaluating blast-furnace briquettes than tests for compression and drum testing. At the same time, it is precisely such a test that most adequately simulates the process of delivering briquettes to the blast furnace, which makes it possible to consider such a test as an obligatory component of a comprehensive assessment of briquettes for strength.

To determine the drop strength, different methods are used, according to which batches of briquettes are dropped onto a metal plate from a height of 1.5–2 m and the output of the formed fines is determined (fineness class less than 5, 10 or 25 mm, depending on the size of the briquettes).

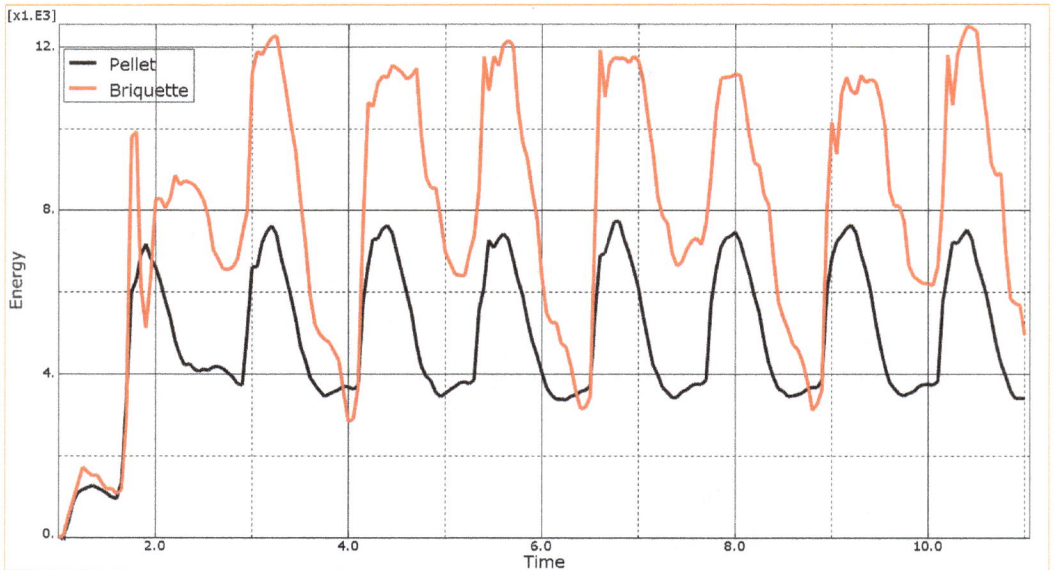

Figure 5.9. Change in kinetic energy of pellets (black line) and briquettes (red line) drum sample.

Large briquettes (up to 100 mm in size) are recommended to be discarded only 1–2 times, and small briquettes (~ 25–30 mm)—from 3 to 5 times or more. It is believed that briquettes meet the drop strength requirements if the number of fines formed does not exceed 5–10 (or even 15%, [9]). This means that the transportation of large briquettes can only be ensured under conditions of "sparing" logistics, which excludes multiple dumps.

In order to study the characteristics of the behavior of cylindrical briquettes during drop tests, we performed modeling within the framework of an elastoplastic model of the behavior of a briquette with the possibility of its destruction within the framework of the SIMULIA Abaqus software package. Further, various options for the fall of a cylindrical briquette are considered during strength tests when dropping (falling on a side surface; on an end surface and on a face).

The physical and mechanical parameters of the material were obtained as a result of processing the experimental data of the samples of experimental briquettes. In a physical experiment on a tabletop single-column electromechanical testing machine Instron 3345 (USA) with a loading capacity of 5 kN, the following material properties were determined: Young's modulus—123 GPa, Poisson's ratio 0.3; tensile strength 2.1 MPa. The simulation results are shown in Figs. 5.10–5.11. To compare the values of the damage parameter, the results for the case of falling briquettes from a height of 2 meters were also presented. Damage parameter is understood as a conditional value characterizing the degree of material damage (0 for intact material and 1 for completely destroyed material).

In the above calculated cases in the body of the briquette, there are areas of irreversible (inelastic) deformations in the zone of contact interactions.

When falling from a height of 2 m, the destruction of thin near-surface layers of the briquette is observed in almost all calculated cases. It is worth noting that when the briquette falls on the butt end (Fig. 5.11), there is a localization of the damage parameter in the center of the briquette (without destruction). This fact is explained by the interference of compression-tension waves in the model upon impact. These results complement the data obtained in [15] on the options for the destruction of roll briquettes.

The results obtained form the basis for revising the acceptability of standard test methods for the strength of sinter and pellets without adjusting the test conditions and rejection limits for the strength of briquettes. When determining the requirements for the compressive strength of briquettes,

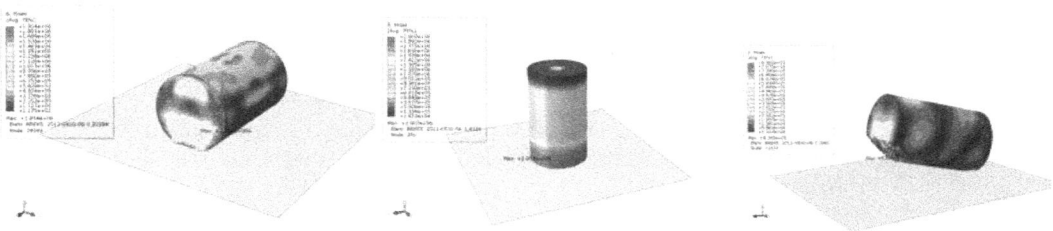

Figure 5.10. Distribution of von Mises equivalent stresses in a briquette. Impact type: left - lateral surface; in the center - butt; on the right is a face. Fall height - 2 m.

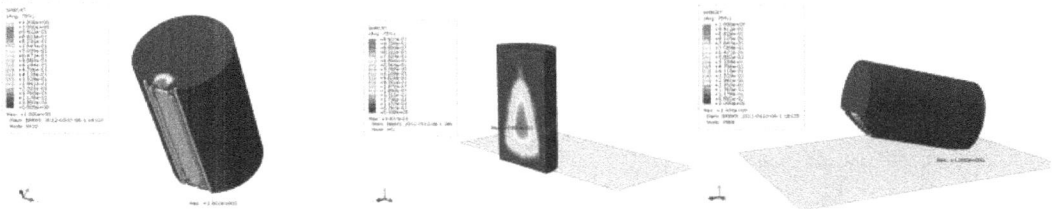

Figure 5.11. Distribution of the damage parameter in the briquette Impact type: left - lateral surface; in the center - butt; on the right is a face. Fall height - 2 m.

it is necessary to take into account the appearance in the body of the briquette in the backfill of a noticeably lower stress in comparison with the pellet, which makes it unreasonable to ensure its value at a level of at least 150–250 kgF/cm². When carrying out a drum test, one should keep in mind the fundamental difference in the behavior of briquettes and pellets due to the difference in the mass and quantity of such agglomerated products, leading to a notably greater impact on the briquette. The rejection limits of a drum sample for a briquette should be revised, or the conditions for conducting a drum test should be changed. To carry out an adequate assessment of the strength of the briquette, reflecting the specifics of its delivery to the furnace, a drop strength test should be provided in the complex of test methods. Improvement of this technique involves a deeper study of the briquette destruction mechanism. The influence of the parameters of ascending gas flows on the strength of briquettes deserves a separate study.

The specification of proposals based on the results obtained and the extension of the conclusions made to other types of briquettes (steel-making, ferroalloy and briquettes for direct iron reactors) will be the subject of future publications.

References

[1] GOST 25471-82 Iron ores, agglomerates and pellets. Method for determining drop strength.
[2] GOST 2787-75 Secondary ferrous metals. General technical conditions.
[3] Geerdes, M., Chaigneau, R., Kurunov, I., Lingiardi, O. and Rikkets, J. 2015. Modern Blast Furnace Ironmaking. IOS Pres BV, Amsterdam.
[4] GOST 10180-90 Concrete. Methods for determining the strength of control samples.
[5] GOST 28570-90. Concrete. Methods for determining strength by samples taken from structures.
[6] Kotenev, V.I., Barsukova, E.Yu. and Murat, S.G. 2005. Production and use of metallurgical briquettes at JSC Tulachermet // Metallurg.Special edition. 6: 33–36.
[7] Titov, V.V., Murat, S.G. and Kiselev, N.I. 2007. Ecology and Industry 1: 16–21 (in Russian).
[8] Kurunov, I.F., Shcheglov, E.M., Kononov, A.I., Bolshakova, O.G. and other. 2008. Research of metallurgical properties of briquettes from technogenic and natural raw materials and assessment of the effectiveness of their use in blast-furnace smelting. Part 1 // Bul. Scientific and Technical and Economical. Inf. "Ferrous metallurgy". Bul. NTIEI 1: 8–16.
[9] Eremin, A.Ya., Babanin, V.I. and Kozlova, S.Ya. 2003. On the formation of requirements for the indicators of mechanical strength of briquettes with a binder. Metallurg 11: 32 (in Russian).
[10] Bizhanov, A.M. and Zagainov, S.A. 2021. Test of briquettes for mechanical strength. Metallurg 3: 11–18 (in Russian).
[11] Sundström, B. 1998. Handbok och formelsamling i hållfasthetslära, KTH, Södertälje.
[12] Bruno, G., Efremov, A.M., Levandovskyi, A.N. and Clausen, B. 2011. Connecting the macro- and microstrain responses in technical porous ceramics: modeling and experimental validations. Journal of Material Science 46: 161–173.
[13] Sokolov, V.N., Yurkovets, D.I. and Razgulina, O.V. 1997. Determination of tortuosity coefficient of pore channels by computer analysis of SEM images. Proceedings of Russian Academy of Sciences, Physical Series 61(10): 1898–1902 (in Russian).
[14] Chicot, D. et al. 2011. Mechanical properties of magnetite (Fe_3O_4), hematite (α-Fe_2O_3) and goethite (α-FeO·OH) by instrumented indentation and molecular dynamics analysis. Materials Chemistry and Physics 129.3: 862–870.
[15] Noskov, V.A. and Petrenko, V.I. 2000. Analytical study of the distribution of compaction of the briquette in the forming elements of the roller press. Metallurgical and Mining Industry (Ukraine) 1: 95–98 (In Russian).

5.2 Reducibility

Reducibility is a value characterizing the ability of iron oxides of iron ore materials to give oxygen to a reducing gas. The reducibility of iron ore material depends on its structure, specific pore surface and mineralogical composition. To determine the reducibility of briquettes, one can use a standard technique (GOST 28657-90 Iron ores Method for determining reducibility, [1]), in accordance with which a sample of a briquette of a certain class is reduced in a fixed bed at a temperature of 950°C in an atmosphere of a reducing gas containing CO and N_2, weighing the sample at certain points in time, calculating the degree of reduction relative to iron (III) and calculating the degree of reduction when the oxygen/iron ratio is 0.9. The weighed portion is placed in the recovery tube so that the surface is flat, the cover of the recovery tube is closed and placed in the oven, suspended in the center above the balance. Then, an inert gas flow is passed through the reducing tube at a rate of about 25 dm³/min and heating is started. When the temperature of the sample reaches 950°C, the flow rate of the inert gas is increased to 50 dm³/min and heating is continued, maintaining the flow of inert gas until the weight of the sample and the temperature are constant. A reducing gas is introduced to replace the inert gas at a rate of 50 dm³/min. The weight of the sample is recorded approximately every 3 minutes in the first 15 minutes, and then every 10 minutes. The recovery is stopped when the oxygen loss is 65%.

The degree of recovery after a lapse of time (R_t) relative to iron (III) in percent is calculated by the formula:

$$R_t = \left[\frac{0.111 W_1}{0.430 W_2} + \frac{m_1 - m_2}{m_0 * 0.430 W_2} * 100 \right] 100 \qquad (1)$$

where: m_0 is the mass of the sample

m_1 - weight of the sample immediately before the start of recovery, g

m_2 - weight of the sample after 4 hours of restoration

W_1 - mass fraction of iron (II) oxide in the control sample before the study and calculated by the mass fraction of iron (II) by multiplying it by a factor of 1.286

W_2 - mass fraction of total iron (II) in the control sample before the study, %

The gas permeability of the layer and shrinkage can be determined in accordance with GOST 21707-76 "Iron ores, agglomerates and pellets. Method for determining gas permeability and shrinkage of a layer during reduction" [2].

To determine the reducibility using the reducibility index according to the ISO 4695: 2015 standard, a 500 grams briquette sample is heated in an inert atmosphere up to 950°C, then its isothermal reduction begins in an atmosphere of gases consisting of CO (40%) and N (60%) N_2. The gas mixture consumption is 50 liters per minute. The tests are continued until the weight loss of the sample stops. Then the sample is cooled to room temperature in an inert atmosphere. The degree of recovery is determined by the formula (1). According to the ISO 11258: 2015 standard, it is possible to determine the degree of reducibility, reduction and metallization of briquettes, which are components of the charge of direct reduction reactors.

The hot strength of the briquettes after low-temperature reduction in a rotating drum is then determined according to ISO 4696-1: 2015 and ISO 4696-2: 2015. According to ISO 4696-1: 2015 [3], a test briquette sample is dried in an oven at a temperature of 105°C ± 5°C for two hours. Further, before testing, it is cooled to room temperature. In accordance with the standard, the size of the pieces of briquettes is from 10 to 12.5 mm.

The test sample (500 g) is placed on porcelain pellets in a reduction tube. A thermocouple is placed in the center of the test sample. The tube is filled with an inert gas (argon flow rate 20 l/min), and the test sample is heated until 500°C. Next, the sample is purged with a reducing gas for 1 hour (composition: CO - 20%, CO_2 - 20%, H_2 - 2%, N_2 - 58%) with a flow rate of 20 l/min, and then cooled to a temperature below 100°C with an inert flow gas.

The test sample is removed from the reduction tube, its mass (m_o) is determined and placed in a tumbling drum, in which the sample is subjected to destructive loads by rotating the drum at a speed of 30 rpm for 10 minutes. After that, the sample is removed from the drum and sieved by hand on sieves with a mesh diameter of 6.3 mm; 3.15 mm and 500 microns. The strength is determined from the results of at least two tests. The hot strength of the pellets measured in this way is considered acceptable if the yield of the + 6.3 mm fraction exceeds 80%. The proportion of the fraction with a particle size of less than 0.5 mm should not exceed 10%. The acceptance criterion for the agglomerate is the yield of the fraction –3.15 mm not exceeding 20%.

The Hot Strength Index (Reconstitution Disintegration Index) RDI-1 is calculated as a quantitative measure of the degree of crushing of a material after reconstitution and mixing. Material larger than 6.3 mm and less than 500 μm refers to the test batch after reconstitution and before mixing.

Thus, the hot strength index RDI-1 is calculated from the following equalities:

$$RDI-1 = \frac{m_1}{m_O}100 \tag{2}$$

$$RDI-1_{-3,15} = \frac{m_O-(m_1+m_2)}{m_O}100 \tag{3}$$

$$RDI-1_{-0,5} = \frac{m_O-(m_1+m_2+m_3)}{m_O}100 \tag{4}$$

where m_O is the mass of the test sample after recovery, but before rotation in the tumbling drum, g

m_1 - weight of the oversize fraction remaining on a sieve with a cell of 6.3 mm, g

m_2 - weight of the oversize fraction remaining on a sieve with a mesh of 3.15 mm, g

m_3 - weight of the oversize fraction remaining on a sieve with a cell of 500 μm, g

According to ISO 4696-2: 2015 [4], preparation of briquette samples is carried out in the same way as in the first part of the standard, with the only difference that the size of briquette samples is 16 to 20 mm. Heating the sample to a temperature of 550°C is carried out in an inert atmosphere at a gas flow rate of 15 l/min. After reaching the specified temperature, the sample is purged with a reducing gas (CO - 30%, N_2 - 70%) for 30 minutes at a flow rate of 15 l/min, and then cooled with an inert gas to a temperature of less than 100°C. The cooled sample is subjected to destructive loads for 30 minutes in a drum rotating at a speed of 30 rpm. After removing the material from the drum, it is sieved by hand on a 2.8 mm sieve. At least two tests are carried out on each sample. The hot strength index RDI - 2.$_{2,8}$, expressed as a percentage of mass, was calculated from the following equation:

$$RDI-2_{-2,8} = 100 - \frac{m_1}{m_O}100 \tag{5}$$

where m_O is the mass of the test portion after reconstitution and before rotation in the drum, g

m_1 is the mass of the oversize product; retained on a sieve 2.8 mm, g

The ISO 4697 standard also serves to define rotary drum low temperature recovery. The method is based on the reduction of a sample at a temperature of 500°C in an atmosphere of CO, CO_2, H_2 and N_2 with a gas flow rate of 20 liters per minute, it is further cooled an hour later to a temperature below 100°C and sieving into fractions +6.30 mm, + 3.15 mm and – 500 microns. The dynamic index of recoverability according to this standard is determined by the ratios:

$$DRDI_{+6.3} = \frac{m_1}{m_0+m_4}*100 \tag{6}$$

$$DRDI_{+3.15} = \frac{m_1+m_2}{m_0+m_4}*100 \tag{7}$$

$$DRDI_{-0.5} = \frac{(m_0+m_1)-(m_1+m_2+m_3)}{m_0+m_4}*100 \tag{8}$$

where m_0 is the mass of the material after reduction, g.

m_1 is the mass of sample particles with dimensions greater than 6.3 mm, g;

m_2 is the mass of sample particles with dimensions greater than 3.15 mm, g;

m_3 is the mass of sample particles with sizes greater than 0.5 mm, g;

m_4 is the mass of material in the dust collector, g.

The ISO 13930 standard describes a methodology for the dynamic determination of the parameters of low-temperature disintegration of iron ores as components of a blast-furnace charge and can be used for testing blast-furnace briquettes. Briquettes are tested in a rotating tube at a temperature of 500°C in an atmosphere of a mixture of gases CO (20%), CO_2 (20%), H_2 (2%) and N_2 (58%). The flow rate of the reducing gas, preheated to 500°C, is maintained at 20 liters per minute. An hour later, the gas supply and the pipe rotation stop, the material is cooled to temperatures below 350°C when an inert gas is fed into the pipe at the same flow rate (20 l/m). Then the gas supply is stopped and the material is cooled to 100°C and dispersed into three fractions in the same way, as well as the method according to the ISO 4696 standard. The indices of low-temperature disintegration are calculated by the formulas:

$$LTD_{+6.3} = \frac{m_1}{m_0} * 100 \tag{9}$$

$$LTD_{-3.15} = \frac{m_0 - (m_1 + m_2)}{m_0} * 100 \tag{10}$$

$$LTD_{-0.5} = \frac{m_0 - (m_1 + m_2 + m_3)}{m_0} * 100 \tag{11}$$

where m_0 is the mass of the material after testing, including the mass of dust in the dust collector, g;
m_1, mass of sample particles with dimensions greater than 6.30 mm, g;
m_2, mass of sample particles with dimensions greater than 3.15 mm, g;
m_3 mass of sample particles with sizes greater than 0.5 mm, g;
$LTD_{+6.3}$ - disintegration strength parameter;
$LTD_{-3.15}$ - disintegration index;
$LTD_{-0.5}$ - disintegration abrasion index.

ISO 7992: 2007 describes a method for testing iron ore materials for blast furnace smelting for load reducibility and can also be used to study the behavior of briquettes under blast furnace conditions. Reducibility under load is determined by applying a static load (50 kPa) to a layer of test materials (1200 g) of a certain size (10–12.5 mm) when heated to 1050°C in an atmosphere of CO (40%), H_2 (2%) and N_2 (58%). During the test, the weight loss of the material layer and the pressure drop of the gas mixture are measured. Gas consumption is maintained at 83 liters per minute. Indicators of recoverability under load are determined by the ratios (9–11).

To test the reducibility of briquettes under load, the Burghardt test (ISO/DOC 3772) can also be used, based on the measurement of gas permeability and shrinkage of a layer of material placed in a heated reaction chamber.

Samples of briquettes are subjected to crushing and a fraction with a grain size of 8–20 mm is selected for testing. The mass of the prepared sample is 1000 g. The sample is tightly packed into a glass, which is placed in the reaction chamber and the load is set to 1.5 kg/cm. The samples are subjected to reduction in an atmosphere of H_2 (with a flow rate of 1 m^3/h) for three hours at a temperature of 900°C and 800°C. During the heating and recovery period, the temperature in the retort, the shrinkage of the layer, and the change in the weight of the sample are continuously recorded. The degree of recovery is calculated by the formula:

$$R = \frac{m_0 - m_f}{m_0(0,429 Fe_{tot} - 0,111 FeO)} \tag{12}$$

where m_0 is the mass of the original sample, g;
m_f is the mass of the recovered sample, g;
Fe_{tot} - mass fraction of total iron in the original sample, %
FeO - mass fraction of iron oxide in the original sample, %.

To study the behavior of the so-called washing briquettes in a blast furnace hearth, the technique described in [5–6] can be used. The technique serves to study the behavior of iron ore materials when heated in a coke layer. The reaction space and at the same time, the heater of the smelting installation is a high-density graphite tube with an inner diameter of 60 mm, placed in the inductor of a high-frequency furnace. The bottom of the reaction space is a grid with 25 holes 5 mm in diameter. The investigated pre-reduced material with a 55–60 mm layer of a 7–12 mm fraction is placed in the reaction space between two layers of coke. A pressure of 1 kg/cm^2 is applied to the material. Preliminary reduction is carried out in a reducing atmosphere of a mixture of gases CO (33%), N$_2$ (65%) and H$_2$ (2%) under pressure per sample (100 kPa) when heated to T = 1050°C for 200 minutes. A sealed chamber is located under the reaction tube, in which there is a graphite receiver in the form of a disk with cells for liquid melting products. During the experiment, the temperature of the reaction space and the amount of layer shrinkage are automatically recorded. The temperature of the beginning of droplet separation is determined visually through the viewing window of the sealed receiver. By the amount of the liquid phase that has flowed into the cells of the receiving disk, one can judge the rate of its drainage through the coke packing. The material from each cell and the remaining melt in the coke layer at a temperature of 1600°C are weighed and separated into slag and metal components, after which they are subjected to chemical analysis.

The reducibility of brex made of the LD sludge and BF flue dust with the addition of the iron ore fines with that of lumpy iron ore (Hematite) from the same deposit can be compared.

Brex were produced by industrial stiff extrusion line manufactured by J.C. Steele & Sons, Inc. and based on the Steele 25A Extruder with productivity 12–15 tons per hour—the smallest among the capacities of Steele Extruders ranges from 12 to 120 tons per hour. The composition of the brex was as follows (wt.%): LD sludge 47.2; BF dust 28.3; Iron ore - 18.9; PC - 4.7 and Bentonite 0.9. The Ordinary Portland cement was used as the binder. Volclay®DC-2 Western (Sodium) Bentonite was used as the binder and plasticizer. The lumpy samples of the same iron ore were used in the testing for comparison.

Typical chemical analysis of the iron ore and the brex components is given in Table 5.3.

Phase analysis is done by X-ray diffractometer. It shows that iron ore is represented by Hematite (minor phases—Goethite, Gibbsite, Kaolin, Pyroxene); BF dust—Magnetite, Hematite, Graphite (Wustite, quartz, Calcium ferrite); LD sludge—Magnetite, Wustite (Wollastonite, Calcite).

Prior to the reduction test the bulk density and apparent porosity by DIN 51067 test procedure were measured. The cold compressive strength of the brex was measured following DIN 51056. These values were measured after 24 hours of curing under the natural conditions and also daily during one week of strengthening. The values are given in Table 5.4.

From Table 5.4, it is apparent that there is the local maximum of the CCS. Such a non-linear behavior can be explained by the evolution of the structures in the brex body bonded by the combination of Portland cement and Bentonite. The effect of the Portland cement and Bentonite combination as a binder and plasticizer should be considered.

Table 5.3. Chemical composition of the substances.

Elements	Iron ore fines	BF dust	LD sludge
Fe$_2$O$_3$/ FeO	78.5	51.5	87.5
SiO$_2$	5.6	6.3	0.6
CaO	-	4.9	9.5
MgO	-	0.2	1.2
Al$_2$O$_3$	5.4	5.1	0.3
TiO$_2$	0.8	-	-
C		30.5	-
Fe$_{total}$	53.5	35.6	64

Table 5.4. Physical properties of brex during strengthening.

	Bulk density (g/cm³)	Apparent porosity %	CCS (kg/cm²)
As such (after 24 hours)	2.42	31.5	24
1 day	2.66	25.4	45
2 days	2.43	32	63
3 days	2.44	27	52
4 days	2.45	27.2	56
5 days	2.45	26.2	57
6 days	2.46	26.8	59
7 days			75
8 days			80

The properties of a mixture to be briquetted that make it suitable for extrusion are most often achieved by the addition of a plasticizing agent, mainly bentonite, to a charge. The fraction of bentonite required to ensure the extrusion of a mixture oscillates in the range 0.25–1.0% of the briquetted material density. An important consequence of the plasticized mixture is an increase in the mechanical strength of brex and a change in the character of its fracture from brittle to viscoplastic fracture.

As an example, the results of axial compression tests of the brex made of a charge of a base composition (%), 67 dust of the aspiration of ferrochrome production, 15 chromium ore concentrate, 15 coal, 3 Portland cement) and the brex of the second series made of the same charge with an addition of 0.5% bentonite that were performed in a laboratory extruder are presented. The axial compression tests of the brex of both series were carried out on a bench-type one-column electromechanical Instron 3345 tensile-testing machine at a load up to 5 kN. Figure 5.12 shows the stress-strain curves that demonstrate the change of the character of fracture of bentonite-containing brex from pronounced brittle to viscoplastic fracture. This change is also illustrated in Fig. 5.13: it is seen that, in the case of fracture of bentonite-containing brex, the compression surfaces of the tensile-testing machine should be much closer than in the case of bentonite-free brex. A comparison of the strengths of the brex of two series showed that the average ultimate compressive strength of the brex made of a bentonite-free mixture is 1.88 MPa and that of bentonite-containing brex is 2.08 MPa (which is higher by 10%).

The change in the properties of brex induced by the addition of bentonite to the composition of the mixture to be briquetted also manifests itself after preliminary homogenization of the mixture to be briquetted during its "souring" (storage of the mixture for 1 day). The effect of pickling on the strength properties of brex of various compositions (Table 5.5) were studied. Table 5.6 gives the results of tensile splitting tests of brex 1 week after their manufacture. These results demonstrate that the use of bentonite in combination with Portland cement improves the strength properties of brex. The strength of brex increases substantially in some cases. In addition, the character of their fracture also changes and signs of viscoplastic fracture appear. In this case, brex can better withstand the impact loads that appear during their transportation to the sites of their application.

Regarding the traditional methods of briquetting, this favorable influence of bentonite or bentonite in combination with Portland cement on the strength properties of briquettes is well known [7]. An attempt to determine the optimum ratio of the fractions of Portland cement and bentonite in a charge to be briquetted was described in patent [8]. It was stated that the addition of the binder consisting of bentonite (15–25%) and Portland cement (85–75%) enhances the mechanical strength of the corresponding briquette. For the briquettes consisting of converter slime (24.3%), scale (23.4%), blast furnace dust (23.4%), iron ore fines (18.9%), and the binder (10%) consisting of 75% Portland cement and 25% bentonite, the result of a drum test (fraction of a size of > 6.3 mm) increased to 76% (it was 67% for the briquettes of the same composition but without bentonite).

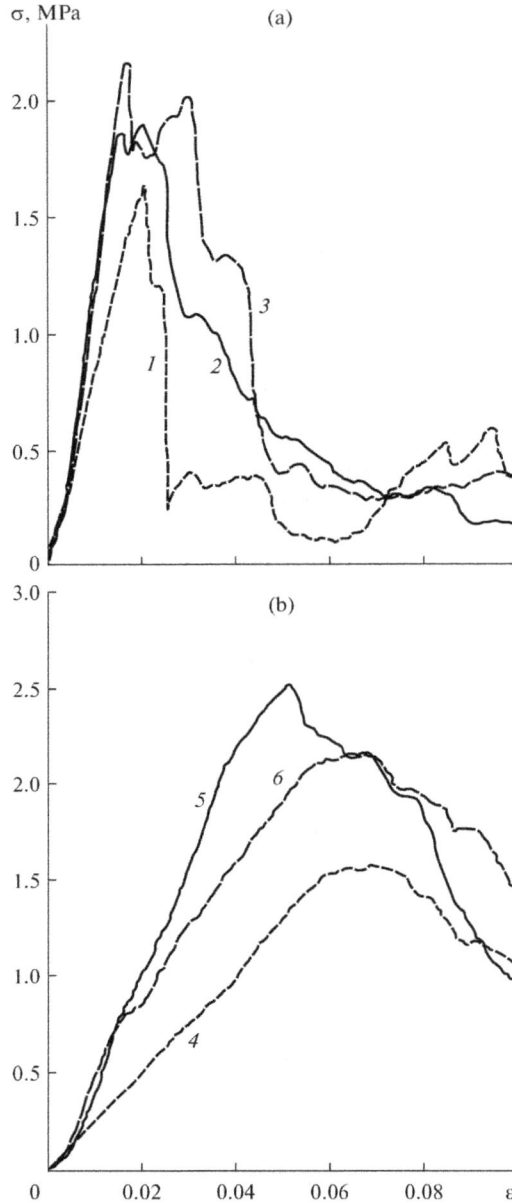

Figure 5.12. Stress (σ)—strain (ε) curves for brex based on aspiration dust (a) without and (b) with 0.5% bentonite.

The influence of this combined binder on both the compressive strength of brex and the rate of increase of strength during strengthening storage under natural conditions were studied. For tests, the brex that were produced in the production line of stiff vacuum extrusion located at Suraj Products Ltd. (India) and contained (%) 47.2 converter slime, 28.3 blast furnace dust, 18.9 iron ore fines, 4.7 Portland cement, and 0.9 bentonite were chosen. Portland cement and bentonite were manually mixed in a dried state and added to a charge mixture before a clay mill with a vacuum lock. For brex samples, the compressive strength (on Tonipact 3000 (Germany) according to standard DIN 51067), open (apparent) porosity (vacuum method of liquid saturation according to standard DIN 51056), and density (on Metier (United States) balance) were measured daily. Table 5.7 gives the results of daily measurements during 9 days.

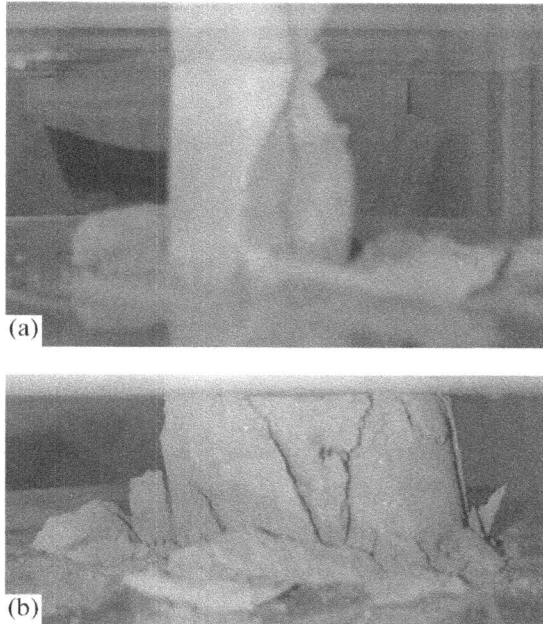

Figure 5.13. Fracture of brex based on aspiration dust: compositions (a) without and (b) with 0.5% bentonite. The initial brex sizes are 50 mm in height and 25 mm in diameter.

Table 5.5. Compositions of the charges of brex 1–6 for souring efficiency (%) tests.

Charge component	1	2	3	4	5	6
Blast-furnace slime	42.8	41.8	41.8	41.2	47.8	28.3
Converter slime	39.8	38.8	38.8	38.2	43.7	25.4
Iron ore concentrate	-	-	-	-	-	29.3
Scale	13.0	13.0	13.0	12.1		10.7
Portland cement	4.0	4.0	6.0	8.0	8.0	5.8
Bentonite	0.4	0.4	0.4	0.5	0.5	0.5
Silica micro granules	-	2.0	-	-	-	-

Table 5.6. Strength* of brex 1–6 (see Table 5.5) in tensile splitting tests (MPa) and the "souring" efficiency** of a briquetted mixture.

1	2	3	4	5	6
$\frac{0.86}{2.45}$	$\frac{1.93}{3.83}$	$\frac{2.08}{5.76}$	$\frac{1}{1.88}$	$\frac{1}{1.29}$	$\frac{0.77}{1.25}$
2.85	1.98	2.76	1.88	1.28	1.64

* Numerator, the strength of brex without pickling of the mixture to be briquetted; denominator, the same after pickling.
** It is estimated as the ratio of the strengths of brex after pickling and without it.

Figure 5.14 shows the brex porosity and strength curves measured during structuring storage. A pronounced local maximum is clearly visible in the brex strength curve on the third day. The next day, it changes into softening. In the course of further storage, the strength increases. It should be noted that, before softening, the brex strength accounts for ~ 84% of the brex strength after strengthening storage for 1 week. The change of open porosity almost repeats the change of strength,

Table 5.7. Changes in the physical properties of the brex produced at Suraj Products Ltd. during strengthening 9-day storage.

т, day	Porosity, %	Density, g/cm³	CCS, MPa
1	31.5	2.42	2.35
2	25.4	2.66	4.4
3	32.0	2.43	6.2
4	27.0	2.44	5.1
5	27.2	2.45	5.5
6	26.2	2.45	5.6
7	26.8	2.46	5.8
8	-	-	7.4
9	-	-	7.8

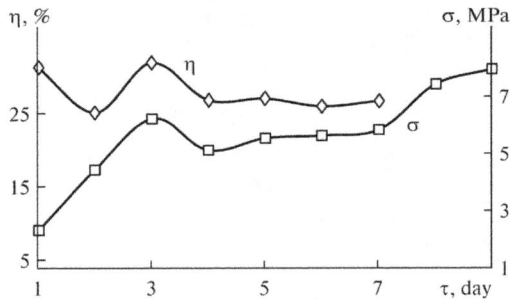

Figure 5.14. Changes in compressive strength σ and porosity η of brex during strengthening 9-day storage.

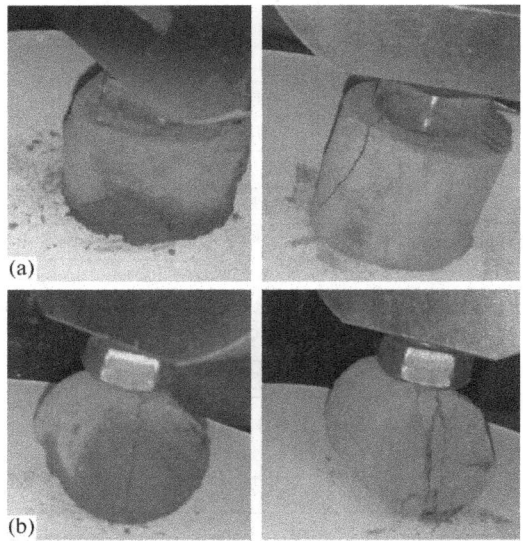

Figure 5.15. (a) Axial compression and (b) tensile splitting tests of the brex produced at Suraj Products Ltd. at the third ((a), (b) on the left) and seventh (on the right) day of strengthening storage. The initial brex sizes are 25 mm in height and 25 mm in diameter.

except for the first day of strengthening. The decrease in the porosity at that time is obviously related to the swelling of bentonite, which fills the pore space [9].

Similar results on the strength of brex during compressive strength and splitting strength tests performed on a strength-testing machine, which consists of a hand-power press, a strain gage and a recording secondary device (Fig. 5.15) were obtained. On the third day of increasing strength, the brex samples retained their integrity up to the end of testing and demonstrated a viscoplastic character of fracture during compressive and tensile splitting tests. On the third day of strengthening, both halves of the brex were connected despite almost complete development of a crack in the tensile splitting test. It is seen that compression and splitting of brex led to moisture release at the bottom. Moisture release was absent during compression at the seventh day of strengthening. When samples were subjected to compressive and tensile splitting tests at the seventh day of strengthening, a viscoplastic character of fracture was substantially lost.

This effect of nonmonotonic strengthening of briquettes with a cement binder, which is characterized by a local strength maximum and is accompanied by viscoplastic behavior at a breaking load, has not been detected and described earlier. The strengthening of monomineral cement stone and concrete is known to have a monotonic character [10]. A nonmonotonic character of increasing strength of brex having an alternative (noncement) binder or having no binder is well known. Ozhogin [11] presents data on the impact fracture (analog of dropping strength) oscillations of briquettes having no binder and having different compositions on drying under natural conditions (at a temperature of 20°C). The detected softening was thought to be caused by recrystallization, and it was recommended to limit the transportation and transfer of briquettes at this time and to store briquettes in closed vessels.

It is necessary to note in the phenomenon described above that the application of a cement—bentonite binder changes the behavior of brex during a braking action, increases its impact strength, and decreases the probability brittle fracture.

This behavior of brex can be explained by the properties of a cement-bentonite binder and related to the formation of coagulation structures in the cement-bentonite-water system, which leads to the modification of the properties of the binder. Such structures are known to form in the gel-cement solutions used for the cementation of boreholes. The properties of the gel-cement solutions and the theoretical and practical aspects of the formation and fracture of cement-bentonite systems are well known and systematized in, e.g., [12]. The driving force of the formation of such structures is the attraction of negatively charged bentonite particles to positively charged Portland cement particles, which results in their rapid coagulation and the formation of a structure with suspended cement particles. Hydrated cement particles are gradually coated with an impermeable shell of flaky bentonite particles. The number of adsorbed bentonite particles is proportional to the activity of cement. During hydration, Portland cement particles grow in size, which leads to tension, a break in the integrity of bentonite shells, and the penetration of water to cement particles (i.e., to further hydration of cement and, apparently, the adsorption of a larger amount of bentonite).

The fracture intensity of such a coagulation structure depends on the cement activity. The application of slag Portland cement is likely to retard this process. The coagulation structure begins to fail gradually as a result of the coagulation effect of calcium ions and changes into the structure of a hardening cement stone. The decrease in the strength after the peak at the third day is explained by the decomposition of the coagulation structure (see Fig. 5.14). The further increase in the strength is completely controlled by the hydration of cement stone.

The possibility of formation of such a stable structure is supported by the conclusions of work [13]. The range of the relative content of bentonite and Portland cement in the binder used for the production of brex (80% Portland cement, 20% bentonite) corresponds to the dark gray region in the water-cement-bentonite phase diagram (Fig. 5.16 [13]). It should be noted that the local moisture near coagulated particles substantially (by 11–12%) exceeds the moisture of the as-prepared brex due to differences in the permeability of minerals. An increase in the fraction of bentonite in the binder up to 30–40% favors the formation of a stable mastic suspension. The changes of the strength of the forming cement-bentonite compositions are also shown in the phase diagram. The strengths of the compositions are seen to decrease as the water content increases or the cement content decreases. Figure 5.17 shows the structure of acicular aggregates in cement.

The effect revealed can be important for practice: brex based on a cement-bentonite binder can be used as a charge component for a metallurgical furnace within 3 days of drying under natural

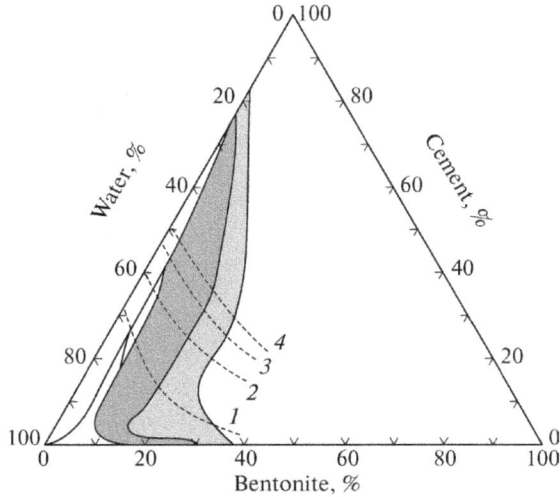

Figure 5.16. Water-cement-bentonite phase diagram [13]. (light gray) Region of formation of a stable mastic suspension and (dark gray) region of a stable and mobile suspension. (1)–(4) Compressive strength of suspension: (1) 69, (2) 345, (3) 689, and (4) 1378 kPa.

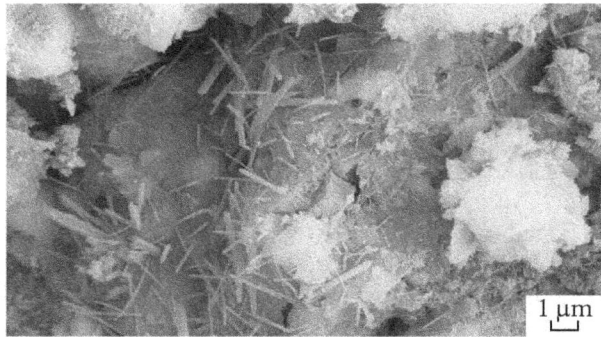

Figure 5.17. SEM micrograph of acicular aggregates (ettringite particles) in cement in the structure of the brex produced at Suraj Products Ltd.

conditions. As a result, the required sizes of brex storage can be decreased. The threshold level of compressive strength can be easily achieved by a simple increase in the fraction of this combined binder. It should be noted that the total fraction of a binder in the brex produced at Suraj Products Ltd. accounts for at most 5.6% of the brex mass, which is significantly lower than the required content of Portland cement used for vibrating pressing (10–12%). The use of bentonite in the binder composition changes the character of fracture of briquettes and makes it viscoplastic, which simplifies the transportation of brex to the site of their application. A quantitative description of the detected effect is the subject of a further investigation.

    The moisture release during the strength tests of briquettes on the third day can be related to restructuring in the agglomerates formed in cement during its hydration, which is accompanied by the release of part of chemically fixed water. It is known that, at a sufficiently high content of volume capillary moisture in cement, intense nucleation and growth of acicular aggregates take place on the third day of hardening [14]. These aggregates are responsible for the connection of cement grains and prevent crack development in primary aggregates, and they were experimentally detected in the cleavages of commercial brex. The experimental investigation was carried out using a high-resolution scanning electron microscopy (SEM) on an Auriga CrossBeam (Carl Zeiss) analytical working station equipped with an INCA X-Max energy dispersive spectrometer. The

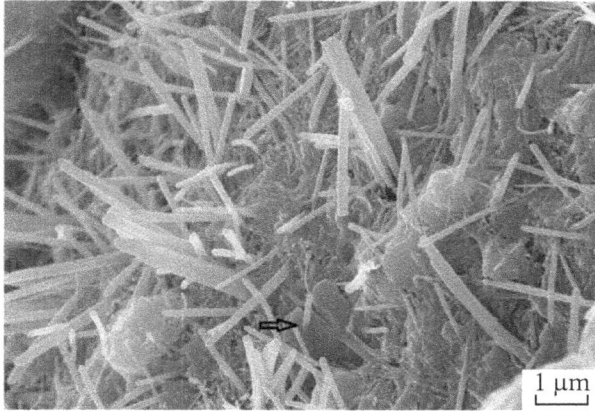

Figure 5.18. SEM micrograph of acicular aggregates (ettringite particles) surrounded by bentonite plates in cement in the structure of the brex produced at Suraj Products Ltd.

accelerating voltage was 20 kV. Figure 5.17 shows micrographs of acicular cement aggregates (ettringite). Ettringite forms during the hydration of cement as a result of the reaction between calcium aluminate and sulfate: $3CaO \cdot Al_2O_3 + 3CaSO_4 \rightarrow$ Ettringite.

Using SEM, we were able to reveal the elements of the decomposed coagulation structure. Ettringite particles surrounded by bentonite plates can be seen in Fig. 5.18.

Based on these results, it can be concluded that:

1) The use of bentonite in a charge for briquetting by stiff vacuum extrusion can substantially improve the service properties of agglomerated products, increasing the compressive strength and the impact strength.

2) The homogenization of a charge using bentonite for 1 day increases its extrusion ability and the strength brex, which ensures the achievement of the required strength of briquettes at a lower binder content in the charge. The effect revealed in this work can be important for practice: brex based on a cement-bentonite binder can be used as a charge component for a metallurgical furnace within 3 days of drying under natural conditions. As a result, the required sizes of brex storage can be decreased.

3) The threshold level of compressive strength can be easily achieved by a simple increase in the fraction of the combined binder. It should be noted that the total fraction of a binder in the brex produced at Suraj Products Ltd. accounts for at most 5.6% of the brex mass, which is significantly lower than the required content of Portland cement used for vibrating pressing (10–12%).

4) The presence of bentonite in the charge composition as a binder changes the character of fracture of briquettes and makes it viscoplastic, which simplifies the transportation of brex to the site of their application.

5) A quantitative description of the detected effect is the subject of a further investigation. In particular, the possibility of using slag Portland cement in combination with bentonite as a binder will be studied.

Industrial brex are air cured under natural conditions during 24 hours. A tumbler test was carried out following ASTM methods and the results were compared with the same for the lumpy iron ore sample. Comparative data are as follows: tumbler index for brex was 96% and for the iron ore – 82%; abrasion index for brex – 1.8%, for the iron ore sample 2.9%.

Comparative reducibility test for iron ore and brex was done in a control atmosphere furnace connected with vacuum pump and constant air supply with a regulator. A hollow alumina tube having length 1000 mm, outside diameter 85 mm and inside diameter 75 mm makes the test chamber. Weights of the brex and iron ore are taken before and after incipient fusion at 1300°C for 30 minutes. Brex and iron ore are placed in graphite crucible inside alumina crucible covered by coke breeze having size 1–5 mm throughout the diameter engulfing the graphite crucible completely. The thickness of the coke layer is 25 mm. Then it was blocked by a perforated plug from both sides and covered by insulating blocks. Water-cooled metallic clamps were fitted in both the side of alumina tube. The vacuum pump is connected in one of the clamps along with air supply connected with the flow meter. Before staring the heating, the inside chamber of the alumina tube was evacuated up to 0.5 mbar. The furnace was started with a heating rate of 10°C/min. When the furnace attained 800°C air supply was started at a rate of 10 liters/min till the completion of the test. Both brex and iron ore are fired from 1000°C to 1500°C for 2 hours in step of 100°C. The schematically test facility is described in Figs. 5.19–5.20.

Both the iron ore samples and brex were chemically evaluated. Chemical analysis data is given in Table 5.8.

Reducibility was determined from oxygen loss calculated from weight. Metallic iron content and metallization was evaluated. Reducibility, metallic iron content and metallization of iron ore and brex are given in Table 5.9.

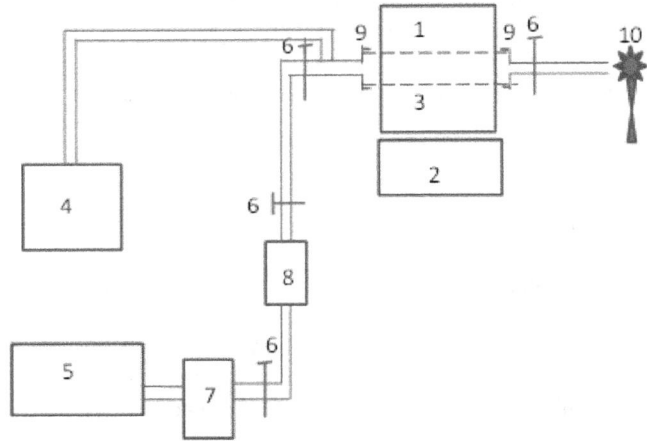

Schematic Diagram for Reducibility Test

1. Furnace
2. Controller
3. Alumina Tube (Sample Carrier)
4. Vacuum Pump
5. Compressor
6. Shut off valve
7. Dehumidifier
8. Flow controller
9. Vacuum seal
10. Burner

Figure 5.19. Schematic diagram of the testing facility.

Detail of Alumina Tube (Sample Carrier)

1. Alumina Tube
2. Perforated block
3. Metallurgical Cake
4. Sample
5. Vacuum Sealent
6. Shut off value

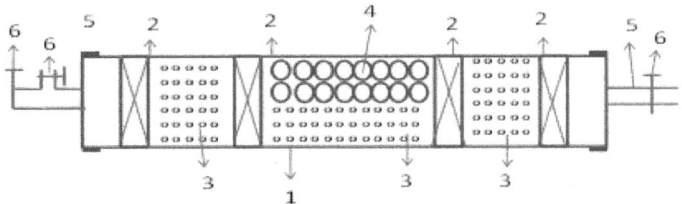

Figure 5.20. Sample arrangement.

Table 5.8. Chemical composition of iron ore and brex, %.

Elements	Iron ore	Brex
CaO	0.2	7.4
SiO$_2$	2.21	4.6
Al$_2$O$_3$	3.11	4.35
Fe$_2$O$_3$	88.53	68.03
MgO	-	0.4
K$_2$O+Na$_2$O	0.2	0.2
TiO$_2$	0.22	0.23
LOI	5.82	15.02

Table 5.9. Metallization and reducibility of iron ore/brex.

T, °C	Fe$_{metallic}$, %	Fe$_{total}$, %	Metalization %	Reducibility %
1000	2.16/14.20	67.48/60.5	3.20/23.5	35/48
1100	12.69/20.96	69.52/61.7	18.25/33.97	52/62
1200	17.65/30.36	70.25/63.7	25.10/47.6	61/72
1300	30.33/44.74	73.3/65.6	41.30/68.2	65/80
1400	53.75/47.04	80.79/67.46	66.53/69.7	82/88
1500	84.25/79.22	90.4/80.87	93.20/97.95	93/99

Phase analysis of iron ore fired at different temperature shows that at 1000°C major phases are represented by Hematite, Magnetite, Wustite (minor phase – Almandine); at 1100°C major phases are represented by Magnetite, Wustite (minor phase – Almandine, iron); at 1200°C major phases are represented by Magnetite (minor phases – Wustite, iron, Almandine); at 1300°C major phases are represented by Magnetite, Wustite (minor phases – iron, Iscorite, Almandine); at 1400°C major phase is iron (minor phases – Wustite, Glass); at 1500°C major phase is iron (minor phase – Glass). X-ray diffractogram of iron ore fired at 1300°C to 1500°C are compared in Fig. 5.23. X-ray diffractogram of brex fired at 1000°C to 1500°C are compared in Figs. 5.21–5.23.

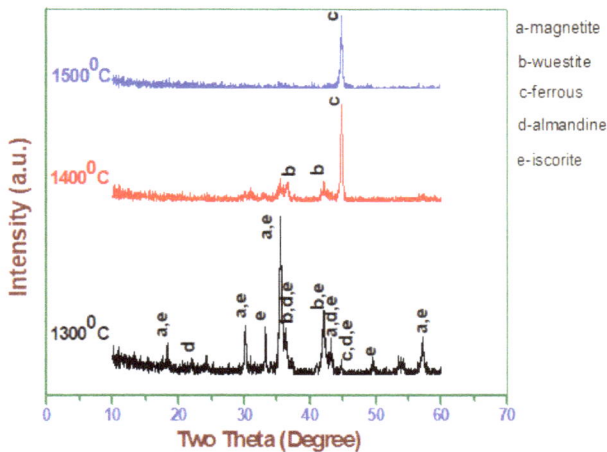

Figure 5.21. X-Ray diffractogram of iron ore fired at 1300°C to 1500°C.

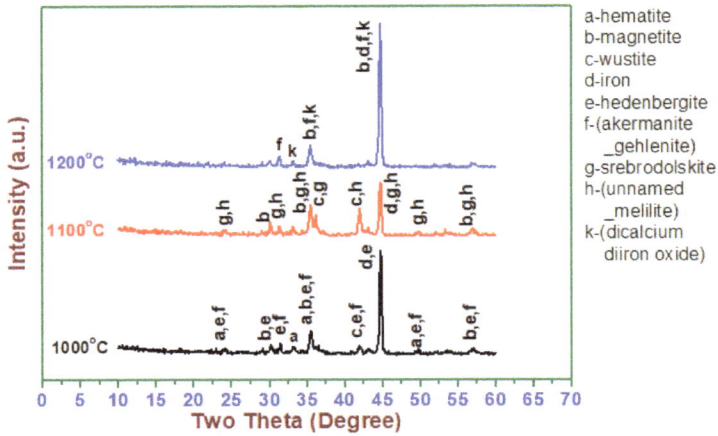

Figure 5.22. X ray diffractogram of brex fired at 1000°C to 1200°C.

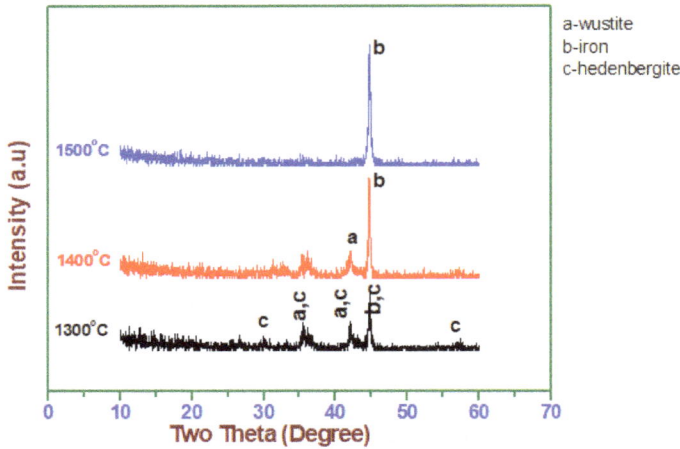

Figure 5.23. X-ray diffractogram of brex fired at 1300°C to 1500°C.

The dynamics of the apparent porosity (DIN 51056) values while firing at different temperatures (for the non-cured samples) were also estimated. At 1000°C the value of the apparent porosity was 37.36%, reaching its maximum value of 46.87% at 1200°C and then falling till 22.6% at 1400°C. The contribution of the high porosity values to good reducibility of the brex is evident. Porosity levels typical for the Hematite iron ores are usually lower.

Microscopic evaluation of iron ore and brex fired at different temperature are evaluated on polished section under reflected light in a universal microscope with image analyzer. The amplification is 200. Photomicrograph of iron ore fired at 1200°C is given in Fig. 5.24. Photo micrograph of brex fired at 1200°C is given in Fig. 5.25.

It follows from the chemical phase and mineralogical analysis that the degree of reducibility is higher in case of brex than for iron ore samples.

The necessity to add the iron ore fines followed from the experience we gained during the early stages of the mentioned above micro-BF. Once we start adding 10–20% of the iron ore fines to the brex we witnessed evidently improved BF operation. This has been attributed to the better sintering properties of the iron ore—the only component of the brex which did not undergo the thermal treatment (such as LD-sludge and flue dust).

Under the particular conditions of the considered test facility better reducibility of brex can be explained by principally different mineralogical structure of the agglomerated product and of

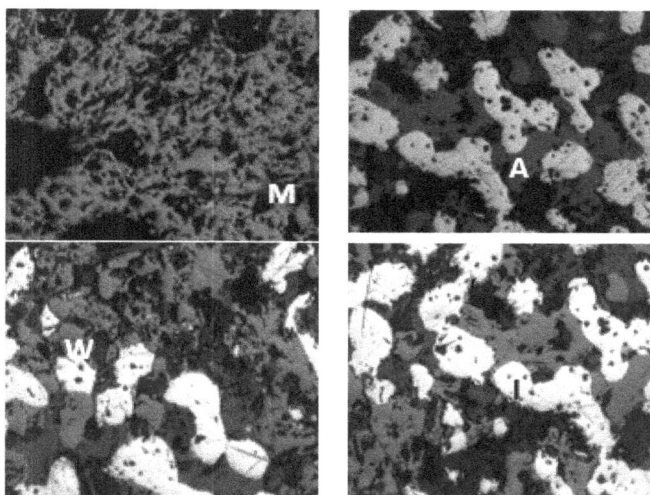

Figure 5.24. Iron ore fired at 1300°C/2 hrs (x200) Magnetite (M); Wustite (W); Iron (I); Almandine (A).

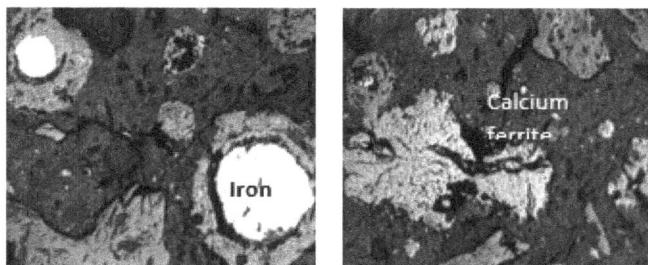

Figure 5.25. Photomicrograph of brex fired at 1200°C/2 hours (x200).

the iron ore at different temperatures. Major silicate phases after firing of iron ore samples at a different temperature are iron silicate and iron aluminium silicate which might envelope the grain and thus retard the reduction. In case of brex presence or formation of ferrite which is easier to reduce than iron silicate or iron aluminium silicate may have facilitated the reduction. Ferrite was brought to the brex by flue dust or might have been additionally generated because of the presence of calcium oxides due to the very high basicity of brex. The latter is possible since the conditions in the test facility partially simulate the sintering process (burning Coke) where the ferrites are very well known to appear.

The contribution of the brex high porosities values will have to be compared with the evolution of the iron ore samples during reduction. This will be done during the next stage of the research. However, for now it is also clear that this feature of the brex favors their utilization as the BF charge components.

The addition of the iron ore fines (10–20% of the brex composition) improves the reducibility (and hot strength) of the brex due to the higher sintering ability of this substance. Porosity of the brex at different temperatures is comparable or better than the porosity of the pellets. brex made of the LD sludge and BF flue dust with the addition of the iron ore fines exhibited better reducibility than the lumpy iron ore of the same deposit. Better reducibility may be related with different mineralogical evolution of the brex and iron ore during reduction. The essential role in more reduction of the brex is due to the ferrites.

In combination with the high capacity of the stiff vacuum extrusion agglomeration lines (up to 120 kty) this technology can be considered as very challenging for the efficient recycling of the fine natural and anthropogenic raw materials of metallurgy.

References

[1] GOST 28657-90 Iron ores. Method for determining reducibility (in Russian).

[2] GOST 21707-76 Iron ores, agglomerates and pellets. Method for determining the gas permeability and shrinkage of the layer during restoration (in Russian).

[3] ISO 4696-1: 1998 Iron ore - Statistical reduction test - degradation at low temperatures. Part 1. Reaction with CO, CO_2, H_2.

[4] ISO 4696-2: 1998 Iron ore - Statistical fracture test after low temperature reduction. Part 2. Reaction with CO.

[5] Knowledge of the processes of blast-furnace smelting. Collective labor ed. Bolshakova, V.I. Dnepropetrovsk: Porogi, 2006. 440 p. (in Russian).

[6] Tovarovskiy, I.G., Gladkov, N.A. and Nesterov, A.S. 1994. Features of the formation of melt in the conditions of low-coke blast furnace melting. Steel 2: 7–12.

[7] Koizumi Hideo, Yamaguchi Arata, Doi Teranobu and Noma Fumio. 1988. Fundamental development of iron ore briquetting technology. ISIJ 74(6): 22–29.

[8] Jpn. Patent S63 196689 (A), 1988.

[9] Bogdan, E.A. and Cole, R.L. 1995. US Patent 5 395 441.

[10] Komine Hideo and Ogata Nobuhide. 2003. Can. Geotech. J. 40: 460–475.

[11] Dvorkin, L.I. and Dvorkin, O.L. 2011. Building Mineral Binding Materials (Infra-Inzheneriya, Moscow) (in Russian).

[12] Ozhogin, V.V. 2010. Fundamentals of the Theory and Technology of Briquetting of Fragmented Metallurgical Raw Materials (PGTU, Marioupol') (in Russian).

[13] Bulatov, A.I. and Danishevskii, V.S. 1987. Grouting Mortars (Nedra, Moscow) (in Russian).

[14] Jones, G.K. 1963. Chemistry and flow properties of bentonite grouts. pp. 22–28. In Proceedings of Symposium on Grouts and Drilling Muds in Engineering Practice (Butteworths, London).

[15] Shmit'ko, E.I., Krylov, A.V. and Shatalova, V.V. 2006. Chemistry of Cement and Binding Substances (Prospekt Nauki, St. Petersburg) (in Russian).

5.3 "Large Sample Method" for Studying the Kinetics of Carbothermal Reduction

An original research technique has been developed and used [1], which makes it possible to study the process of recovery of industrial-size samples weighing 30 g using gas chromatography. The object of the study was ore-coal briquettes consisting of poor oxidized nickel ore of the Buruktal deposit, a reducing agent and a flux.

The experiments were carried out on the basis of an electric resistance furnace (Fig. 5.26) with a graphite heater with a diameter of 65 mm. In the isothermal zone of the heater, a tall alundum crucible with a diameter of 60 mm (6) was placed to protect the atmosphere of the working crucible from the graphite heater, inside which a replaceable alundum working crucible (7) was installed, and an ore-coal briquette was dropped into it (1). Before the start of the experiments, the furnace with the working replaceable alundum crucibles installed in it was evacuated with a foreline pump to a residual pressure of 10^{-1} Pa and filled with high purity argon. Then, the gas discharge to the atmosphere was opened and the argon flow rate through the furnace was set at 0.4 L/min. The furnace was heated, the preset temperature was reached and stabilized, and the tested ore-coal briquette was dropped into the crucible through the dosing lock (2). The moment of discharge was considered the beginning of the experiment.

The temperature in the furnace was controlled and stabilized by a control thermocouple, located in the isothermal zone of the furnace on the outside of the heater. The true temperature of the process was measured with a second measuring thermocouple with an accuracy of ± 5 degrees, located inside the heater and lowered from above into a working alundum crucible; the temperature was recorded.

The briquette surface temperature was measured using a Pyrovision M9000 thermal imager. The temperature distribution over the melting briquette is uneven, therefore, the instrument option of instrumental temperature averaging over the area selected in the frame that was used, is shown in Fig. 5.27. Such averaged temperature of the briquette surface more accurately reflects the

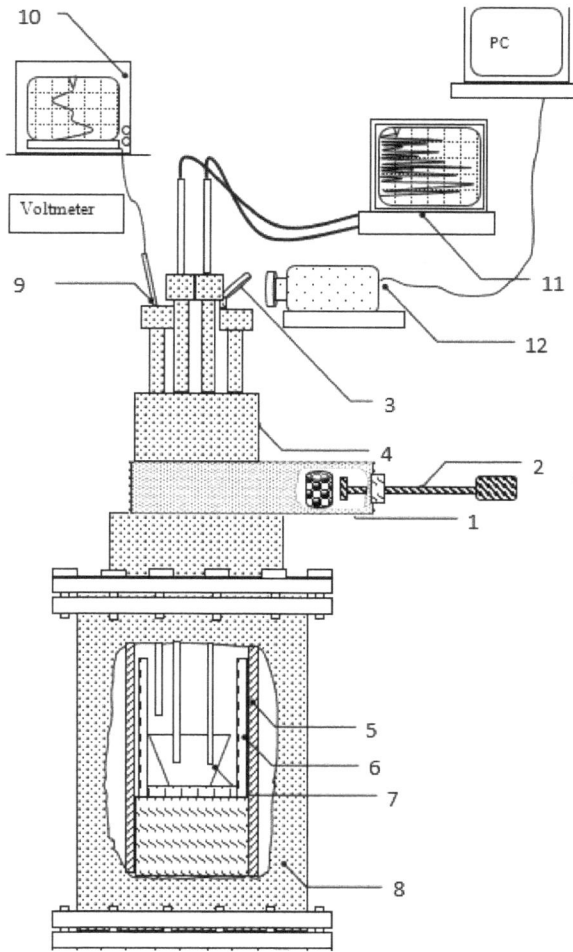

Figure 5.26. Scheme of the installation for research in isothermal conditions (temperature 1400–1550°C): 1 - briquette; 2 - dispenser sluice; 3 - mirror; 4 - furnace cover; 5 - graphite heater; 6 - high alundum crucible; 7 - working alundum crucible; 8 - furnace body; 9 - thermocouple BP 5/20; 10 - KSP-4 recorder; 11 - chromatograph "Gazochrome-3101"; 12 - thermal imager "Pyrovision M9000".

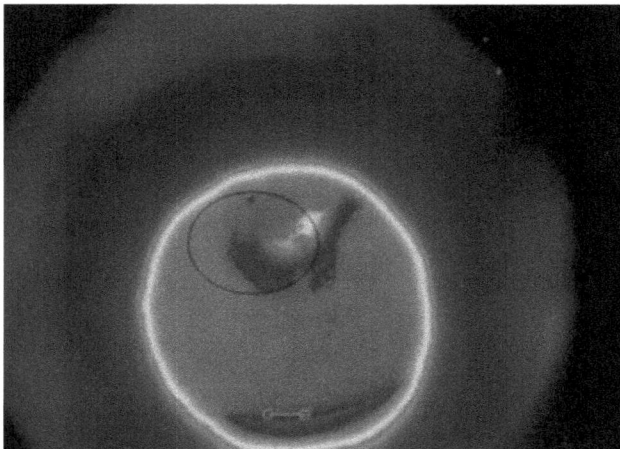

Figure 5.27. Averaging the briquette temperature using the instrument option.

integral consequences of the physicochemical processes occurring in the briquette than the internal measuring thermocouple.

During the experiment, the composition of the gas phase was monitored for the content of carbon monoxide and hydrogen using a Gasochrome-3101 chromatograph (13), the sampling frequency was one sample per minute (determined during blank experiments).

The charge materials for research were ground in a mortar to a fraction of 0.63 mm (a common fraction of grinding in industrial conditions is less than 1 mm, this provides an increase in the reaction rate due to the large contact area of the charge components), mixed according to the selected compositions and briquetted. The production of briquettes was carried out by the method of shock-vibration pressing in a steel mold. The experiment lasted 15–20 minutes at a process temperature of 1500°C [2].

After completion of the experiment, the furnace was depressurized, the crucible with the melting products was removed and quenched in air. The furnace was closed, rinsed with argon, and after the required temperature had been established, the experimental cycle was repeated.

The cooled smelting products (metal droplets) were analyzed to determine the chemical composition on an optical emission analyzer (OBLF, Switzerland), the slag phase was analyzed on a Magnesium-1 X-ray fluorescence analyzer (RF).

Before the start of experiments with ore-coal briquettes, a series of blank experiments was carried out with sampling and analysis of off-gases when briquettes from individual pure components used in the experiments were introduced into the installation: briquettes only from nickel ore and briquettes only from reducing agents—charcoal (ChC), Semi-Coke (SC) and Shubarkul coal (ShC). The lag of the gas system was checked by injecting a reference gas into the system to introduce corrections when processing the obtained experimental data.

Blank experiments were carried out according to the method described above: the briquette was dropped into the hot zone of the furnace at a temperature of 1500°C and held for 15 minutes, during the experiment, the exhaust gas phase was taken from the working space of the furnace, after which the obtained data were processed.

Experiments with briquettes from pure nickel ore made it possible to establish the content of hydration water in the ore by the difference in briquette mass before and after the experiment. Hydration water in the ore contains about 35%, this figure was taken into account when calculating the degree of recovery of base metals from ore.

Experiments with briquettes of pure reducing agents (charcoal, Shubarkul coal, semi-coke) made it possible to introduce a correction for the released carbon monoxide in subsequent experiments with ore-coal briquettes when calculating the degree of conversion for each type of reducing agent.

The lag of the response of the gas system was checked using a reference gas with a known content of carbon monoxide $\{CO\} = 28.45$ [%, vol.], as shown in Fig. 5.28.

It was found that at a flow rate of 400 ml/min (the used argon flow rate in the experiments), the reference gas was quickly washed out of the furnace space, which made it impossible to determine the concentration of the reference gas in the argon flow. To determine the inertia of the gas system, the argon flow rate was reduced to 100 ml/min and the reference gas was "injected", sampling the gas phase every minute, the results of the experiment are shown by triangles on the graph (see Fig. 5.28).

To check the authenticity of the data obtained in the experiment, the change in the concentration of carbon monoxide was calculated under the same condition—the rate of the argon purge was set equal to 100 ml/min. After that, the rate of change in the concentration of carbon monoxide was calculated from real 50 to 0% at an argon flow rate of 400 ml/min, which was used in experiments with ore-coal briquettes.

In the course of this experiment, the volume of the reaction cell was determined—82 ml from the physical 500 ml of the reaction volume, it can be interpreted as the apparent volume of the reaction cell in the approximation of an ideal mixing reactor. The calculated data completely coincided with the experimental data, which is confirmed by the graph (see Fig. 5.28).

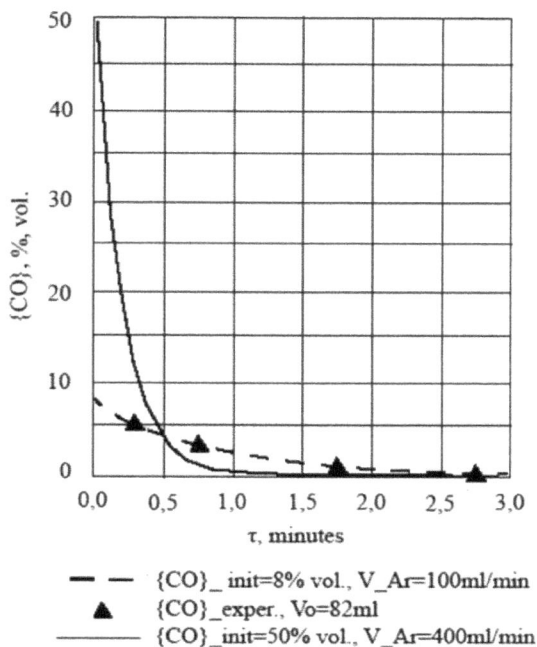

Figure 5.28. The rate of {CO} exit from the system.

As can be seen from the graph (see Fig. 5.28), carbon monoxide {CO}, [% vol.] in the system with a concentration of 50% is completely washed out of the working space of the furnace within one minute without leaving any traces in the working space. The results of processing the obtained data made it possible to determine the optimal time for sampling in experiments with briquettes— every minute.

Model experiments have shown that the constant installation time is much shorter than for the ideal mixing model, in other words, the displacement of gases in the installation occurs in an intermediate mode between ideal mixing and the piston mode [3].

The choice of the method for processing the obtained experimental data depended on the rate of change in the temperature of the briquette during the experiment, which had to be determined.

The widely used thermogravimetric analysis method (TGA) of studies for calculating the kinetic parameters of the reduction process is not applicable in this case [4, 5]. In TGA, a small sample of 0.01–1 g is used, the temperature of which during the experiment, due to the small mass of the sample, is equal to the temperature of the furnace; it usually rises automatically according to a linear law. The kinetic data obtained on samples of low mass (the mass of the samples differs by several orders of magnitude) is difficult to apply to industrial aggregates, which is the main disadvantage of the method [6].

It was experimentally established that the temperature of a sample with a mass of about 30 g when it is dropped into the furnace always changes according to the same logarithmic law (Fig. 5.29), the coefficients of the equation of which depend on the furnace temperature and the mass of the sample (equation (1)).

$$T = C \cdot \ln \tau + d, \frac{dT}{d\tau} = \frac{C}{\tau} \tag{1}$$

where: T – temperature, K
τ – time, min;
C and d – coefficients.

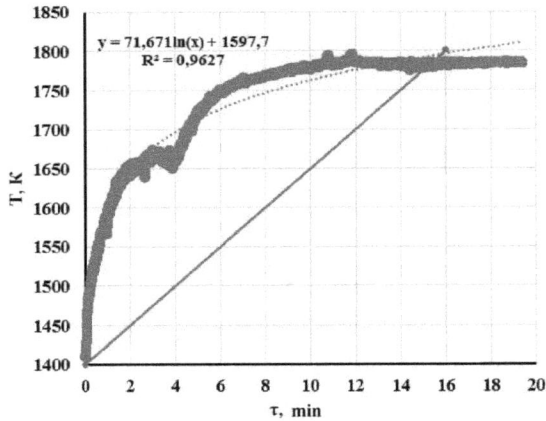

Figure 5.29. Logarithmic dependence of the temperature of the briquette on time.

After determining the main parameters of the experimental technique, such as: the logarithmic law of the rate of change in the temperature of the sample, the rate of sampling by a gas chromatograph in experiments with ore-coal briquettes—every minute; and also assuming that—the formation of only carbon monoxide as a result of the reduction of the briquette in the gas phase (carbon dioxide is neglected due to its small volume in the gas phase and rapid oxidation to carbon monoxide), which was taken into account when processing the experimental data, the method for processing the obtained experimental data was chosen.

To check the adequacy of the experimental technique and the method of processing the obtained primary experimental data, two independent methods were used: calculating the degree of conversion by gas analysis (α_CO) and calculating the degree of conversion by the mass of the obtained metal (α_Me).

The expression for the degree of conversion in general representation has the form:

$$\alpha_O = \frac{m_o(\tau)}{\sum m_o} \tag{2}$$

To calculate the degree of conversion of α_CO from the results of gas analysis, the following considerations were used: when 1 mole of oxygen ($\mu O = 15.999$ g) interacts with the carbon of the reducing agent, 1 mole of carbon monoxide ($V\mu = 22\ 414$ ml) is released, which can be further oxidized to carbon dioxide, but the number of moles of gaseous reaction products will not change in this case:

$$NiO + C = Ni + CO$$
$$Fe_2O_3 + 3C = 2Fe + 3CO \tag{3}$$
$$CO + 1/2O_2 = CO_2$$

Consequently, when the mass of oxygen changes by an amount dm_O, dV_{CO} ml of carbon monoxide will be released, hence

$$dm_O = \frac{\mu_O}{V_\mu} dV_{CO} \tag{4}$$

Dividing the left and right sides by $d\tau$, we get:

$$\frac{dm_O}{d\tau} = \frac{\mu_O}{V_\mu} \frac{dV_{CO}}{d\tau} \tag{5}$$

Carbon monoxide concentration {CO}, [%, vol.] is related to the instantaneous volumetric flow rate of carbon monoxide υ_{CO}, [%/min] by the following obvious relationship:

$$\upsilon_{CO} = V_{Ar} \frac{\{CO\} + \{CO_2\}}{100 - \{CO\} - \{CO_2\}} \tag{6}$$

where V_{Ar} – instantaneous consumption of argon through the measuring system, ml/min;

Using these equations (2)–(6), an expression relating the rate of loss of oxygen mass due to the formation of carbon monoxide with the concentration {CO} in the gas phase were obtained. The amount of carbon dioxide gas {CO_2} was neglected:

$$\frac{dm_O}{d\tau} = \frac{\mu_O}{V_\mu} V_{Ar} \frac{\{CO\}}{100 - \{CO\}} d\tau \tag{7}$$

Integrating expression (7) from 0 to τ and substituting it into the original expression (6), the final equation relating the degree of conversion to the composition of the exhaust gases was obtained:

$$\alpha_{(O)} = \frac{\mu_O}{V_\mu} \frac{V_{Ar}}{\sum m_O} \int_0^\tau \frac{\{CO\}}{100 - \{CO\}} d\tau \tag{8}$$

where μ_O – oxygen atomic mass;

V_μ – molar volume of ideal gas at normal pressure (22 414 ml);

Σm_O – total mass of oxygen in ore oxides;

{CO} – volumetric concentration of carbon monoxide in the gas phase, % vol.

To calculate the degree of conversion from the composition of the metal phase, the following formula was derived, based on the formal expression of the degree of conversion (2):

$$\alpha_{Me}(\tau) = \frac{m_O^{Me}(NiO) + m_O^{Me}(Fe_2O_3) + 0,05 m_O^{Me}(SiO_2)}{m_O^p(NiO) + m_O^p(Fe_2O_3) + 0,05 m_O^p(SiO_2)} \tag{9}$$

where: $m_O^{Me}(NiO)$, $m_O^{Fe}(Fe_2O_3)$ and $m_O^{Si}(SiO_2)$ – the mass of oxygen taken from nickel oxide, iron (III) oxide and silicon (IV) oxide, respectively, at the time τ;

$m_O^p(NiO)$, $m_O^p(Fe_2O_3)$ and $m_O^p(SiO_2)$ – the amount of oxygen contained in the ore in nickel oxide, in iron oxide (III) and in silicon oxide (IV).

The numerator of equation (9) is calculated by the chemical composition and mass of the obtained metal, the denominator—by the chemical composition and mass of the original ore and ash of the reducing agent.

The reliability of the method is verified by comparing the calculated values of the degree of recovery obtained by two independent methods. With the correct calculation of the degree of recovery, the values of **α_CO** and **α_Me** should coincide and on the graph lie on the control line drawn at an angle of 450 through the origin.

To calculate the kinetic constants of the process: the apparent activation energy **E_act** (kJ/mol), the rate constant K and the order of the reaction n, followed by the determination of the limiting stage of the reduction process, the direct differential method was used, the advantage of which is good linearity of the graphs in the case of correctly selected parameters of the equation [7].

Three basic equations should be considered. Formal-mathematical expression of the degree of transformation:

$$\frac{d\alpha}{d\tau} = Kf(\alpha); \quad f'(\alpha) = (1 - \alpha)^n \tag{10}$$

where n – reaction order.

Arrhenius equation:

$$K = A\exp\{-E/RT\} \tag{11}$$

where: A – rate constant;

R – universal gas constant equal to 8.31 J/(mol*K);

E – activation energy, J/mol;

The equation for the heating rate according to the logarithmic law for the case of a large sample (see equation (1)):

$$T = C \cdot \ln\tau + d, \quad \frac{dT}{d\tau} = \frac{C}{\tau}$$

Combining these equations and taking into account that $\frac{d\alpha}{dT} = \frac{d\alpha}{d\tau}\frac{d\tau}{dT}$, we get the basic kinetic equation:

$$\frac{d\alpha}{dT} = \frac{\tau}{C}(1-\alpha)^n \exp\{\frac{-E}{RT}\} \tag{12}$$

To apply this basic kinetic equation to the experimental results, it is necessary that:

- the reaction rate was described by one kinetic equation in the entire range of values;
- the rate constant satisfies the Arrhenius equation.

Direct application of the differential equation is considered to be the simplest way to obtain kinetic parameters from non-isothermal data [8]. The equation in logarithmic form, taking into account the parameters of our experiment can be represented by:

$$\frac{\ln\frac{d\alpha}{dT}}{f'(\alpha)} = \ln\frac{A}{b} - \frac{E}{RT} \tag{13}$$

Since the function f'(α) is known - in our case $f(a) = \frac{C}{\tau}(1-\alpha)^n$, the apparent activation energy and the order of the reaction can be found from the graph.

To use a direct differential equation, a preliminary value of the order of reaction n is required, which is unknown and is itself determined from the same measurements.

To select the reaction order n that best describes the experimental data, we used the following technique, which consists in determining the equation coefficient of the reaction order (n) for a graph plotted in coordinates $\left(\frac{\frac{d\alpha}{dT}}{f'(\alpha)}, \frac{1}{T}\right)$. The reaction order n was varied from 0 to 2 with a step

of 0.1 and for n from 2 to 3 with a step of 0.25. The value of the apparent activation energy **E_act** is determined from the tangent of the slope of the dependence l gk = f(1/T), obtained by taking the logarithm of the Arrhenius equation.

References

[1] Bizhanov, A.M., Pavlov, A.V. and But, E.A. 2018. The study of the reduction process parameters of ore-coal briquettes for ferroalloys production. The Fifteenth International Ferro-Alloys Congress. Infacon XV (Cape Town, South Africa). pp. 139–151.

[2] Pavlov, A.V. and But, E.A. 2018. Study of solid-liquid carbothermal reduction of nickel from ore-coal briquettes. Izv. Ferrous Metall. 61(2): S. 120–127 (in Russian).

[3] Ioffe, I. and Pismen, L. 1965. Engineering chemistry of heterogeneous catalysis. - M.: Chemistry. 456 p. (in Russian).

[4] Arsentiev, P.P., Paderin, S.N., Serov, G.V. and other. 1989. Experimental work on the theory of metallurgical processes. - M.: Metallurgy. 288 p. (in Russian).

[5] Baysanov, A.S., Takenov, T.D., Tolymbekov, M.Zh. and other, 2005. Determination of the magnitude of the activation energy of phase transformations in ferromanganese ores. Materials Int. Scientific-Practical Conference, Dedicated. To the 80th Anniversary of E.A. Buketova. Academician E.A, Buketov - Scientist, Teacher, thinker. Karaganda. 3: 76–81 (in Russian).

[6] Fialko, M.V. 1981. Non-isothermal kinetics in thermal analysis. Tomsk: Publishing House of Tomsk University. 110 p. (in Russian).

[7] Piloyan, G.O. and Novikov, O.S. 1967. Thermographic and thermogravimetric methods for determining the activation energy of dissociation processes. Journal of Inorganic Chemistry. - T. 12(3): 602–604 (in Russian).

[8] Paderin, S.N. and Serov, G.V. 2008. Thermodynamics and kinetics of metallurgical processes. Study guide number 14. Part 2. - MISiS. Publisher: OJSC "VyksaPoligrafIzdat". 112 p. (in Russian).

5.4 Porosity

The porosity of briquettes has a significant effect on their recoverability. In accordance with the universal classification, pores are distinguished by size into three classes. Large and super-large pores have a diameter of 1 to 100 microns and more, micropores—from – 0.1 to 1 micron, submicropores and mesopores have an average diameter of less than 0.1 microns. The total porosity of a briquette is the sum of open and closed porosity. Open or apparent porosity is the ratio of the total volume of pores communicating with the atmosphere to the volume of the briquette. Closed porosity is the fraction of pores isolated from the atmosphere.

Porosity can be measured in a number of ways. We will not go into detail on the description of traditional methods for determining porosity, which are described in sufficient detail in the reference literature. To identify macropores and determine the number, volume and distribution of pores, methods of direct visual-optical observation and observation using light and electron microscopy can be used. Visual-optical observation allows to determine porosity in the range of macropore sizes from 10 to 75 microns. Light microscopy is applicable for pore sizes in the range of 0.5–100 microns, and electron microscopy allows the study of pores ranging in size from 0.002 to 0.5 microns. Capillary methods for determining porosity (capillary permeability) are used to study porosity with pore sizes from 0.01 to 100 microns. Mercury porosimetry methods are used to determine pore sizes, pore size distribution and specific surface area in a wide range of pore sizes from 0.0015 to 800 microns. The methods for determining porosity by this method are described in the standards ISO 15901-1: 2006, ISO/NP 15901-2 and ISO 15901-1: 2016. Pycnometric methods (gas and liquid pycnometry) allow the determination of the total porosity, size and distribution of submicropores with sizes from 0.002 to 0.001 microns. In this way, the porosity is determined in accordance with the standards of GOST ISO 5017-2014. Open porosity is measured pycnometrically based on liquid saturation under vacuum in accordance with DIN 51056.

One of the most accurate and informative methods for determining porosity, which makes it possible to study not only its quantitative parameters, but also pore morphology, is the STIMAN method developed at Moscow State University [1] and used for the first time to study the porosity of briquettes. The method is based on the combined application of Scanning Electron Microscopy (SEM) and computed tomography. X-ray computed microtomography (μCT) allows visualizing the three-dimensional internal structure of an object without violating its integrity. The principle of operation of a computed tomograph is based on transillumination of the object under study with a thin beam of X-rays. When passing through an object, X-rays are absorbed by various structural elements to varying degrees. The X-rays are then recorded by a system of detectors, resulting in a shadow projection that resembles a medical X-ray. By rotating the sample around the axis, a series of shadow projections are collected, according to the totality of which the reconstruction of the internal structure of the object is carried out. As a result of the reconstruction, a cubic volume of data is obtained, divided, as a rule, into 512 voxels in each direction. The data presentation method can be different: orthogonal halftone sections, which have been carried out anywhere in this volume, it is possible to obtain a volumetric representation, etc. The resolution or size of a pixel (voxel) is determined by the sample diameter reduced by the number of pixels (voxels) of the image (as a rule, 512 (x512x512)). The shades of gray in such images correspond to the relative linear X-ray extinction coefficient, which is a function of the density, atomic number and energy of the X-rays. When evaluating a complex structure, measuring its "X-ray density" does not always allow one to state with precision what substance is being visualized. Usually, one can only highlight the denser (shown in a light tone) and looser (shown in a dark tone) areas of the sample. Thus, the boundaries, for example, between pore space and matter, are determined by the operator by setting a gray level threshold. After binarization of the image at the selected level of gray shade, quantitative operations can be performed, for example, to estimate the percentage of any component of interest. A fresh cleavage of the sample surface is used for research.

For morphological studies of the microstructure, the secondary electron mode is used, which makes it possible to obtain high-quality grayscale images in a wide range of magnifications.

The quantitative analysis of the microstructure is carried out using the STIMAN software by two methods. (1) Based on a series of SEM images obtained in the reflected electron mode. This mode allows one to obtain correct images with sharper boundaries between pores and particles. (2) complex analysis of a series of SEM images and μCT images of cross sections of the sample obtained at different magnifications.

1. n_{im} is the total porosity calculated from the SEM image.
2. D_1, D_2, D_3, D_4, D_5 - different size categories of pores.
3. D_{max} is the maximum equivalent pore diameter.
4. K_f - pore shape factor. It is calculated as the ratio of the minor and major semi axes of an ellipse inscribed in the pore. For isometric pores $K_f = 0.66$–1.00, for anisometric pores - $K_f = 0.1$–0.66, for slit-like pores $K_f < 0.1$.

The quantitative morphological analysis of the images obtained in SEM is carried out using the STIMAN software. A series of different-scale SEM images covering the entire range of sizes of structural elements are used as input information for the STIMAN software. Usually, the analysis is carried out in series containing at least eight images obtained at magnifications, changing exponentially with a coefficient equal to 2. Quantitative analysis is carried out in stages. Each stage corresponds to a special software module that makes up the STIMAN software.

The results of applying the STIMAN procedure to measuring the porosity of various briquettes and brex can be illustrated.

Using the STIMAN technique we measured the porosity of the vibropressed briquette of the following composition—iron ore fines (61%), coke breeze (14.5%), flue dust (14.5%), and cement (10%). The porosity values obtained using micro tomograph characterize a fraction of the total porosity, represented by macropores with a pore size (equivalent diameter) greater than 100 microns. For the considered vibropressed briquette, this value was 2.5%. The porosity determined by the electronic microscopy was 13.37%. The total porosity of the briquette is 15.87%. The morphological parameters of the structure of the briquette porosity are shown in Table 5.10.

Table 5.11 shows the compositions of the brex we have chosen for the utilization as the charge components for the direct solid-state reduction in the industrial reactors.

Morphological microstructure investigation has been conducted using the regime of secondary electrons, which allows obtaining high-quality halftone images in a wide range of magnifications. The STIMAN procedure allows getting correct images with clear boundaries between the pores and particles. Quantitative analysis of microstructure was carried out using the software "STIMAN" by two methods. (1) Based on a set of SEM-mode images in the regime of the reflected electrons, (2) complex analysis of the set of SEM images and X-ray computed tomography with various amplifications.

Measured this way, the morphological parameters of the extrusion briquettes and pore characteristics are listed in Table 5.12. The values of the apparent porosities measured by DIN 51056 were at the level of 24.5% for the cured briquettes and 33.5% after reduction.

Table 5.10. Morphological parameters of the porosity of the vibropressed briquette.

Sample	Micro-morphological parameters	Pores category						n, %	K_a, %
		D_1 < 0.1 μm	D_2 0.1–1.0 μm	D_3 1.0–10 μm	D_4 10–100 μm	D_5 > 100 μm	D_{max} μm		
Vibropressed briquette	N, %	1.9	15,9	35,8	46,5	-	91.4	13.37	19.07
	K_f	0.25–0.33; 0.42–0.50							
	K_f	0.33–0.42; 0.50–0.58							

where

1. N_{im} – total porosity calculated by SEM image.
2. N – the contribution of pores of different size categories to the total porosity (n).
3. D_1, D_2, D_3, D_4, D_5 – various dimensional pore categories.
4. D_{max} – maximum equivalent pore diameter.
5. K_f – pore shape factor. It is calculated as the ratio of the minor and major axes of the ellipse inscribed at the time. For isometric pores, $K_f = 0.66$–1.00, for anisometric pores, $K_f = 0.1$–0.66, for slit-like pores, $K_f < 0.1$.
6. K_a – anisotropy coefficient, or the degree of orientation of solid structural elements. Evaluated using the software "STIMAN". For the samples studied, the presence of a medium-oriented microstructure is characteristic.

Table 5.11. Brex compositions for the solid-state reduction.

Briquette components	01–01	01–02	01–03
Pellets fines	50,0	50,0	50,0
Sludge	25.0	25,0	25,0
Mill scale	15,0	15,0	15,0
EAF dust	5,0	4,75	5,0
Slaked lime	5,0		
Portland cement		5,0	
Magnesium binder			5,0
Bentonite		0,25	

Table 5.12. Porosity and pores micromorphology measured by SEM (STIMAN) +X-ray tomography.

Sample	Porosity SEM/SEM+ X-ray tomography	Sizes distribution (n_{im}), %					Maximum diameter	Shape factor		
	n, %	D_1 < 0.1*	D_2 0.1–1.0	D_3 1.0–10	D_4 10–100	D_5 > 100	D_{max}	K_f		
01–01	31,6	0,6	8,6	25,6	65,2	0,0	57,6	0,25– 0,33	0,42– 0,50	
	37,3	0,5	7,2	21,4	66,8	4,1	407,2			
01–01-reduced	32,9	0,2	9,0	38,4	52,4	0,0	70,94	0,25– 0,33	0,42– 0,50	0,75– 0,83
	38,3	0,1	7,7	32,4	49,8	10,0	504,49			
01–02	31,2	0,6	8,0	30,1	61,3	0,0	45,78	0,33– 0,42		
	35,7	0,5	7,0	26,2	61,3	5,0	299,96			
01–02-reduced	37,9	0,2	9,3	31,1	59,1	0,3	100,51	0,33– 0,42	0,50– 0,58	
	40,4	0,2	8,9	29,5	55,2	6,2	583,77			
01–03	31,8	0,6	5,8	28,4	65,2	0,0	62,92	0,42– 0,50		
	32,9	0,5	5,5	27,2	65,3	1,5	191,89			
01–03-reduced	32,8	0,2	3,2	35,7	60,9	0,0	71,75	0,42– 0,50	0,58– 0,67	0,75– 0,83
	39,3	0,2	2,6	28,9	58,0	10,3	736,41			

*- in μm

It follows from these data that the largest growth in the porosity values after reduction exhibited the briquette bonded by a magnesium-based binder—19.5% against 13% for the lime bonded one and only 2.6% for the Portland cement bonded briquette. The difference in the maximum sizes of these briquettes after their reduction can also be seen. The reasons for this will be explained in the next paper (under preparation).

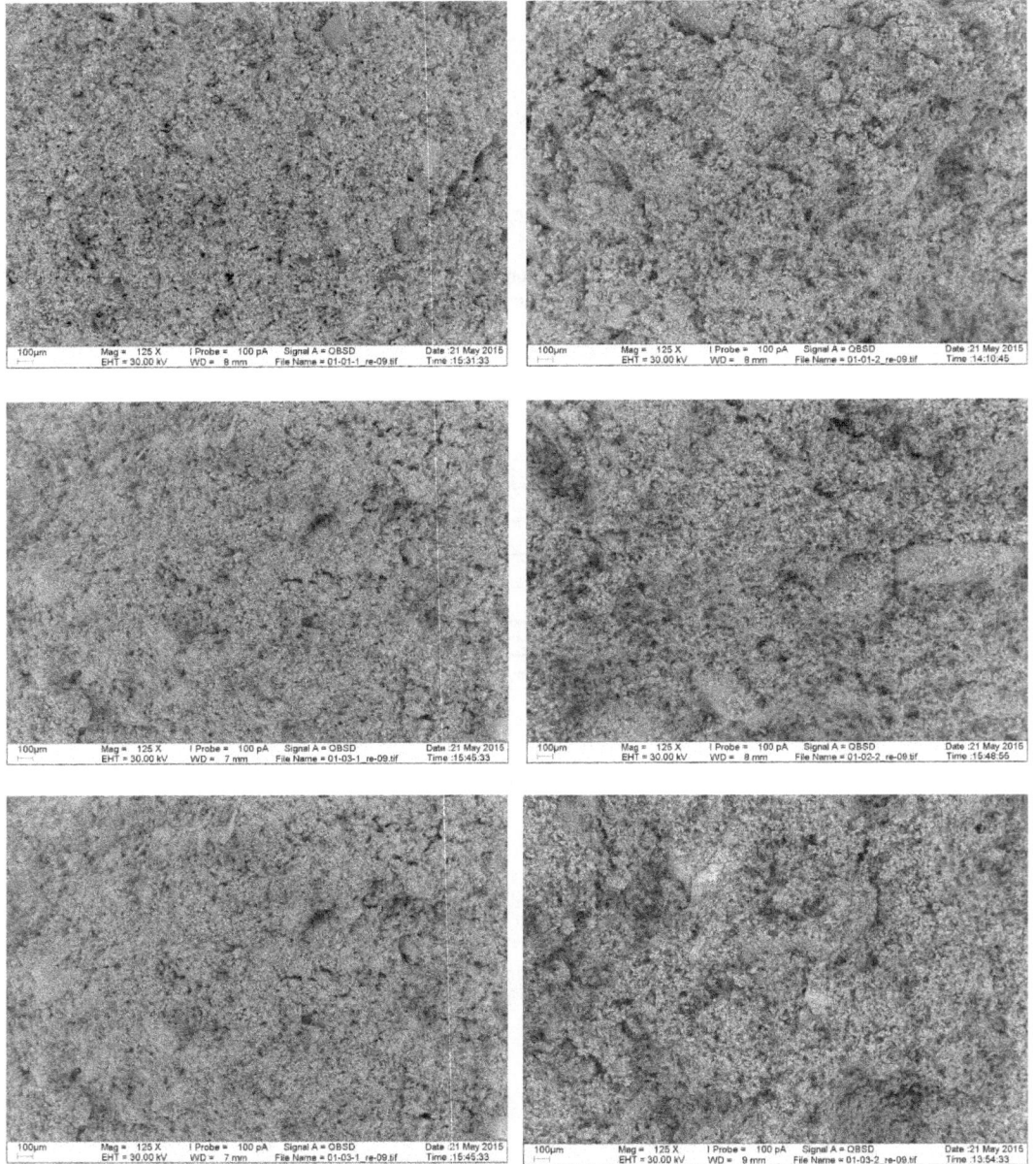

Figure 5.30. Comparison of the pore distribution in the cured and reduced extrusion briquettes (top row – briquette 01–01, lime bonded; middle row – 01–02, cement bonded; bottom row – 01–03, magnesium-based binder).

Figure 5.30 shows the difference in the pore structure and distribution for the considered briquettes. These are SEM images processed by the STIMAN procedure.

References

[1] Sokolov V.N., Yurkovets D.I., Razgulina O.V. Determination of Tortuosity Coefficient of Pore Channels by Computer Analysis of SEM Images. // Proceedings of Russian Academy of Sciences, Physical Series. 1997. V. 61. No. 10. P. 1898-1902.

CHAPTER 6

Basic Industrial Technologies of Briquetting in Ferrous Metallurgy

6.1 Briquetting Using Roller-Presses

Before the advent of vibropressing and the technology of stiff vacuum extrusion, roll presses were most widely used for briquetting in ferrous metallurgy.

6.1.1 Physical processes and designs of roll briquetting presses

The principle of operation of such presses is that the briquetted mixture is fed into the gap between two rolls rotating towards each other, on the surface of which symmetrically located cells in the form of semi-briquettes are arranged in a checkerboard pattern. During the rotation of the rolls, there is a convergence of cells, the capture of the material and its sealing compression. The briquetted material is subjected to bilateral compression, which contributes to a more uniform distribution of its density by volume. Then, as the rolls rotate, the cells diverge, and the briquette drops out of the cell under its own gravity. In some cases, in the designs of roller presses, pre-compressing dispensers are used for preliminary agglomeration of the briquetting charge in order to increase its bulk weight, which allows for an increase in the compressive load and the overall performance of the press.

There are several types of press roll design: solid rolls, sleeve rings and segmented rolls (Fig. 6.1). The main advantage of segment rolls (Fig. 6.1c), assembled from separate sections or segments, is the ability to quickly replace worn segments with minimal labor and time, which simplifies the operation of briquette equipment and increases its efficiency.
A variant of the sleeve ring is shown in Fig. 6.2.

The processes occurring in the briquetted mass in a roller press can, to a certain degree, can be described by methods of the general pressing theory. The compaction process can be divided into three stages. The first stage is the reorientation of particles without changing their shape and size. In the second stage, the deformation of the soft and the destruction of brittle particles of material take place. At the last stage, plastic changes in the structure of the material occur, and its compaction occurs.

Figure 6.1. Designs of rolls of roller presses: a - one piece solid roll; b - sleeve rings; c - segmented roll.

Figure 6.2. Roller press sleeve.

The strength of briquettes produced in this way depends on a number of factors: the particle size and component composition of the briquetted material, the magnitude of the applied pressing force, the moisture of the charge, the duration of the pressing process, the shape and size of the cells of the sleeves and friction coefficients. The influence of these factors on the effectiveness of roller briquetting will be described.

The factor limiting the effectiveness of compaction of the material to be pressed is the presence of air, which is "trapped" inside the material during compression and is released after the external pressure ceases, thus leading to a softening of the briquette. The amount of pressed air can be 30–70% of the initial volume of air in the material. It is this that causes the frequent reduction in the strength of raw roll briquettes. Air displacement can also be difficult due to the presence of a binder that occupies a significant proportion of the pores. To avoid this phenomenon, it is necessary to ensure maximum removal of air from the briquetted mixture. Air removal can be facilitated by a reduction in the rate of material compression itself, facilitating the release of air from a decreasing pore space. It should be noted that during vibropressing and in stiff vacuum extrusion, as will be seen later, the removal of air from the briquetted mass is an important component of the technology and is achieved by displacing the air as a result of high-frequency vibrations in the vibropress and the complete removal of air from the stiff extruder's working chamber.

The maximum particle size for roller briquetting usually does not exceed 5–6 mm. At the same time, multi-fractionality is a factor favoring briquetting efficiency, since in this case the space between large particles is filled with smaller particles, which contributes to the displacement of air from the material.

When briquetting multicomponent mixtures, one should also take into account the difference in the manifestation of the elastic-plastic properties of the individual components, leading to a significant difference in the degrees of compaction. Figure 6.3 [1] illustrates the degree of compaction of roll briquettes based on manganese ore fines with the addition of different carbon-containing materials (coal or peat). It is seen how largely the degree of compaction of the mixture varies depending on the physicomechanical properties of the components of the mixture.

When briquetting loose mixtures that exhibit significant shrinkage (40 ... 50% or more), it may be necessary to use a pre-compressing that allows to provide a specific preliminary pressure to 0.007–0.7 MPa, and the actual pressure to 68.9–690 MPa [1].

The growth of the applied pressure, which means an increase in the energy consumption for briquetting, leads to a decrease in the porosity of the briquette, which may reduce its metallurgical value in the reduction processes. Moreover, such a measure makes sense only for briquetting without a binder, since the application of high pressure can lead, for example, in the case of using organic binders, to squeezing the binder onto the surface of briquettes and sticking them together.

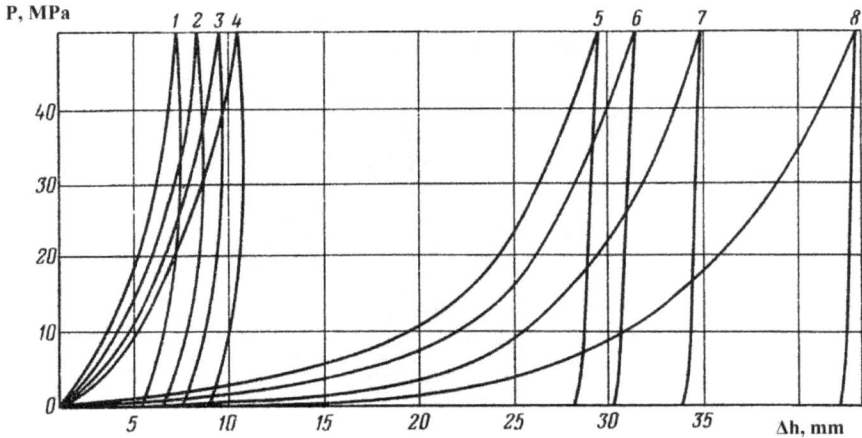

Figure 6.3. The degree of compaction of the mixture with coal and peat components depending on the compression load (1–4 - manganese ore fines and coal in proportions of 70:30, 60:40, 50:50 and 40:40 respectively; 5–8 - manganese ore fines and peat in proportions of 70:30, 60:40, 50:50, and 40:40, respectively [1].

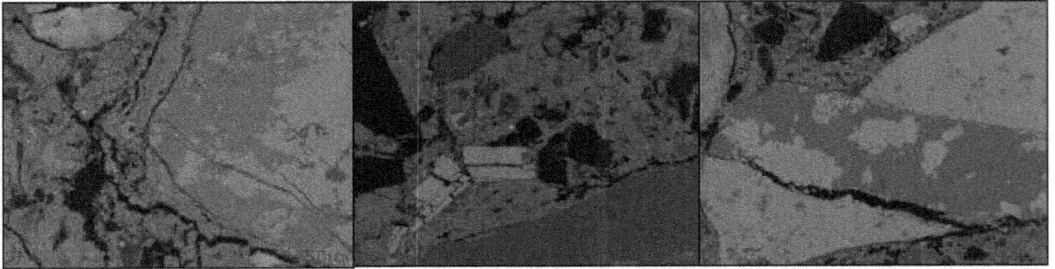

Figure 6.4. Structure of the roller-pressed briquette.

The structure of the material pressed in the rollers undergoes changes due to elastic and irreversible deformation, destruction of the particles of the pressed material and the formation of cracks in it (at pressures above 10 MPa). Figure 6.4 shows the structure of a roller-pressed briquette (manganese ore fines - 47.6%, dust of the gas cleaning system - 38.1%, coal - 9.5% and lignosulfonate as a binder - 4.8%) in comparison with the structure of extrusion briquette (manganese ore fines - 66%, gas cleaning dust - 28%, cement - 5% and bentonite - 1%).

It can be seen that as a result of high pressure (up to 100 MPa), the briquette structure is characterized by the presence of many large cracks and low porosity. The occurrence of such cracks may be due to elastic expansion after the finished briquette leaves the cells, which leads to a decrease in its strength.

A feature of roller briquetting is the limitation on the moisture content of the charge material (not higher than 5–10%). The required low moisture content of the charge limits the use of a binder that hardens with hydration, for example, cement, or requires the creation of combined binders with an increased content of cement in their composition.

The strength of the briquettes is significantly affected by the duration of the pressing process. For example, the duration of pressing for rolls with a diameter of 1.1 m at a frequency of rotation of 8–11 revolutions per minute is only 0.29–0.38 seconds. Exposure of the briquette under pressure allows not only more completely to force out the air from the narrowing pore space without the formation of pressed "air pockets", but also to reduce the number of elastic deformations that can lead to its softening. The increase in pressing time is achieved by restrictions on the speed of rotation of the rolls. For iron ore briquettes, usually no more than 6–8 revolutions of rolls per minute are recommended [1].

In 2003, a patent was obtained for the design of a roller press with a so-called Extended Compaction Zone (ECZ) [2].

The developed design allows to intensify the process of removing the gaseous phase from the volume of the pressed material and to achieve high degrees of compaction of the material by a combination of static compaction of the material during its movement along the narrowing channel into the lower part of the space between the rollers and the subsequent dynamic compaction in the zone of action of the vibrating compactor pumping the material into the pressing zone. The scheme of a roller press with ECZ is shown in Fig. 6.5, [2]. The strength of briquettes increases by 10–15%, and the productivity of such a press increases by 15–20%.

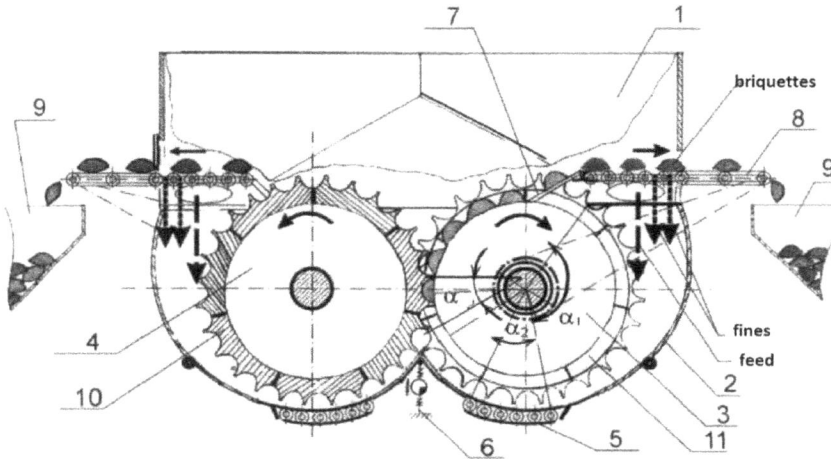

Figure 6.5. The scheme of a roller press with an extended compaction zone: 1 - loading hopper; 2 - arcuate plate of the extended sealing zone; 3, 4 - rolls; 5 - endless tape; 6 - vibrating device; 7 - cleaning comb; 8 - device for spillage recycling and removal of briquettes; 9 - finished product bins; 10 - removable forming elements of the gear type; 11 - removable forming elements of the grooved type [2].

The principle of the roller press with ECZ is as follows. Briquetted charge from the hopper (1) is fed into the extended sealing zone (2) through the loading openings made in the side walls of the press. In the channel of constant cross section formed by the working surface of the rolls, the side parts and the surface of the tapes (5), the internal stresses relax in the charge layer and are pre-stabilized. Next, the mixture is forcibly injected into the lower space between the rolls due to the rotation of the rolls (3 and 4) and the movement of the belts and under the influence of the vibrating compaction device (6). The mixture is pressed by rollers into briquettes, which remain in the annular depressions-grooves, from which they are removed by cleaning combs (7). Next, the briquettes come into the spillage selection device (8). The spillage and fines of the briquettes through the gaps between the rollers are returned to the loading zone of the roller press, and quality briquettes are transported to the bunkers of the finished product (9). To improve the performance characteristics, the roller press is equipped with removable forming elements (10 and 11), the use of which allows for the rapid restoration of the working surface of the rolls due to wear, as well as to obtain briquettes of various geometric shapes and sizes.

Figure 6.6 ([3]) shows the most common forms of the geometry of the cells of the roller press sleeves.

Operating experience of roller presses with traditional cell shapes showed that the displacement of the contacting cells resulted in the asymmetry of the briquettes formed, and the strength of the contact area of the halves of such a briquette is only 10–20% of the strength of the rest of the briquette mass. Due to the incomplete closure of cells of this type, a significant amount of return (up to 30%) may be formed when the roller press operates with the traditional form of cells.

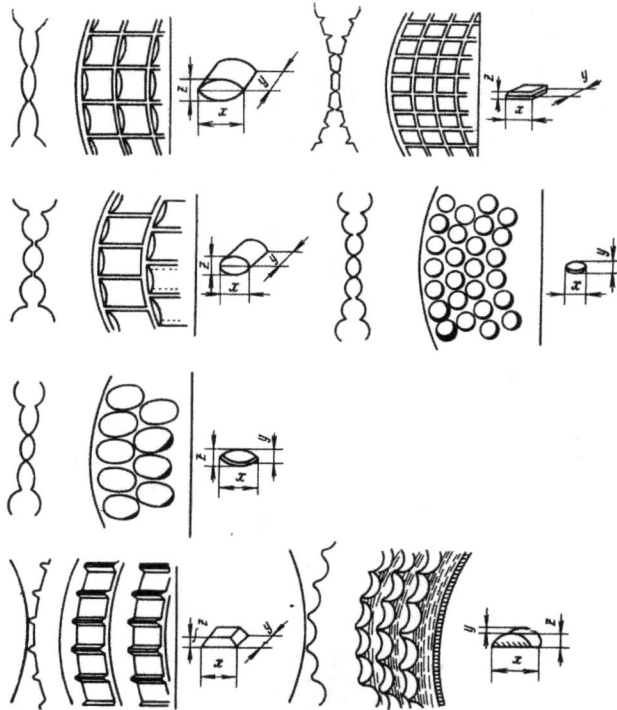

Figure 6.6. Different types of cells of the roller press sleeves [3].

Figure. 6.7. Grooved toothed cell shape [2].

In the design of a roller press, an open cell shape can be used, as, for example, in the ECZ press described above, the so-called grooved-toothed cell of the bandage (Fig. 6.7, [2]).

This form of cells does not require combining half-molds, allows doubling the volume of the charge captured by the cells in the compaction zone and refuse from additional pressing, allows achieving high workability in the manufacture of interchangeable sleeves in the conditions of briquetting factories in the metallurgical enterprises.

The sizes of the briquette manufactured by roller presses depend on the type of the metallurgical unit in which they are used as a charge component. One of the largest roller presses producers Köppern uses the following briquette sizes: $33 \times 30 \times 20$ mm for blast furnace briquettes and briquettes for direct iron production reactors, $46 \times 34 \times 20$ mm for electric arc furnaces, ore-smelting furnaces and basic oxygen converters, $62.5 \times 53 \times 34$ mm for cupola furnaces.

In roller briquetting, energy is expended on overcoming the friction forces between the particles of the material and the walls of the cells of the bandage and on the friction between the particles of the material themselves. The frictional forces in briquetting without a binder are important. If the magnitude of the adhesion force of the briquette with the surface of the cell exceeds the value of its tensile strength, the briquette is broken into two halves that remain in the cells.

6.1.2 Methods for modeling the processes of roll briquetting

The choice of the most optimal design of a roller press, which ensures the cost-effective production of strong briquettes, its rapid adaptation to various charge mixtures and binders in the factory may require simple and convenient methods for calculating the density and strength of briquettes, depending on the design parameters of the press itself and physical and mechanical properties of briquetted materials. The construction of a realistic mathematical model of roller briquetting, even with the use of modern computational software packages, is an extremely difficult task, since it requires taking into account the features of the geometry of the tire cells, the multicomponent nature of the briquetted mixture, its moisture content and the presence of a binder material.

Results satisfactory in terms of accuracy can be obtained using known calculation methods from the theory of rolling. The most widespread in solving problems of compaction by rolling was the Johansson method [4], who built a one-dimensional analytical model of rolling granular powder (Fig. 6.8). The process in the Johansson model is divided into the following zones:

1. The entry zone, in which the material to be pressed moves between the rollers at constant pressure due to the force of gravity. At this stage, the density of the material is assumed to be constant.

2. The sliding zone in which the material is still not compressed enough to prevent it from sliding over the roll surface. In this area, the material is subjected to shear stresses. The shift of the deformable charge takes place along the sliding lines, as a result of which the charge is pushed towards the supplied layers of material, slides along the surface of the rolls, thereby forming the so-called lagging zone. In the area under the sliding lines, the briquetted charge enters the compaction zone.

3. The compaction zone that starts when the material has the same speed as the roll surface. In this area, the pressure is at its highest and the powder is continuously deformed until it reaches maximum compaction with a minimum roll gap. The seal area is shaded in Fig. 6.9. It corresponds to the capture angle α. In aggregate, the first three zones have been referred to in the literature as the "deformation zone".

4. A discharge zone in which material exits at a speed that exceeds the speed of the roll surface.

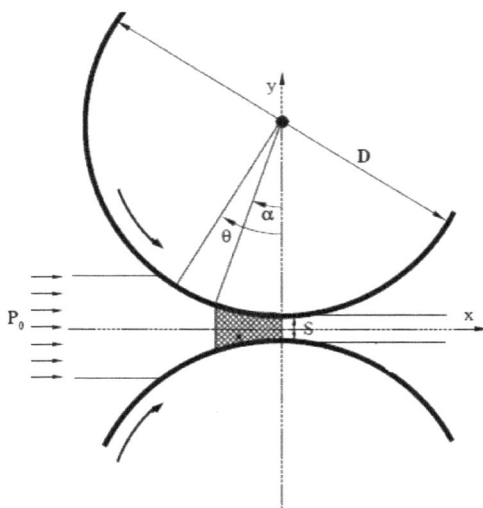

Figure 6.8. Calculation area in the Johansson method.

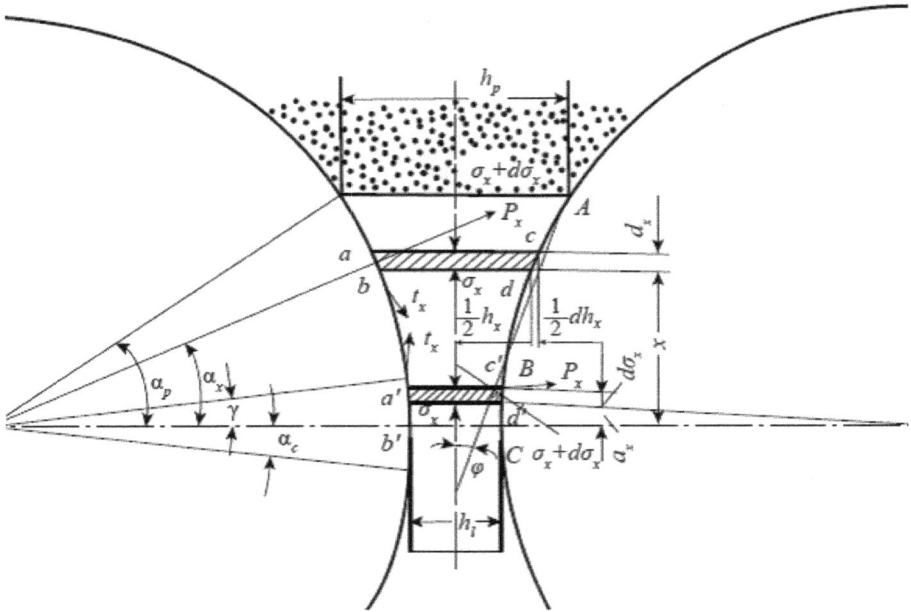

Figure 6.9. Scheme for calculating the rolling parameters by the Katashinsky method [4].

In the model, it is also assumed that the material is isotropic and deformable, the surface of the rolls is solid. Coulomb's law of friction is valid for the pressed mass. Taking into account these assumptions, it is possible to obtain the dependence of pressure on density and geometric parameters of the process (capture angle, gap width, etc.) in the form:

$$\sigma_\theta = \sigma_\alpha \left| \frac{\left(1 + \dfrac{S}{D} - \cos\alpha\right)\cos\alpha}{\left(1 + \dfrac{S}{D} - \cos\theta\right)\cos\theta} \right|^k \quad for \ \theta \leq \alpha \tag{1}$$

k - coefficient of compressibility, determined from the ratio:

$$\frac{\sigma_\alpha}{\sigma_\theta} = \left(\frac{\rho_\alpha}{\rho_\theta}\right)^k \tag{2}$$

The distribution of pressure gradients in the sliding zone is determined by the following differential equation:

$$\frac{d\sigma}{dx} = \frac{4\sigma\left(\dfrac{\pi}{2} - \theta - v\right)tg\delta}{\dfrac{D}{2}\left[\left(1 + \dfrac{S}{D} - \cos\theta\right)\right][ctg(A-U) - ctg(A+U)]} \tag{3}$$

outside the sliding zone is true:

$$\frac{d\sigma}{dx} = \frac{k\sigma_\theta\left(2\cos\theta - 1 - \dfrac{S}{D}\right)tg\theta}{\dfrac{D}{2}\left[\left(1 + \dfrac{S}{D} - \cos\theta\right)\right]\cos\theta} \tag{4}$$

where

$$A = 0.5\left(\theta + \mu + \frac{\pi}{2}\right) \tag{5}$$

$$U = \frac{\pi}{4} - \frac{\delta}{2} \tag{6}$$

$$2v = \pi - arcsin\frac{sin\phi_w}{sin\phi_e} - \phi_w \tag{7}$$

ϕ_w - wall friction angle, ϕ_e - effective angle of friction.

The analytical method for plotting specific pressure plots developed in 1966 by V.P. Katashinsky was also widely used for calculating the parameters of compaction of material by rolls (Fig. 6.9). In literature, it is called the Slab method.

When the material is compacted with rolls, the length of the deformation zone is much greater than the average thickness of the rolled stock, which means that it can be assumed that the deformation is uniformly distributed over the section. In Fig. 6.9, the rolling angle α_p determines the position of the beginning of the deformation zone and the thickness of the initial section h_p. The angle of elastic compression of the rolls α_c decides the cross-section of the exit of the rolled product from the rolls. In accordance with the accepted provision on the constancy of the density and thickness of the rolled product after passing through the neutral section, the deformed arc BC is replaced by a chord parallel to the rolling axis. The arc AB is replaced by a chord passing through points A and B. Such a replacement is permissible, since when rolling bulk materials, the angle α_p usually does not exceed 0.2 rad (11° 25'). It is also assumed that during rolling the density change occurs only in the so-called lagging zone on the α_p—γ arc. After the neutral section, there is no change in the density and thickness of the rolled stock. Elastic deformation in the Katashinsky model is also neglected, since its value does not exceed 2%. In the lagging zone, an infinitely small element *abdc* is distinguished, bounded by the surfaces of the rolls and by two planes perpendicular to the rolling direction and located one from another at an infinitely small distance *dx*. From the equilibrium conditions of this element, by projecting all the forces acting on it on the direction of rolling, the following equation can be obtained:

$$h_x d\sigma_x + \sigma_x dh_x - p_x dh_x + t_x \frac{dh_x}{\tan\alpha_x} = 0 \tag{8}$$

where σ_x – average normal compressive stress occurring in the rolled material over the section *bd*, p_x – specific pressure of rolls on rolled material; t_x – contact frictional force acting on the material from the side of the rolls; α_x – center angle that defines the position of the element *abdc*. When replacing the arc AB with a chord, the angle α_x can be considered constant and equal to the angle ϕ, which is equal to $0.5(\alpha_p + \gamma)$. To solve equation (8), it is necessary to establish a relationship between p_x and σ_x and between p_x and t_x. The model assumes that the contact friction force t_x is proportional to specific pressure. Then $t_x = \mu p_x$. The relationship between specific pressure p_x and tension σ_x can be obtained from the plasticity equation for two-dimensional deformation. The plasticity equation for the plane deformation scheme has the form:

$$\sigma_1 - \sigma_3 = K \tag{9}$$

where σ_1 and σ_3 - the largest and the smallest major normal stresses; K - resistance to plastic deformation.

It can be assumed with satisfactory accuracy that $\sigma_1 = p_x$ and $\sigma_3 = \sigma_x$. Taking this into account, omitting intermediate calculations, it can be seen that equation (8) is reduced to the following form:

$$\frac{dp_x}{dh_x} + p_x \frac{\mu \cot\varphi}{h_x} = \frac{K}{\Delta h}\left(\frac{h_p}{h_x} - 2\right) \tag{10}$$

Simplifying the solution of the ordinary differential equation obtained in general form, it is possible to obtain an analytical expression for the dependence of pressure on rolling parameters and material properties in the form:

$$P_x = \frac{K}{R(\alpha_p - \gamma)} \left[\frac{h_p}{2\mu} - \frac{h_x}{\mu + \varphi} + \frac{\mu - \varphi}{2\mu(\mu + \varphi)} h_p \left(\frac{h_p}{h_x} \right)^{\frac{\mu}{\varphi}} \right]$$ (11)

The slab method was further developed in the work of Musikhin [5], who developed a one-dimensional model of a roller compaction that takes into account the difference between the maximum and minimum stresses determining that there is slip in the entire rolling area except for the neutral angle zone. It is assumed that as the particulate material enters this area, it begins to exhibit the properties of a solid porous body. Just as in Katashinsky's work, the approach is based on considering the equilibrium conditions for the elementary volume of a deformed body. The differential equation determines the dependence of the pressure distribution depending on the geometric parameters and physical and mechanical properties of the material. The solution describes the pressure distribution in two areas—the areas of lagging and sliding. The disadvantage of Musikhin's model is the loss of shear stresses. It is assumed that only slippage occurs on the roll surface. In 1991, Dec published a paper in which the possibility of plastic deformation was additionally considered in the slab model [6].

The development of numerical modeling methods and the emergence of the ABAQUS application package led to the widespread use of the finite difference method and the finite element method, and later the discrete element method, for modeling the compaction of materials in roller presses [7]. Comparison of the results of numerical modeling and calculations based on the simplified analytical models of Johansson and Katashinsky with the experimentally obtained data showed that the finite element method leads to a higher accuracy in calculating the distributions of pressure, neutral angle, angle of capture and other important compaction parameters during rolling [8]. An important advantage of the finite element method is the ability to tune the model to take into account as much information as possible, such as press geometry, roll surface properties, strain energy, shear stress, etc., which allows this method to obtain a realistic picture of the process to optimize the design of the roll press. As an illustration, Fig. 6.10 shows an image of the velocity field of the rolled material obtained using the ABAQUS computational package [9]. The results obtained in this way showed that the analytical model of Johansson can lead to exaggerated and often physically meaningless values of the pressure values in the compacted mass.

Comparison of the simulation results using the ABAQUs package with the calculations based on the slab method (Katashinsky) showed a very satisfactory coincidence in the localization of the point at which the speeds of the material and rolls coincide (neutral point), which is illustrated in Fig. 6.11 [10].

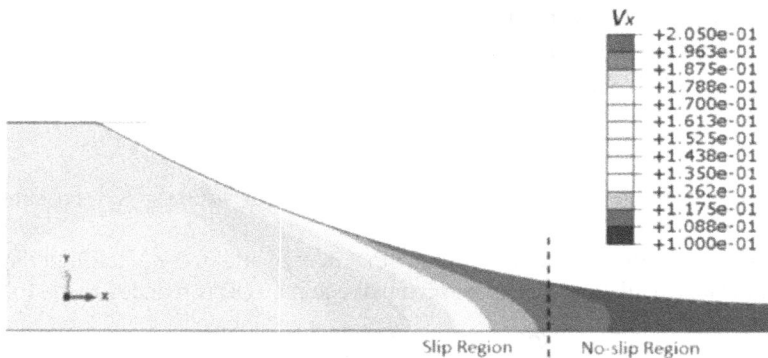

Figure 6.10. The field of velocities in the body of the rolled powder obtained using the ABAQUS package [10].

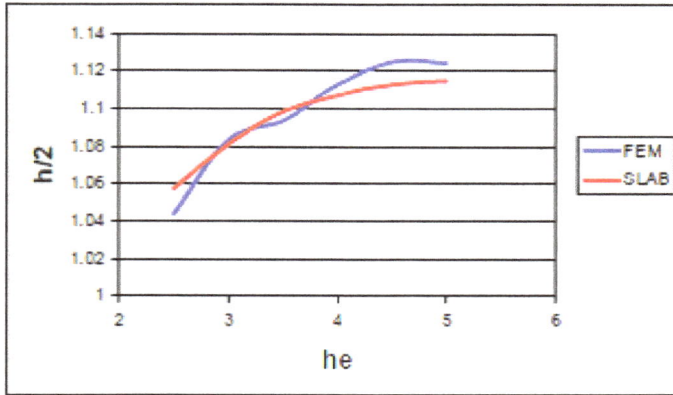

Figure 6.11. Dependence of the position of the beginning of the compaction zone (h) on the width of the entry zone (h_e). Blue FEM curve, red - slab method [10].

Thus, the methods based on analytical simplified calculations and powerful computational packages used in the theory of compaction of materials by rolling are currently the only way to predict the properties of roll briquettes. In literature, there are practically no mathematical models that would describe the process of obtaining precisely roll briquettes, and not rolled products. In the absence of such specialized calculation methods, empirical and semi-empirical methods for predicting the strength properties of such briquettes, based on the approximation of experimental results by power polynomials, began to gain wide acceptance. Such results do not have a universal meaning and must be recalculated every time for a new charge, but they can be successfully used to optimize the operation of a particular roller press.

References

[1] Ravich, B.M., Vasiliev, V.I. and Myakinkov, Yu.I. 1977. Investigation of pressing parameters from the diagrams of plastic and elastic deformations. In Collection: Enrichment and Processing of Coal, TsNIIEugol pp. 12–13 (in Russian).

[2] Patent of the Russian Federation No. 2204486 Application dated 06.08.2001 Published on 20.05.03 Bulletin No. 14 (in Russian).

[3] Ravich, B.M. 1975. Briquetting in Non-Ferrous and Ferrous Metallurgy. Moscow: Metallurgy, 232 p. (in Russian).

[4] Vinogradov, G.A., Cartush, O.A. and Katashinsky, V.P. 1969. Rolling metal powders. M.: Metallurgy, p. 382 (in Russian).

[5] Musikhin, A.M. 1979. Compaction in the lead zone when rolling powders. Powder Metallurgy 2: 1–6 (in Russian).

[6] Dec, R.T. 1991. Study of Compaction Process in Roll Press, IBA Proceedings, Vol. 22, November 1991, 22nd Biennial Conference, San Antonio, TX, pp. 207–217.

[7] Electronic resource. https://www.3ds.com/support/hardware-and-software/.

[8] Zavaliangos, A., Dec, R.T. and Komarek, R.K. 2003. Analysis of powder processing in the roller press using finite element modeling, XXII International Mineral Processing Congress, Cape Town, South Africa.

[9] Liu, Y. and Wassgren, C. 2016. Modifications to Johanson's roll compaction model for improved relative density predictions. Powder Technology 297: 294–302.

[10] Reyero, J., León, J., Luri, R. and Luis, C.J. 2006. Comparative Analysis between FEM and Slab Method for Flat Rolling Process. 10th International Research/Expert Conference. Trends in the Development of Machinery and Associated Technology TMT 2006, Barcelona-Lloret de Mar, Spain, 11–15 September, 2006.

6.1.3 Major manufacturers of roller presses for briquetting in the iron and steel industry

One of the largest manufacturers of roller presses used for briquetting natural and industrial raw materials in the iron and steel industry is Köppern, which has supplied 450 cold briquetting roll presses and 143 hot briquetted iron presses since 1950. Figure 6.12 shows one of the first roller presses of the company, which was used in the 20s of the last century for coal briquetting and a modern briquetting press.

An important design feature of the company's roller presses is that the gap between the rolls is controlled automatically by a hydraulic station, depending on the requirements of a particular process, which guarantees the passage of all the feed material through the gap between the rollers under the same technological conditions, thereby ensuring a stable quality of the final product. Depending on the characteristics of the processed material and its ability to fill the forming cells of the press rolls, loading into the roll press is carried out by means of either a screw or gravity feeder. The maximum pressing force of the Köppern equipment is 20,500 kN. Equipment weight—up to 400 tons. The productivity of Köppern briquette roller presses is up to 100 tons of briquettes per hour. It is clear that in this case one is describing the briquetting of metallized raw materials.

Köppern is also renowned for its pioneering work in improving the strength and wear resistance of materials used in roller press designs, including tires. The cells of the bandages for the formation of briquettes have been manufactured using the technology of electrochemical machining (ECM – Electro Chemical Machining) since 1965, which has gained wide popularity and distribution in industry, energy and astronautics. Electrochemical treatment consists in removing metal from the surface of the workpiece being processed by electrolytic dissolution until the required shape and size is achieved (Fig. 6.13).

Figure 6.12. Köppern briquette roller presses. Left - a press for coal briquetting (20s of the last century). On the right is a modern briquette press.

Figure 6.13. ECM surface treatment of sleeves on a Köppern roller press.

The development of another original development of the Köppern Company—the HEXADUR® wear protection system, developed for the cement industry, is the patented RESIDUR® technology for the production of steel tires for briquetting with a highly wear-resistant metal-powder surface. The coating is applied to the base ring using a special technology, for which the base ring is first placed in a special capsule. The distance between the outer walls of the capsule and the base ring is filled with powder, which interacts with the surface of the base ring. The capsule is sealed hermetically, having being evacuated before, subjected to sintering at a temperature of 1000°C and a pressure of about 1000 bar, as a result of which a non-porous wear-resistant coating is formed on the surface of the base ring. Next, the mold is heat treated to achieve the required level of strength and ductility, after which RESIDUR® bands are processed using ECM technology to form cells of the required shapes and sizes.

The roll structure consists of two parts: a roll base (core) and a wear-resistant band, fixed on the base by a shrink-fit method, which allows the roll core to be reused at the end of the service life of the band. Figure 6.14 shows the comparison of the degree of wear for the same time of operation of fragments of a cast band made of tool steel and a band made using the technology RESIDUR®.

These technological developments have significantly reduced equipment downtime caused by wear and related repairs.

Köppern briquetting presses in the iron and steel industry enable the briquetting of a wide range of natural and man-made materials with a range of binders. However, the restrictions on the moisture content of the briquetted charge create difficulties in using Portland cement as a binder, which has become widespread. Roller briquettes on cement have low strength at the exit from the press, and the exit itself can be difficult. The required proportion of cement in the briquette mass can reach 8–12%, which is comparable to the content of this binder in vibropressed briquettes, but twice as high as required in stiff extrusion briquettes.

The company's experience in briquetting heated plastic materials, high productivity of presses (up to 100 tons per hour), their reliability and high wear resistance of the tires have led to the fact that today Köppern is the world leader in hot briquetting technology. The Köppern roll briquetting technology transforms sponge iron immediately after it leaves the direct reduction reactor at a temperature of 750°C into a form more convenient for transportation and storage—into hot briquetted iron (HBI, Fig. 6.15). Such pressing makes it possible to effectively prevent manifestations of pyrophoricity of sponge iron during its transportation and storage, due to compaction and reduction of porosity.

Hot briquetting roll presses and other ancillary equipment are especially designed by Köppern for high temperature applications. The design of the equipment is made of heat-resistant steel, water cooling of the different parts of the equipment and inertization of machine bodies are provided.

Another major manufacturer of roller briquetting presses is Komarek, part of the Köppern group of companies. The company has been in the briquetting market for over a century, starting, like Köppern, with coal briquetting. In 1962, the company was headed by Richard Komarek, that

Figure 6.14. Degrees of wear of sleeves with the same duration of operation. On the left is a cast tool steel sleeve. On the right is a sleeve made using the technology RESIDUR®.

Figure 6.15. Hot briquetted iron produced by Köppern roller press.

Figure. 6.16. Komarek segmented roller press sleeves.

Figure 6.17. Komarek roller press DH500-28 × 2.

was described earlier. Since 1968, the company has marketed over 600 briquetting, pelletizing and compacting presses in more than 40 countries around the world. Komarek was the first to use segmentable bandages with replaceable elements (Fig. 6.16).

The maximum productivity of Komarek presses is 54 tons of briquettes per hour (Model DH 500-28 × 2, Fig. 6.17). This model is designed for coal briquetting.

Figure 6.18. Roller press Komarek DH500.

For briquetting of metal oxides, ore fines and sludge, a Komarek DH500 press (Fig. 6.18) is used with a capacity of up to 45 tons per hour.

The design features of this press are in the use of solid or segmental tires, in the vertical feeding of the charge by gravity or using a screw feeder (if pre-pressing is necessary), in the vacuum deaeration of the fine powder components of the charge, which, as noted above, is very important for roller briquetting, in the use of motors with the ability to control the speed of rotation of rolls and auger, in the use of materials from wear-resistant alloys, etc. Roll diameter 710 mm, roll width 229–508 mm, pressing force up to 3000 kN. Roll drive power 200 kW, feeder drive 22 kW. Press weight—up to 33 tons.

In 2001, Komarek acquired a major stake in EURAGGLO, now the European division of Komarek. The maximum productivity of EURAGGLO roller presses reaches 65 tons per hour (model E92, Fig. 6.19). The model is designed for briquetting at medium and high pressures. Pressing force - 4200 kN, roll diameter 1200 mm, width - 300–460 mm, roll drive power 500 kW.

Hosokawa-Bepex is one of the largest manufacturers of roller briquetting presses. For briquetting using high pressure, roller presses of the MS series are used, which are also suitable for processing abrasive and hot materials (Fig. 6.20). A design feature of Hosokawa-Bepex presses is the possibility of their operation in an inert gas atmosphere, as, for example, when working with materials that require exclusion of contact with oxygen. In this case, the bodies of the MS series roller presses are made in a gas-tight version. The overpressure can reach over 140 mbar. Of all roll press options, Hosokawa Bepex veneers the inner surfaces of the pressing device to minimize contact with the workpiece. The entire pressing device can be made entirely of stainless steel or other alloys to meet the requirements of specific applications. The surface of the rolls can be smooth, profiled or with grooves. For MS series presses there are rolls suitable for a wide variety of applications (segmented, for highly abrasive materials and high temperatures; with an outer band, for moderately abrasive materials and temperatures up to 450°C; solid rolls, Fig. 6.21). The pressing force of the MS series presses ranges from 360 to 6000 kN. Roll diameters—from 300 to 1100 mm.

An important distinguishing advantage of Hosokawa-Bepex roller presses is their high productivity—up to 150 tons per hour (for metallized raw materials).

The largest French manufacturer of roller presses for briquetting coal and ore concentrates is Sahut-Conreur, one of the oldest briquetting companies in the world. The first coal briquette factory was built by this company in 1859. Roller presses from this company have been used for

Figure 6.19. EURAGGLO E92 roll press.

Figure 6.20. Hosokawa-Bepex MS series roller press.

briquetting since 1860. The productivity of modern roller presses of the company is from 500 kg to 100 tons of briquettes per hour (for metallized raw materials). The pressing force reaches values of 10–50 kN per linear centimeter of roll width. Roll diameter - 250–1400 mm (Fig. 6.22). The presses are equipped with a hydraulic roll compression system and an automatic roll rotation speed control system.

Figure 6.21. Hosokawa-Bepex roll options.

Figure 6.22. Sahut-Conreur roller press.

Figure 6.23. Sahut-Conreur roll press sleeves for hot briquetting.

Like Köppern, Sahut-Conreur produces briquette roll presses for hot briquetting (Fig. 6.23).

The company is also the developer of the patented Cold Briquetted Iron and Carbon (CBIC) concept. The raw material for such briquetting is the so-called cold iron of direct reduction (Fig. 6.24) and carbon. The largest producer of such iron is Iran. The Sahut-Conreur process serves the same purpose as Köppern hot briquetting presses—passivation of pyrophoric sponge iron.

Figure 6.24. Cold briquetted direct reduced iron with carbon (CBIC).

Briquettes from cold direct reduced iron with added carbon (CBIC) have some advantages over the original form of this material—higher bulk density (2.5 versus 1.9 t/m³), higher oxidation resistance, lower hygroscopicity and, of course, higher carbon content (up to 10%). These factors not only make storage and shipping easier and more economical, but also have a significant impact on the steelmaking process. Pilot tests have shown that when using such briquettes for steel smelting instead of cold direct reduced iron, it led to a decrease in dust emission and energy consumption and to an increase in furnace productivity. In early 2018, a commercial CBIC plant was commissioned in Iran.

6.2 Vibropressing for Briquetting

6.2.1 *The physical essence of vibropressing and the structure of the briquette*

The possibility of applying vibration for the agglomeration of natural and anthropogenic raw materials in metallurgy is because of a number of reasons. The very first experience of vibratory molding in powder metallurgy in the late 40s of the last century showed that the use of vibration when filling powder into a mold or in the process of compaction allows significantly reducing the required compressive load and homogenizing the distribution of its density over the volume. It was found that at vibrations with a frequency exceeding 50 Hz, the bonds between the particles in the compacted dry powder are destroyed and the internal friction in the compressible mass sharply decreases, which facilitates the convergence of particles and compaction of the mixture. In this case, a higher degree of compaction is achieved lower than during compression, the values of the applied pressing pressures, which is well illustrated in Fig. 6.25 [1].

Another important physical process that occurs when vibration is applied to a formable mass containing gel phases is the so-called "thixotropy" (The word comes from Ancient Greek *thixis* "touch" and *tropous* "to turn")—a decrease in viscosity (liquefaction) under mechanical action and thickening in state of rest. Manifestations of thixotropy are central to the process of vibration compaction of concrete [2]. Under the influence of vibration, the cement gel transforms into a sol, simplifying the movement (convergence) of solid aggregate particles under the action of its own gravity, which leads to compaction of concrete. Similar processes take place during briquetting with the use of a cement binder, when, due to the reversible transformation of a cement gel into sol when exposed to vibration, at the stages of its liquefaction, particles of the briquetted charge approach each other under the action of their own gravity, which contributes to compaction of the briquette, and the air displaced by the approaching particles is released on the surface of the compressible mass in the form of bubbles. The thixotropy of the gel and the reduction of internal friction due to the oscillatory movements of the particles caused by vibration lead to the fact that the briquetted mass acquires some properties of the liquid, which simplifies the molding.

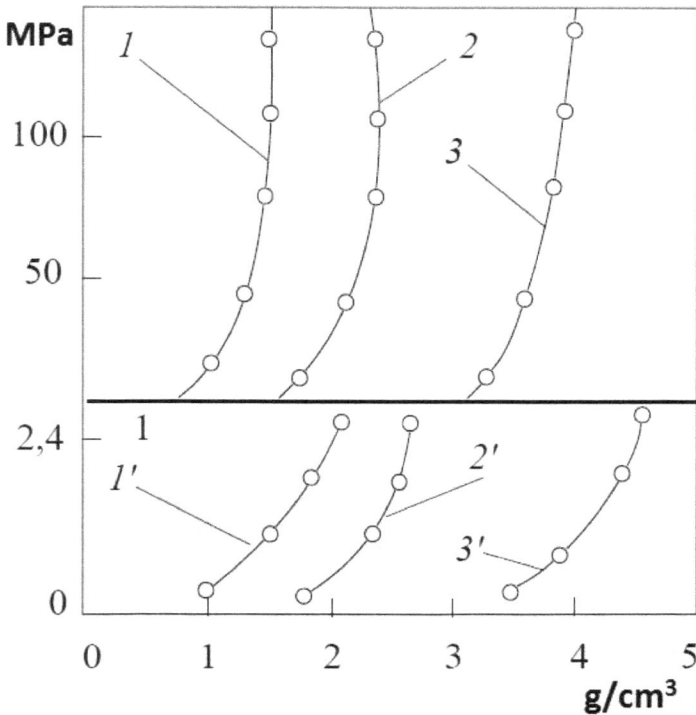

Figure 6.25. Comparison of the degree of compaction of the material during static pressing (1–3) and vibropressing (1'–3').
1, 1' - boron carbide; 2, 2' - silicon carbide; 3,3' - titanium carbide.

It is clear that thixotropy plays a key role in vibropress briquetting. It is no coincidence that almost all known vibropress briquette factories use cement as a binder. That is why the moisture content of the charge components plays an important role in briquetting with vibropressing. Its quantity should be sufficient to preserve the properties of the cement gel and for further hydration hardening of the cement. Usually, its content in the briquetted mixture is limited to 5–8% by weight of the briquette. The cycle of vibratory compaction lasts less than 30–40 seconds, which is obviously not enough for the cement to "set". As a result, newly formed vibropress briquettes have a very low mechanical strength, which does not allow them to be transported and stacked like roll briquettes. Therefore, the equipment of the vibropress briquetting factories include special mechanisms to transport and accumulate finished products on pallets. In addition, to speed up the curing of briquettes, "steaming" is used—heat and humidity treatment at a temperature of 70–95°C in an atmosphere of saturated steam. Movement of moisture and steam in a briquette that has not yet become strong may lead to its softening.

The requirements for the granulometric composition of the briquetted charge during vibropressing are not as rigid as for roller briquetting (less than 5–6 mm). An important advantage of this method of briquetting is the possibility of agglomeration of materials with a particle size of up to 10 mm. Figure 6.26 shows the briquette obtained by the method of vibropressing, with the presence of large particles of material in the structure of the briquette.

The structure of the briquette produced by the method of vibropressing is highly homogeneous. Figure 6.27 shows an image of the structure of an industrial briquette sample (center and periphery), produced by vibropressing. The composition of the briquette is iron ore fines (61%), coke breeze (14.5%), flue dust (14.5%) and cement (10%). The uniform distribution of the main components of the briquette mixture (iron oxides, coke breeze and cement) attracts attention. It can be seen that the largest grains are concentrated in the central part of the briquette.

Figure 6.26. Vibropressed briquette.

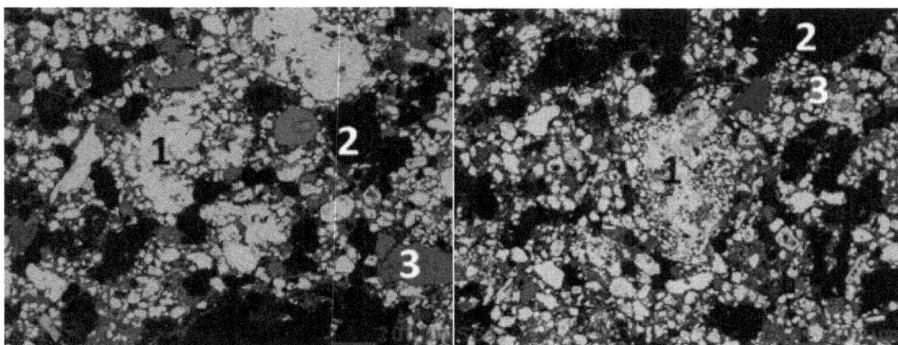

Figure 6.27. SEM image of the vibropress briquette structure. On the left - the center of the briquette, on the right - the periphery (1 - iron oxide, 2 - coke breeze, 3 - cement).

The properties of briquettes produced by the method of vibropressing are significantly influenced by the vibration parameters (duration, amplitude and frequency). Insufficient intensity of vibration may not lead to the manifestation of thixotropy, which will not allow achieving the required values of the density of the briquetted mass. Too intense vibration can lead to stratification of the briquetted mass and also degrade the density characteristics of the briquette. There is no theoretical basis for choosing a vibration mode. In each case, one is describing the experimental testing of such modes on the bench models. To some extent, recommendations on the vibration modes developed in the technology of concrete vibration compaction may be useful. According to [3], the whole process of vibro compaction can be formally divided into two stages. The first is the rearrangement of large particles of a flexible mixture and the formation of a macrostructure, the second is a manifestation of the thixotropy of the cement gel and the formation of a microstructure. For the first stage, it is recommended to apply vibration, which causes low-frequency oscillations with large displacement amplitude to overcome the friction forces of the non-compacted particles. At the second stage, higher frequencies of vibration are advisable.

The methods of the theory of similarity and dimension allow determining which factors influence the process of compaction of the vibrating mass. The main parameters that determine the process of vibro compaction are 10 values: N is the power of vibration exposure, $[Fl/t]$; A is the amplitude of oscillation, $[l]$; ω is the angular frequency of oscillations, $[1/t]$; t is the duration of the vibration; ρ is the density during vibro forming, $[Ft^2/l^4]$; τ_0 - shear resistance of the briquetted mixture, $[F/l^2]$; ν - mixture viscosity, $[Ft/l^2]$; l - the specific size of the forming zone, $[l]$; β is the linear damping coefficient of oscillations; g - gravitational acceleration. Here, F is force, t is time, and l is length. In accordance with the well-known π-theorem, the number of dimensionless combinations of defining parameters in the case of vibropressing is equal to 7. These combinations can be represented as:

$$\rho l^3 A^2\ \omega^3 l^3/N^2;\ \rho A^2\omega^2/\tau_0;\ \rho A\omega l/\beta v;\ \rho A\omega l/v;\ Nt/\tau_0\ l^3;\ Nt^2/vl^3;\ A\omega^2/g$$

The dependence of the density of the mixture during vibropressing can be obtained from the equation in general:

$$\rho(A^2\omega^3l^3/N^2;\ A^2\omega^2/\tau_0;\ A\omega l/\beta v) = f(Nt/\tau_0\ l^3;\ Nt^2/v\ l^3;\ A\omega^2/g;\ A\beta)$$

The qualitative conclusions from the dependencies obtained in this way are reduced to the fact that the process of compaction of the mass during vibropressing is influenced by the acceleration of oscillations ($A\omega^2$), the duration of the process is associated with the specific power of the vibration effect and with the properties of the briquetted mixture. The higher the vibration amplitude, the more effective the compaction.

6.2.2 Technology of vibropressing, transportation, heat treatment and storage of briquettes

The process of vibropressing consists of several stages. The pallet is mounted on the vibrating table. The briquetted mixture prepared in the mixer with the addition of a binder is poured into a replaceable mold tooling—a matrix. Next, the mixture is compressed by the punch, a kind of "mirror" reflection of the matrix, ideally entering it exactly like a piston in a cylinder, and the vibration of the whole unit is turned on (Fig. 6.28).

The duration of the vibration cycle is 15–40 seconds. At the end of the molding cycle, the vibration is automatically turned off, the pressure in the hydraulic system decreases, the matrix rises, and the formed briquettes remain on the process pallet. The appearance of the briquetting vibropress is shown in Fig. 6.29.

The low strength of freshly formed briquettes does not allow them to be delivered to the zone of curing by the conveyor. Preserving the integrity of freshly formed briquettes requires special measures. Briquettes are transported, remaining on technological pallets.

Pallets with freshly formed briquettes along the conveyor (Fig. 6.30) are fed to a special hoist drive (Fig. 6.31), from where a stack of pallets is delivered to the heat treatment zone.

After accumulating the required number of pallets on the pallet stacker, they are removed and further transportation to the heat treatment chamber is carried out by an automatic stacked cart (Fig. 6.32) moving along the rails.

It is possible to transport a stack of pallets to the heat treatment zone and with the help of a transborder consisting of transfer and dispensing carts, which ensures the accuracy of the installation of pallets in the heating chambers. The appearance of the heat treatment chambers is shown in Fig. 6.33.

Figure 6.28. Matrix (left) and punch (right).

Figure 6.29. Vibropress for briquette production.

Figure 6.30. Transporting briquette pallets by conveyor.

The duration of heat treatment can reach 24 hours or more. The temperature in the chamber is 70–95°C. The cost of heat treatment chambers can exceed 20% of the cost of equipment and the engineering costs.

After heat treatment, the briquettes on the pallets are moved to the stacker and further using a hydraulic pusher to the finished product conveyor for further unloading to the warehouse or loading onto a vehicle. The pallets freed from briquettes are returned to the vibropress. With a briquette line capacity of up to 130 thousand tons of briquettes per year, the number of formable pallets can exceed one million pieces. The cost of technological pallets may exceed the cost of the vibropress itself. In general, the cost of special mechanisms and devices for automatic lines of vibropressing, allows ensuring the preservation of raw briquettes intact until delivery to the heat treatment chambers and

Figure 6.31. Pallet stacker.

Figure 6.32. Stacked carts for transporting briquette pallets.

the return of pallets can double the cost of vibropress. It is clear that in the absence of automation of these procedures, the capacity of the briquetting plant will be significantly lower. The cost of automation can be up to 50% of the cost of the vibropress.

With regard to the specifics of vibropressing, the productivity of a vibropress is usually understood as the number of units of products produced during an hour or eight-hour shift. First, productivity depends on the degree of automation, which affects the rate of removal of pallets with raw briquettes from the working area. Equipment productivity essentially depends on the sizes and volume of briquettes. Therefore, the performance is sometimes used to compare the number of pallets filled in a certain period of time. According to the results of operation of the known automated vibropress briquette factories, their productivity does not exceed 20–30 tons of briquettes per hour.

Figure 6.33. Heat treatment chambers.

Figure 6.34. Unloading of vibropressed briquettes from the car.

The size of briquettes is 60–100 mm; the most widely used form is a hexagonal prism. Such a size of vibropressed briquettes can lead to their "bridging" when unloaded into the bunker (Fig. 6.34). The smaller the size of the briquette produced, the lower the performance of the vibropress.

6.2.3 Major manufacturers of vibropresses for briquetting in ferrous metallurgy

The largest manufacturers of equipment for the production of briquettes by vibrocompression are Hess and Masa (Germany). Hess supplies to the concrete vibratory markets the MULTIMAT series of concrete forming machines, which can also be used for briquetting in the iron and steel industry. The MULTIMAT RH 2000-3 MA has the highest productivity (Fig. 6.35).

The dimensions of the technological pallet are 1400 × 1300 mm, the forming area is 1300 × 1250 mm, the cycle time is 10 seconds. The productivity of the machine for the production of paving stones (10 × 20 × 6 cm) is 352 m² of this coating per hour. A feature of Hess vibropresses is the patented VARIO TRONIC vibration system, which allows setting vibration parameters (frequency and amplitude) individually for each type of vibration and achieving optimal compaction

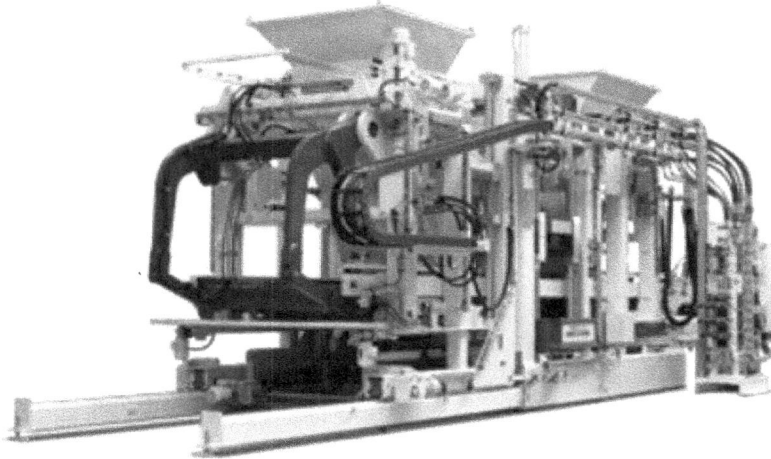

Figure 6.35. Hess MULTIMATRH 2000-3 MA concrete forming machine.

Figure 6.36. Transporting pallets with products to the heat-moisture treatment zone.

using eight vibrators, with minimal wear of equipment components. The company supplies the market with automatic briquetting lines, which, in addition to vibrating presses, also include special devices for transporting pallets with raw briquettes to the heat-moisture treatment zone and finished briquettes to the places of their shipment to consumers (pallet conveyors, storage hoists, transborders, multiformes, cameras heat treatment, briquette ejectors, etc.). Figures 6.36 and 6.37 show the components of Hess vibratory compaction lines for transporting and collecting pallets for heat treatment.

Masa manufactures high-performance stone forming machines XL 9.1 and XL 9.2 (Fig. 6.38).

The productivity of the XL 9.2 machine reaches 2938 m³ of rectangular paving slabs with dimensions of 200 × 100 × 80 mm, with a standard technological pallet size of 1400 × 1300 mm. The company supplies complete briquetting lines on a turnkey basis. The main equipment, like that of Hess, includes special devices and mechanisms that ensure the safety of raw briquettes up to delivery to the heat treatment chambers for curing (Figs. 6.39–6.41).

Figure 6.37. Accumulation of pallets in the heat-moisture treatment area.

Figure 6.38. Masa XL series stone forming machine.

Figure 6.39. Transporting pallets with briquettes.

Figure 6.40. Products in curing chambers.

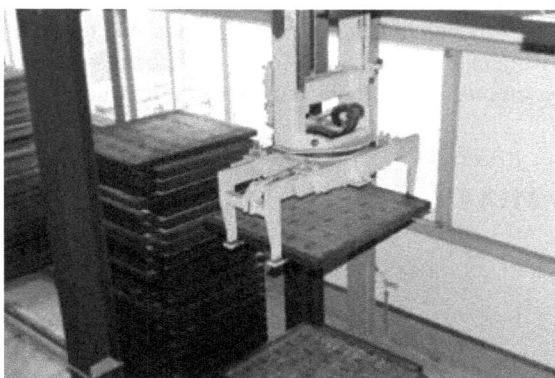

Figure 6.41. Masa automatic production pallet accumulator.

References

[1] Guzman, I.Ya. (Ed.). 2003. Chemical technology of ceramics. Textbook Manual for Universities - M.: OOO RIF "Stroimaterialy", 496 p. (in Russian).

[2] Akhverdov, I.N. 1981. Fundamentals of Concrete Physics. Moscow: Stroyizdat, 464 p. (in Russian).

[3] Gusev, B.V., Olenich, D.I. and Nuryeva, M. 2017. Application of methods of the theory of similarity and analysis of dimensions in the study of wave phenomena. Building Materials, Equipment, Technologies of the XXI Century. 9-10(224-225): S. 50–51 (in Russian).

6.3 Stiff Vacuum Extrusion Briquetting Technology

Extrusion is a process used to create objects with a fixed cross section profile. The material is pushed through the die of the desired cross section. This method has found a great deal of distribution in the industry of the production of ceramic bricks. Vacuum is maintained in the working chambers of modern brick-making extruders, which contributes to achieving greater uniformity and density of the product. Stiff Vacuum Extrusion (SVE) technology is applied in the production of ceramic bricks in 64 countries around the world, including the United States, Britain, Germany, South Korea and South Africa. The world's largest brick factory in Saudi Arabia produces a million bricks per day using SVE technology.

In accordance with brick industry terminology, the word "stiff" is used to describe the process of extrusion, which is carried out at pressures ranging from 2.5 to 4.5 MPa and moisture contents ranging from 12–18% (Table 6.1, [1]).

Table **6.1.** Types of extrusion.

Type of extrusion	Low-pressure extrusion	Medium-pressure extrusion	High-pressure extrusion	
Designation used in structural ceramic industry	Soft extrusion	Semi-stiff extrusion	Stiff extrusion	
Extrusion moisture, % on dry	10–27	15–22	12–18	10–15
Extrusion pressure, MPa,	0,4–1,2	1,5–2,2	2,5–4,5	Up to 30

6.3.1 Technological process of briquetting by method of stiff vacuum extrusion

Typical layouts of the SVE briquetting line are shown in Fig. 6.42.

The mixture of raw materials is fed by a front loader to a J.C. Steele E Series Even feeder (Fig. 6.43), equipped with wear-resistant spiral cast-iron auger elements made of a chromium alloy.

Next, the prepared mixture with added binder and plasticizer is fed for mixing in the pug sealer. The line can also contain primary open pug-mill. The pug sealer consists of a large open part and the sealing node. The open part (Fig. 6.44) consists of a trough and blades for mixing. The blades are fastened to the steel rod shaft by bolted clamps, making it possible to rotate the blades to adjust the angle at which the processing takes place and, thereby, change the machine's performance.

The pug sealer is combined in a single unit with the extruder and is positioned above it (Fig. 6.45).

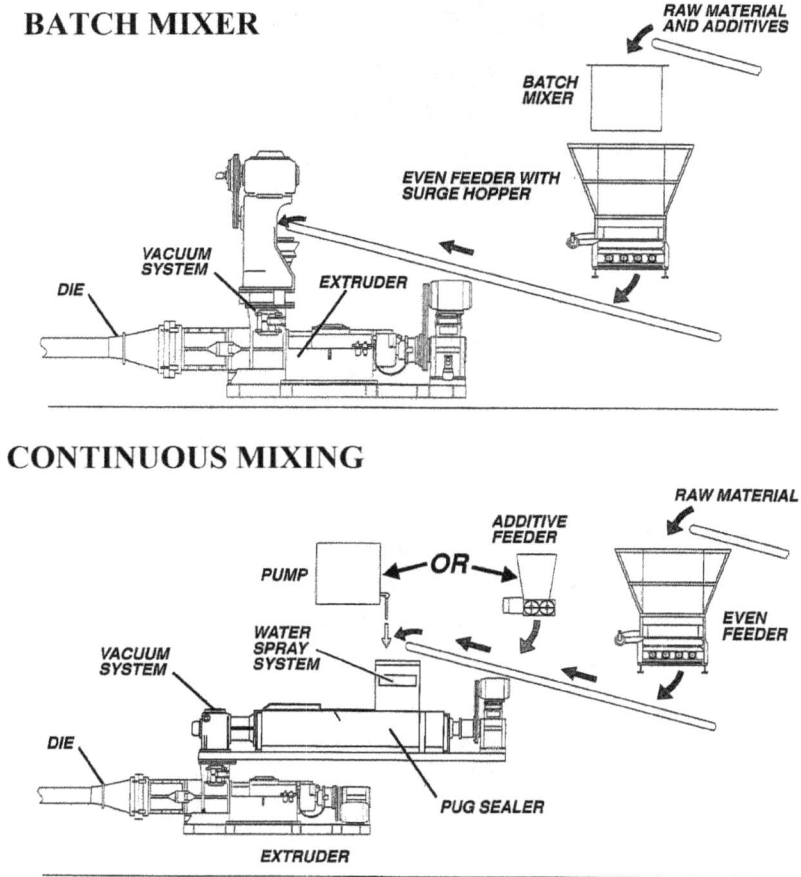

Figure **6.42.** Typical layouts of SVE briquetting line.

Figure 6.43. J.C. Steele E series even feeder.

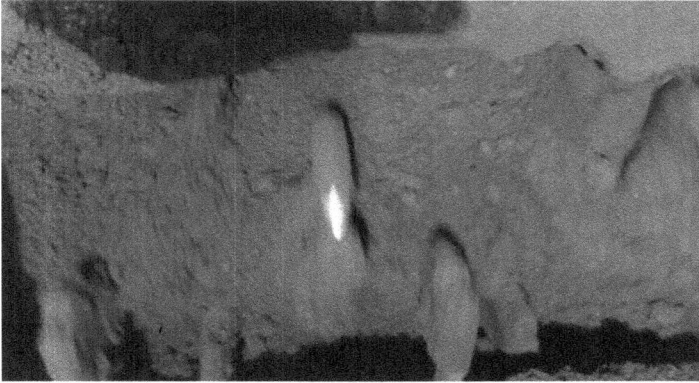

Figure 6.44. Mixing in the open part of the pug sealer.

Figure 6.45. The appearance of extruder (bottom) and pug sealer.

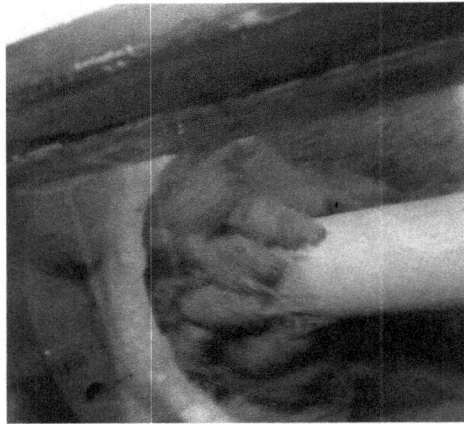

Figure 6.46. Partial agglomeration of the mix in pug sealer.

The mixture enters the vacuum chamber partially agglomerated (Fig. 6.46) and due to the high vacuum inside the chamber the pieces of the mixture immediately crumble into isolated particles, which fall down on the blades of the auger. It is known [2] that air adsorbed by the surface of particles of plastic material in the form of polymolecular layers held by van der Waals forces slows down their wetting with water, prevents the mass from being evenly compacted, contributes to an increase in elastic deformations during plastic molding, forming delamination as well as micro-cracks. Filling the pores, the air also prevents the penetration of moisture into them, separates the particles of the mass, acting as a leaner. Evacuation leads to the removal of air from the pores and promotes closer contact of the particles.

The vacuum is maintained throughout the working volume of the extruder up to the die. The pressure of the vacuum is at least 100 mm Hg (in absolute value). The area of the vacuum in the working chamber of the extruder and pug sealer is shown in Fig. 6.47. The combination of mechanical pressure and vacuum in the working extruder chamber helps to remove almost all compressible air from the material before densifying, which allows them to be immediately transported by conveyors and stacked nearly without the formation of a fine fraction. In addition, as is well known, the vacuum slightly decreases the viscosity of the cement paste, which facilitates its uniform distribution in the briquetting mass and improves its interaction with water [3]. This situation in combination with a

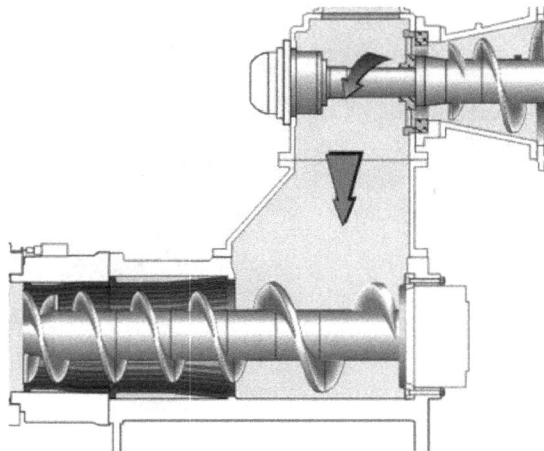

Figure 6.47. Vacuum areas in extruder and pug sealer.

higher density of the briquetting mass, due to the removal of air from it, leads to a decrease in the consumption of cement binder.

6.3.2 The movement of the briquetted mass in the extruder

Due to the rotation of the auger blades in the working chamber of the extruder, the briquetted mass performs translational and rotational motion, which is slowed down by the walls of the extruder's working chamber. The extruder can be represented in the form of two coaxial cylinders and a auger rotated by the inner cylinder around the axis Z, or along the orthogonal axis of radius r with speed $\hat{v}_- = r\hat{\omega}_-$ of angular velocity $\hat{\omega}_-$ to transport the mass along the same axis Z with speed w_- and the stationary outer cylinder of $w_+ = \hat{v}_+ = 0$ (Fig. 6.48).

In zone 1, the mixture is supplied to the working blades of the auger and is moved without compaction. In zone 2, the mixture is compacted. As the mixture moves toward the die, its rotation slows down, while the peripheral layers move at a faster speed. In zone 3, density irregularities that arise in zone 2 due to uneven movement of the molding mass are leveled. Homogenization is achieved due to the special geometry of the point auger blades and their mutual arrangement. In zone 4, further alignment of motion inhomogeneities occurs.

Mathematical modeling of the movement of the briquetted mass in an extruder in its full formulation continues to remain an extremely difficult task, that requires considering the rheological properties of the loaded materials. Most well-known studies use a simplified approach that combines mathematical and physical modeling. A review of the current state of modeling the formation of cast mass in an extruder is given in [1].

The movement of the briquetted mass in the working chamber of the extruder from the standpoint of the basic laws of continuum mechanics and with the intention of obtaining simplified qualitative dependencies that can serve as a basis for approximate calculations of the main parameters of the extrusion agglomeration process should be noted. The rotating mass of a wet, continuous medium has an isotropic molecular pressure field [1, 4] and is subject to the fundamental laws of conservation of mass, momentum and energy [4–6] with an appropriate rheology for the coefficient of dynamic viscosity μ [7–9].

First, as in [10], the motion of a particle, or point $\mathbf{r} = x\mathbf{i} + y\mathbf{j} + z\mathbf{k}$, will be measured in cylindrical coordinates r, θ, z,

$$r = \left(x^2 + y^2\right)^{1/2} \text{ and } \theta = \arctan\frac{y}{x}, \text{ or } x = r\cos\theta \text{ and } y = r\sin\theta,$$

or in polar orts

$$\mathbf{I} = \cos\theta = \mathbf{i}\cos\theta + \mathbf{j}\sin\theta \text{ and } \mathbf{J} = -\sin\theta = \mathbf{j}\cos\theta - \mathbf{i}\sin\theta \text{ and } \mathbf{k}$$

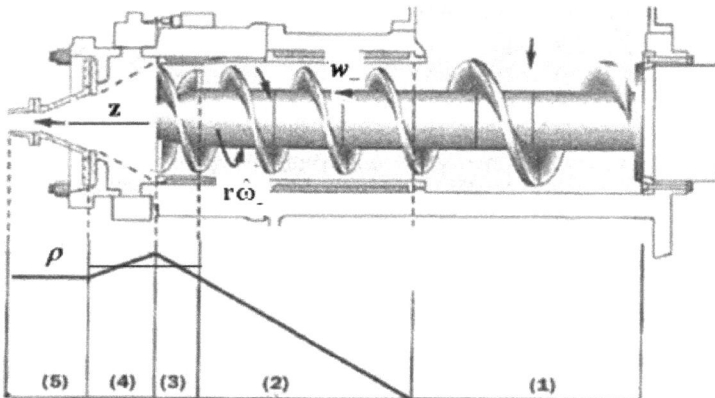

Figure 6.48. Stages of compaction in the working area of the extruder. 1 - feed zone without compaction, 2 - compression zone, 3 - homogenization zone, 4 - molding zone, 5 - exit from the die.

preserving the sum of the double orts of the direct product and

$$\mathbf{ii} + \mathbf{jj} = \mathbf{II} + \mathbf{JJ},$$

"inheriting" the properties of the sine and cosine functions,

$$\mathbf{I}_\theta = \partial_\theta \mathbf{I} = \frac{\partial \cos\theta}{\partial\theta} = -\sin\theta = \mathbf{J} \; and \; \mathbf{J}_\theta = -\partial_\theta \sin\theta = -\cos\theta = -\mathbf{I},$$

but otherwise remaining the same as the Cartesian orts **i, j, k**:

$$\mathbf{I} \times \mathbf{J} = \mathbf{i} \times \mathbf{j} = \mathbf{k}, \; \mathbf{J} \times \mathbf{K} = \mathbf{I}, \; \mathbf{k} \times \mathbf{I} = \mathbf{J},$$
$$\mathbf{I} \cdot \mathbf{J} = \mathbf{J} \cdot \mathbf{k} = \mathbf{k} \cdot \mathbf{I} = 0 \; and \; \mathbf{I} \cdot \mathbf{I} = \mathbf{J} \cdot \mathbf{J} = \mathbf{k} \cdot \mathbf{k} = 1,$$

so:

$$\mathbf{r} = x\mathbf{i} + y\mathbf{j} + z\mathbf{k} = r\mathbf{I}(\theta) + z\mathbf{k}.$$

In addition, the specified field mass accelerations,

$$\mathbf{g}(t, \mathbf{r}) = g^x\mathbf{i} + g^y\mathbf{j} + g^z\mathbf{k} = g^r\mathbf{I} + g^\theta\mathbf{J} + g^z\mathbf{k},$$

field velocities or flow

$$\mathbf{u}(t, \mathbf{r}) = u^x\mathbf{i} + u^y\mathbf{j} + u^z\mathbf{k} = u\mathbf{I} + v\mathbf{J} + w\mathbf{k} = \mathbf{r}_t,$$

receives three components of velocity, axial $w = z_t$, radial, $u = r_t$, and azimuthal, $v = r\theta_t$ from the polar orts

$$u^z = w, \; u^x = u\cos\theta - v\sin\theta, \; u^z = u\sin\theta + v\cos\theta,$$

will be considered further to be axially symmetric, or independent of the polar angle θ:

$$u_\theta = v_\theta = w_\theta = \rho_\theta = p_\theta = g^r_\theta = g^z_\theta = 0 \; for \; g^\theta = 0.$$

In this case, the gradient

$$\nabla = \begin{pmatrix} \mathbf{i}\partial_x \\ +\mathbf{j}\partial_y \\ +\mathbf{k}\partial_z \end{pmatrix} = \begin{pmatrix} \mathbf{I}\partial_r + \mathbf{k}\partial_z \\ +\mathbf{J}\dfrac{\partial_\theta}{r} \end{pmatrix}, \partial_x = \cos\theta\,\partial_r - \frac{\sin\theta}{r}\partial_\theta, \partial_y = \sin\theta\,\partial_r + \frac{\cos\theta}{r}\partial_\theta,$$

and divergences

$$\nabla \cdot \rho\mathbf{u} = \begin{pmatrix} \mathbf{I}\partial_r + \mathbf{k}\partial_z \\ +\mathbf{J}\dfrac{\partial_\theta}{r} \end{pmatrix} \cdot \begin{pmatrix} \mathbf{I}\rho u \\ +\mathbf{J}\rho v \\ +\mathbf{k}\rho w \end{pmatrix} = \begin{pmatrix} (\rho u)_r \\ +(\rho w)_z \end{pmatrix} + \frac{1}{r}\mathbf{J} \cdot \begin{pmatrix} \rho u \mathbf{I}_\theta \\ +\rho v \mathbf{J}_\theta \end{pmatrix}, \mathbf{I}_\theta = \mathbf{J}, \mathbf{J}_\theta = -\mathbf{I},$$

and

$$\nabla \cdot \mu\nabla u = \begin{pmatrix} \mathbf{I}\partial_r + \mathbf{k}\partial_z \\ +\mathbf{J}\dfrac{\partial_\theta}{r} \end{pmatrix} \cdot \begin{pmatrix} \mathbf{I}\mu u_r \\ +\mathbf{k}\mu u_z \end{pmatrix} = \begin{pmatrix} (\mu u_r)_r \\ +(\mu u_z)_z \end{pmatrix} + \frac{1}{r}\mathbf{J} \cdot \mathbf{I}_\theta \mu u_r, \ldots,$$

lead to the following mass, momentum and viscous volumetric densities:

$$\nabla \cdot \rho\mathbf{u} = \frac{1}{r}(r\rho u)_r + (\rho w)_z, \nabla \cdot \rho u\mathbf{u} = \frac{1}{r}(r\rho uu)_r + (\rho uw)_z$$

$$and \; \nabla \cdot \mu\nabla u = \frac{1}{r}(r\mu u_r)_r + (\mu u_z)_z, \ldots for \; \mu = \mu(t, \mathbf{r}),$$

respectively.

Furthermore, in order to save the record the matrices by double vectors, or more specifically by divectors: an identity matrix is the sum of the direct squares of the orts need to be represented

$$\vec{e} = \begin{pmatrix} 1 & 0 & 0 \\ 0 & 1 & 0 \\ 0 & 0 & 1 \end{pmatrix} = \mathbf{ii} + \mathbf{jj} + \mathbf{kk} = \mathbf{II} + \mathbf{JJ} + \mathbf{kk},$$

further, the matrix of momentum flow by the proportion $\rho\mathbf{uu}$ of $\mathbf{uu} = (u\mathbf{I} + v\mathbf{J} + w\mathbf{k})\mathbf{u}$ density of momentum flow

$$\nabla \cdot \rho\mathbf{uu} = \left(\nabla \cdot \rho u\mathbf{u} - \frac{\rho v^2}{r}\right)\mathbf{I} + \left(\nabla \cdot \rho v\mathbf{u} + \frac{\rho vu}{r}\right)\mathbf{J} + (\nabla \cdot \rho w\mathbf{u})\mathbf{k},$$

finally, the original for fluid mechanics the "matrix of the deformation rates" [11] by the matrix of the rate deformations, or fluid deformations $\mathbf{u_r}$ to be the sum of $\mathbf{u}_x\mathbf{i}+..$ of direct products $\mathbf{u}_x\mathbf{i},...$:

$$\mathbf{u_r} = \begin{pmatrix} u_x^x & u_y^x & u_z^x \\ u_x^y & u_y^y & u_z^y \\ u_x^z & u_y^z & u_z^z \end{pmatrix} = \begin{pmatrix} u_x^x\mathbf{ii}+ & u_y^x\mathbf{ij}+ & u_z^x\mathbf{ik}+ \\ u_x^y\mathbf{ji}+ & u_y^y\mathbf{jj}+ & u_z^y\mathbf{jk}+ \\ u_x^z\mathbf{ki}+ & u_y^z\mathbf{kj}+ & u_z^z\mathbf{kk} \end{pmatrix} = \mathbf{u}_x\mathbf{i} + \mathbf{u}_y\mathbf{j} + \mathbf{u}_z\mathbf{k},$$

with

$$\mathbf{u}_\theta = (u\mathbf{I} + v\mathbf{J} + w\mathbf{k})_\theta = u\mathbf{I}_\theta + v\mathbf{J}_\theta = u\mathbf{J} - v\mathbf{I} = \mathbf{k} \times \mathbf{u}$$
$$\mathbf{u}_{\theta\theta} = \mathbf{k} \times \mathbf{u}_\theta = \mathbf{k} \times (\mathbf{k} \times \mathbf{u}) = \mathbf{kk} \cdot \mathbf{u} - \mathbf{k} \cdot \mathbf{ku} = \mathbf{k}w - \mathbf{u},$$

or

$$\mathbf{u_r} = \mathbf{u}_r\mathbf{I} + \frac{1}{r}\mathbf{u}_\theta\mathbf{J} + \mathbf{u}_z\mathbf{k},$$

with divergence

$$\nabla \cdot \mathbf{u_r} = \left(\frac{1}{r}(ru)_r + w_z\right)_r\mathbf{I} + \left(\frac{1}{r}(ru)_r + w_z\right)_z\mathbf{k} = (\nabla \cdot \mathbf{u})_r\mathbf{I} + (\nabla \cdot \mathbf{u})_z\mathbf{k} = \nabla\nabla \cdot \mathbf{u}$$

convolution or point product

$$\nabla\mu \cdot \mathbf{u_r} = (\nabla\mu \cdot \mathbf{u}_r)\mathbf{I} - \frac{\mu_r v}{r}\mathbf{J} + (\nabla\mu \cdot \mathbf{u}_z)\mathbf{k},$$

with divergence

$$\nabla \cdot \mu\mathbf{u_r} = \mu\nabla \cdot \mathbf{u_r} + \nabla\mu \cdot \mathbf{u_r} = \mu\nabla\nabla \cdot \mathbf{u} + \nabla\mu \cdot \mathbf{u_r} =$$
$$= (\mu(\nabla \cdot \mathbf{u})_r + \nabla\mu \cdot \mathbf{u}_r)\mathbf{I} - \frac{\mu_r v}{r}\mathbf{J} + (\mu(\nabla \cdot \mathbf{u})_z + \nabla\mu \cdot \mathbf{u}_z)\mathbf{k},$$

and the conjugate matrix is the sum of the corresponding reverse multiplications

$$\nabla\mathbf{u} = \begin{pmatrix} u_x^x & u_x^y & u_x^z \\ u_y^x & u_y^y & u_y^z \\ u_z^x & u_z^y & u_z^z \end{pmatrix} = (\nabla u^x)\mathbf{i} + (\nabla u^y)\mathbf{j} + (\nabla u^z)\mathbf{k} = \mathbf{i}\mathbf{u}_x + \mathbf{j}\mathbf{u}_y + \mathbf{k}\mathbf{u}_z,$$

or

$$\nabla\mathbf{u} = (\mathbf{u_r})_* = \mathbf{I}\mathbf{u}_r + \frac{1}{r}\mathbf{J}\mathbf{u}_\theta + \mathbf{k}\mathbf{u}_z, \mathbf{u_r} = (\nabla\mathbf{u})_*,$$

with divergence

$$\nabla \cdot \mu\nabla\mathbf{u} = \left(\nabla \cdot \mu\nabla u - \frac{\mu u}{r^2}\right)\mathbf{I} + \left(\nabla \cdot \mu\nabla v - \frac{\mu v}{r^2}\right)\mathbf{J} + (\nabla \cdot \mu\nabla w)\mathbf{k}$$

so that

$$\nabla \cdot \mu(\mathbf{u}_r + \nabla \mathbf{u}) = \begin{pmatrix} \nabla \cdot \mu \nabla u - \dfrac{\mu u}{r^2} \\ +\mu(\nabla \cdot \mathbf{u})_r + \nabla \mu \cdot \mathbf{u}_r \end{pmatrix}\mathbf{I} + \begin{pmatrix} \nabla \cdot \mu \nabla \upsilon \\ -\dfrac{(r\mu)_r \upsilon}{r^2} \end{pmatrix}\mathbf{J} + \begin{pmatrix} \nabla \cdot \mu \nabla w \\ +\mu(\nabla \cdot \mathbf{u})_z \\ +\nabla \mu \cdot \mathbf{u}_z \end{pmatrix}\mathbf{k},$$

It was shown earlier that the movement of the briquetted mass in the extruder can be satisfactorily described on the basis of the concept of an ideal auger. The corresponding homogeneous continuous medium to be of mass conservative, or submitted to both *continuity equation* will be considered

$$\rho_t + \nabla \cdot \rho \mathbf{u} = \rho_t + (\rho w)_z + \frac{(r\rho\hat{u})_r}{r} = 0,$$

and Newton's second law of translational motion under the Earth *specific* (i.e., divided by ρ) *gravity acceleration* **g**, or velocity change per second (1687),

$$\frac{d\mathbf{u}}{dt} = \mathbf{g} = 0 \quad assumed \ to \ be \ negligible$$

extended at one time by Euler (1757) to the *hydrodynamic equilibrium*

$$\frac{d\mathbf{u}}{dt} = -\frac{1}{\rho}\nabla p = -\frac{1}{\rho}\nabla \cdot p\vec{\mathbf{e}} \quad for \quad \frac{d\mathbf{u}}{dt} = \mathbf{u}_t + \mathbf{u} \cdot \nabla \mathbf{u}$$

that has been developed further by Navier (1823) and Stokes (1845) to their equations

$$(\rho \mathbf{u})_t + \nabla \cdot \vec{\mathbf{P}} = 0 \ at \ the \ stress \ \vec{\mathbf{P}} = \rho \mathbf{uu} + p\vec{\mathbf{e}} - \mu\vec{\tau}$$

of dynamic and pressure stresses, $\rho\mathbf{uu}$ and $\rho\vec{\mathbf{e}}$, subtracted by the Stokesian proportion $\mu\vec{\tau}$ of either constant Newtonian (1687) or variable Bingham [7] viscosities $\mu = \mu(t, \mathbf{r}) > 0$ in the so called *shear* $\vec{\tau} = \mathbf{u}_r + \nabla \mathbf{u}$ of velocity *strain*

$$\mathbf{u}_r = \mathbf{u}_x\mathbf{i} + \mathbf{u}_y\mathbf{j} + \mathbf{u}_z\mathbf{k} = \begin{pmatrix} w_z\mathbf{k} + \\ \hat{u}_z\mathbf{I} + r\hat{\omega}_z\mathbf{J} \end{pmatrix}\mathbf{k} + \begin{pmatrix} w_r\mathbf{k} + \\ \hat{u}_r\mathbf{I} + (r\hat{\omega})_r\mathbf{J} \end{pmatrix}\mathbf{I} + \left(\frac{\hat{u}}{r}\mathbf{J} - \hat{\omega}\mathbf{I}\right)\mathbf{J}$$

and *co–strain* (*velocity gradient*)

$$\mathbf{u}_r^* = \mathbf{i}\mathbf{u}_x + \mathbf{j}\mathbf{u}_y + \mathbf{k}\mathbf{u}_z = \mathbf{k}\begin{pmatrix} w_z\mathbf{k} + \\ \hat{u}_z\mathbf{I} + r\hat{\omega}_z\mathbf{J} \end{pmatrix} + \mathbf{I}\begin{pmatrix} w_r\mathbf{k} + \\ \hat{u}_r\mathbf{I} + (r\hat{\omega})_r\mathbf{J} \end{pmatrix} + \mathbf{J}\left(\frac{\hat{u}}{r}\mathbf{J} - \hat{\omega}\mathbf{I}\right) = \nabla \mathbf{u},$$

to form the shear matrix

$$\vec{\tau} = \begin{pmatrix} 2w_z\mathbf{kk} + (w_r + \hat{u}_z)\mathbf{kI} + r\hat{\omega}_z\mathbf{kJ} \\ (w_r + \hat{u}_z)\mathbf{Ik} + 2\hat{u}_r\mathbf{II} + r\hat{\omega}_r\mathbf{IJ} \\ r\hat{\omega}_z\mathbf{Jk} + r\hat{\omega}_r\mathbf{JI} + 2r^{-1}\hat{u}\mathbf{JJ} \end{pmatrix} = \begin{pmatrix} 2w_z & w_r + \hat{u}_z & r\hat{\omega}_z \\ w_r + \hat{u}_z & 2\hat{u}_r & r\hat{\omega}_r \\ r\hat{\omega}_z & r\hat{\omega}_r & 2r^{-1}\hat{u} \end{pmatrix}$$

with *polar shear actions*

$$\tau^z = \begin{pmatrix} 2w_z\mathbf{k} + \\ (\hat{u}_z + w_r)\mathbf{I} + r\hat{\omega}_z\mathbf{J} \end{pmatrix}, \ \tau^r = \begin{pmatrix} (\hat{u}_z + w_r)\mathbf{k} \\ +2\hat{u}_r\mathbf{I} + r\hat{\omega}_r\mathbf{J} \end{pmatrix} \ and \ \tau^\theta = \begin{pmatrix} r\hat{\omega}_z\mathbf{k} + r\hat{\omega}_r\mathbf{I} \\ +2r^{-1}\hat{u}\mathbf{J} \end{pmatrix},$$

constitute the same symmetric strain $\vec{\tau} = \tau^z\mathbf{k} + \tau^r\mathbf{I} + \tau^\theta\mathbf{J} = \mathbf{k}\tau^z + \mathbf{I}\tau^r + \mathbf{J}\tau^\theta = \vec{\tau}^*$ and stress

$$\vec{\mathbf{P}} = \rho \mathbf{uu} + \rho\vec{\mathbf{e}} - \mu\vec{\tau} = \mathbf{k}\mathbf{P}^z + \mathbf{I}\mathbf{P}^r + \mathbf{J}\mathbf{P}^\theta = \mathbf{P}^z\mathbf{k} + \mathbf{P}^r\mathbf{I} + \mathbf{P}^\theta\mathbf{J} = \vec{\mathbf{P}}^*$$

of its own *polar actions*

$$\mathbf{P}^z = \rho w\mathbf{u} + \rho\mathbf{k} - \mu\tau^z, \ \mathbf{P}^r = \rho\hat{u}\mathbf{u} + \rho\mathbf{I} - \mu\tau^r \ and \ \mathbf{P}^\theta = \rho r\hat{\omega}\mathbf{u} + \rho\mathbf{J} - \mu\tau^\theta$$

in the resultant *polar Navier–Stokes equations* of

$$(\rho \mathbf{u})_t + \nabla \cdot \vec{\mathbf{P}} = 0 \ and \ \rho_t + \nabla \cdot \rho \mathbf{u} = 0 \tag{1}$$

for the *divergence*

$$\nabla \cdot \vec{\mathbf{P}} = \begin{pmatrix} (\rho ww)_z \\ + \dfrac{(r\rho \hat{u}w)_r}{r} \\ + p_z - 2(\mu w_z)_z \\ - \dfrac{(r\mu(\hat{u}_z + w_r))_r}{r} \end{pmatrix} \mathbf{k} + \begin{pmatrix} (\rho w \hat{u})_z + \dfrac{(r\rho \hat{u}^2)_r}{r} \\ -\rho r \hat{\omega}^2 + p_r \\ -(\mu(\hat{u}_z + w_r))_z \\ -2\dfrac{(r\mu \hat{u}_r)_r}{r} + 2\dfrac{\mu \hat{u}}{r^2} \end{pmatrix} \mathbf{I} + \begin{pmatrix} r(\rho w \hat{\omega})_z \\ + \dfrac{(r^3 \rho \hat{u}\hat{\omega})_r}{r^2} \\ -r(\mu \hat{\omega}_z)_z \\ - \dfrac{(r^3 \mu \hat{\omega}_r)_r}{r^2} \end{pmatrix} \mathbf{J}$$

Those are to be supplied with *no–slip conditions*

$$w\big|_{r=\varepsilon a,a} = w_-, w_+, \hat{\omega}\big|_{r=\varepsilon a,a} = \hat{\omega}_-, \hat{\omega}_+, w_+, \hat{\omega}_+ = 0, w_-, \hat{\omega}_- = const > 0, \tag{2}$$

interpreted here as the **ideal auger** that for zeroth radial velocity

$$\hat{u} = 0,$$

and constant pressure difference $p_+ - p_-$, *stationary*,

$$\partial_t = 0,$$

rotates the medium between two cylinders $\varepsilon a < r < a$ of rigid walls $r = \varepsilon a$ and $r = a$ when pushed additionally with the *medium pressure constantly dropping* $\sigma > 0$ from one side $z = 0$ to another one, $z = l$, along the symmetry axis $z(r = 0)$,

$$-p_z = \sigma = \frac{p_+ - p_-}{b} = const > 0, 0 < z < b, p_- = p\big|_{z=0}, p_+ = p\big|_{z=b}. \tag{3}$$

Exact solution. Now the stationary motion of an *incompressible Newtonian medium* with density, viscosity and axial pressure drop are supposed to be constant, radial velocity is zero for axial and angular velocities depending only on the radius r will be considered:

$$\rho, \mu, \sigma = const > 0, \hat{u} = 0, w = w(r) \ and \ \hat{\omega} = \hat{\omega}(r).$$

Then, with necessary main relations of (1),

$$\nabla \cdot \vec{\mathbf{P}} = \begin{pmatrix} p_z - \\ \dfrac{(r\mu w_r)_r}{r} \end{pmatrix} \mathbf{k} + (p_r - \rho r \hat{\omega}^2)\mathbf{I} - \frac{\mu(r^3 \hat{\omega}_r)_r}{r^2}\mathbf{J} = \mathbf{0} \ and \ \nabla \cdot \rho \mathbf{u} = \rho w_z = 0 \ in \ V.$$

the problem on ideal auger (1–3) takes the form of

$$-\frac{(rw_r)_r}{r} = \frac{\sigma}{\mu}, p_r = \rho r \hat{\omega}^2, (r^3 \hat{\omega}_r)_r = 0, -p_z = \sigma, \varepsilon a < r < a, 0 < z < b, \tag{4}$$

$$w\big|_{r=\varepsilon a} = w_-, w\big|_{r=a} = w_+, \hat{\omega}\big|_{r=\varepsilon a} = \hat{\omega}_- = \Omega, \hat{\omega}\big|_{r=a} = \hat{\omega}_+ = 0, \varepsilon, a = const > 0, \varepsilon < 1,$$

The obtained hydrodynamic equilibrium (1–3), or (4), can readily be determined uniquely [12] by the *auger swirling flow*

$$\mathbf{u} = w(r)\mathbf{k} + r\hat{\omega}(r)\mathbf{J}, \tag{5}$$

combined for $\hat{u} = 0$ the well-known Hagen–Poiseuille and Couette flows [4–5, 13], of axial velocity

$$w(r) = w_+ + \frac{\sigma(a^2 - r^2)}{4\mu} + \left(w_- - w_+ - \frac{\sigma a^2(1 - \varepsilon^2)}{4\mu}\right) \frac{\ln \dfrac{r}{a}}{\ln \varepsilon} \tag{6}$$

angular twist

$$\hat{\omega}(r) = \frac{\hat{\omega}_+ - \varepsilon^2 \hat{\omega}_-}{1-\varepsilon^2} + \frac{\varepsilon^2 a^2}{r^2}\frac{\hat{\omega}_- - \hat{\omega}_+}{1-\varepsilon^2} \ for \ \hat{\omega}_- = \Omega \ and \ \hat{\omega}_+ = 0 \tag{7}$$

and pressure

$$p(r,z) = p_+ - \frac{z}{b}(p_+ - p_-) - \int_r^a \rho r' \hat{\omega}^2(r')dr', \varepsilon a < r < a, 0 < z < b. \tag{8}$$

For the *fixed external cylinder*, $w_+ = \hat{\omega}_+ = 0$, both character *velocity of pressure drop*,

$$w_p = \frac{\sigma a^2}{4\mu} \ for \ \sigma = \frac{p_+ - p_-}{b} > 0, \tag{9}$$

and dimensionless *displacement factor*

$$\delta = \frac{w_-}{w_p} \ for \ w_- > 0 \ and \ \hat{\omega}_- = \Omega > 0 \tag{10}$$

reveal dimensionless axial and angular velocity profiles induced by the internal cylinder in Fig. 6.49,

$$Z = \frac{w}{w_p} = 1 - X^2 + \left(\delta - (1-\varepsilon^2)\right)\frac{\ln X}{\ln \varepsilon} \ and \ \frac{\hat{\omega}}{\hat{\omega}_-} = Y = \frac{\varepsilon^2}{1-\varepsilon^2}\left(\frac{1}{X^2}-1\right),$$

$$for \ 0 < \varepsilon \le X = \bar{r} = \frac{r}{a} < 1,$$

which further proves to be (i) *subcritical*, $\delta < 1 - \varepsilon^2$, and (ii) *supercritical*, $\delta \ge 1 - \varepsilon^2$, as shown in Fig. 6.49.

It can be readily shown as well that for the case of a stationary, where there is a critical rotation speed of the inner cylinder that, with the appropriate subcritical or supercritical mode, the maximum transport speed $w(r_*)$ is reached between the cylinders or on the inner cylinder and is less than or equal to w_p, respectively:

$$w(r_*) < w_p \ and \ \varepsilon a < r_* < a \ at \ \delta < 1 - \varepsilon^2,$$
$$w(r_*) = w_p \ and \ r_* = \varepsilon a \ at \ \delta \ge 1 - \varepsilon^2.$$

Braking factor. The pressure p from (8) and flow $\mathbf{u}$ from (5) allow equilibrium (1),

$$-\frac{(rw_r)_r}{r} = \frac{\sigma}{\mu}, \ p_r = \rho r \hat{\omega}^2, (r^3 \hat{\omega}_r)_r = 0, -p_z = \sigma, \ \varepsilon < \bar{r} = \frac{r}{a} < 1, \ 0 < z < b,$$

$$w|_{r=\varepsilon a} = w_-, \ w|_{r=a} = w_+ = w_-(1-\alpha), \ \hat{\omega}|_{r=\varepsilon a} = \hat{\omega}_- = \Omega, \ \hat{\omega}|_{r=a} = \hat{\omega}_+ = 0, \tag{11}$$

$$\Omega, w_\pm, \varepsilon, a, \alpha = const > 0, \ \varepsilon < 1,$$

in the space $\varepsilon < \bar{r} < 1$ between solid coaxial cylinders, of which the inner one $\bar{r} = \varepsilon$, rotates around a common axis of symmetry $\bar{r} = 0$ with a constant angular velocity Ω and simultaneously displaces along the specified axis with a certain and usually proportional to Ω transport speed w_- becoming

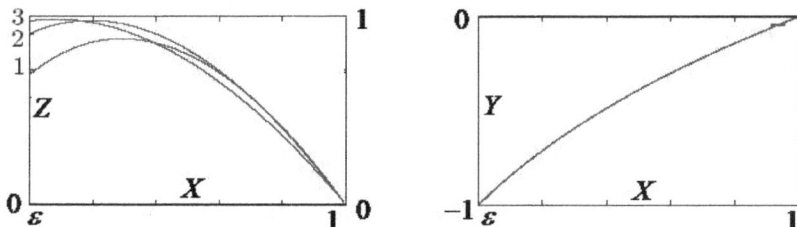

Figure 6.49. Profiles of axial (left) and azimuthal (right) components of the considered spiral flow with two subcritical (1 and 2) and one supercritical (3) factor δ values.

$$\hat{w} = w_- - \alpha w_- \frac{\ln\left(\bar{r}/\varepsilon\right)}{\ln\left(1/\varepsilon\right)},\ 0 < \varepsilon < \bar{r} = \frac{r}{b} < 1,\ \left(r\hat{w}_r\right)_r = 0,\ \hat{w}\big|_{\bar{r}=\varepsilon,1} = w_-,\ w_-\left(1-\alpha\right),$$

dragging the mass with the axial velocity behind it by sticking alone

$$w = \hat{w} + W\ or\ W = w - \hat{w} = w_*\left(\frac{1-\bar{r}^2}{1-\varepsilon^2} - \frac{\ln\bar{r}}{\ln\varepsilon}\right),\ w_* = \frac{W_*\left(1-\varepsilon^2\right)}{2},$$

the term of the speed of the *viscous pressure head*

$$\frac{W}{W_*} = \frac{1-\bar{r}^2}{2} + \kappa^2\ln\bar{r},\ W_* = \frac{\sigma a^2}{2\mu} = 2w_p,\ \kappa = \bar{r}_{max} = \frac{\sqrt{1-\varepsilon^2}}{\sqrt{2\ln\left(1/\varepsilon\right)}}$$

or

$$W = \frac{W_*}{2}\left(1 - \bar{r}^2 - \frac{1-\varepsilon^2}{\ln\varepsilon}\ln\bar{r}\right),\ 0 < \varepsilon < \bar{r} = \frac{r}{b} < 1$$

satisfying the same equation as w, however, boundary vanishing as

$$-\frac{\left(rW_r\right)_r}{r} = \frac{\sigma}{\mu},\ \varepsilon b < r < b,\ W\big|_{r=\varepsilon b} = 0\ and\ W\big|_{r=b} = 0,$$

and reaching its maximum W_* at the radius $\bar{r}_{max} = \kappa$, as in Fig. 6.50,

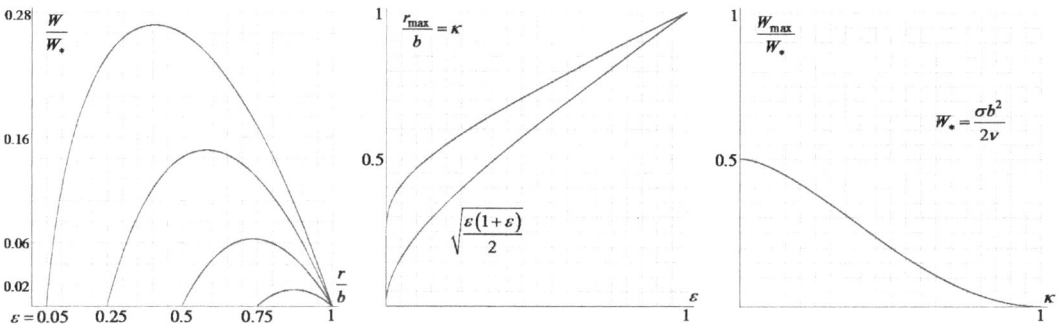

Figure 6.50. Velocity of viscous head (left) between the cylinders and its highest value (right) depending on the radius of the maximum of the first (middle).

with angular twist

$$\hat{\omega} = \frac{\varepsilon^2}{\bar{r}^2}\frac{1-\bar{r}^2}{1-\varepsilon^2}\Omega,\ \hat{\omega} = \Omega\,(\hat{\omega}=0)\ at\ \bar{r} = \varepsilon\,(\bar{r}=1).$$

On a stationary outer cylinder $\bar{r} = 1$, the speed $w = \hat{w} + W$ is reduced to a value $w_-(1-\alpha)$ with the *braking coefficient* $0 < \alpha < 1$.

Using both function

$$\chi\left(\varepsilon\right) = 1 - \varepsilon^2 + \varepsilon^2\ln\varepsilon^2,\ 0 < \varepsilon < 1,$$

and related inequalities

$$0 < \chi\left(\varepsilon\right) < 1,$$

it can be seen that with *strong,* $\beta \geq \chi(\varepsilon)$, or *weak braking* $\beta = 2\alpha A < \chi(\varepsilon)$, $0 < \alpha < 1$, by the outer cylinder $\bar{r} = 1$, dimensionless, as related to the maximum value of the viscous head, the speed of *boundary mass transportation*

$$A = w_-/W_*$$

by the inner cylinder $\bar{r} = \varepsilon$, i.e., when *braking factor*

$$\beta = 2\alpha A = 2\alpha\,w_- / W_* = 2\alpha\,w_- / 2w_p = \alpha\,w_- / w_p = \alpha\delta,$$

as α–proportional to the above displacement factor δ, is not less than or strictly $\chi(\varepsilon)$, provided by the auger dimensionless axial speed (6),

$$\bar{w} = \frac{w}{W_*} = A(1-\alpha) + \frac{1-\bar{r}^2}{2} + \frac{1-\varepsilon^2-\beta}{\ln\left(1/\varepsilon^2\right)}\ln\bar{r}\ \ for\ W_* = \frac{\sigma a^2}{2\mu}\ and\ w_+ = w_- - \alpha w_-,$$

reaches a maximum $\bar{w}_{max} = \bar{w}(\bar{r}_{max}) > 0$ on the inner cylinder $\bar{r}_{max} = \varepsilon$, or on a cylinder with a larger radius, as in Fig. 6.50, to the left, respectively.

For the $\beta \geq \chi(\varepsilon)$ derivative

$$\bar{w}_{\bar{r}} = \frac{1-\varepsilon^2-\beta}{\ln\left(1/\varepsilon^2\right)}\frac{1}{\bar{r}} - \bar{r} \leq \frac{1-\varepsilon^2-\chi(\varepsilon)}{\ln\left(1/\varepsilon^2\right)}\frac{1}{\bar{r}} - \bar{r} < -\bar{r} < 0,\ \varepsilon < \bar{r} < 1,$$

and at $0 \leq \beta < \chi(\varepsilon)$ it is equal to zero between the cylinders,

$$\bar{w}_{\bar{r}} = 0\ for\ 1 > \bar{r} = \bar{r}_{max} = \kappa = \sqrt{\frac{1-\varepsilon^2-\beta}{\ln\left(1/\varepsilon^2\right)}} > \sqrt{\frac{1-\varepsilon^2-\chi(\varepsilon)}{\ln\left(1/\varepsilon^2\right)}} = \varepsilon,$$

as required to show.

Auger capacity. As a consequence, the translational rate of the mass flow

$$Q[\hat{w}] = 2\pi\rho\int_{\varepsilon b}^{b}\hat{w}r\,dr\ at\ \frac{\hat{w}}{W_*} = A - \frac{\beta\ln\bar{r}}{2\ln\left(1/\varepsilon\right)} - \frac{\beta}{2},\ A = \frac{w_-}{W_*}\ and\ \beta = 2\alpha A,$$

or

$$\frac{Q[\hat{w}]}{\pi\rho b^2 W_*} - A\left(1-\varepsilon^2\right) = \frac{\beta}{2}\left(\frac{1-\varepsilon^2}{\ln\left(1/\varepsilon^2\right)} - 1\right) = \frac{\beta}{2}\left(\frac{\chi(\varepsilon)}{\ln\left(1/\varepsilon^2\right)} + \varepsilon^2 - 1\right) \geq \frac{\beta}{2}\left(\varepsilon^2 - 1\right),$$

in total with the same speed from the pressure head,

$$\frac{Q[W]}{\pi\rho b^2 W_*} = \frac{1-\varepsilon^2}{2}\left(\frac{1+\varepsilon^2}{2} - \frac{1-\varepsilon^2}{\ln\left(1/\varepsilon^2\right)}\right),$$

gives the required performance of the ideal auger, or the capacity of an ideal auger,

$$Aq = \frac{Q[\hat{w}+W]}{\pi\rho b^2 W_*} = A\left(1-\varepsilon^2\right) + \frac{\beta}{2}\left(\frac{1-\varepsilon^2}{\ln\left(1/\varepsilon^2\right)} - 1\right) + \frac{1-\varepsilon^2}{2}\left(\frac{1+\varepsilon^2}{2} - \frac{1-\varepsilon^2}{\ln\left(1/\varepsilon^2\right)}\right).$$

Taken in relation to $A > 0$, it reduces to formula

$$q = q_A\left(\varepsilon,\alpha\right) = 1 - \varepsilon^2 - \frac{1}{A}\left(\frac{1-\varepsilon^2}{2}\right)^2 - \left(\frac{1-\varepsilon^2}{2A} - \alpha\right)\left(\frac{1-\varepsilon^2}{\ln\left(1/\varepsilon^2\right)} - 1\right) = q_\infty + O\left(\frac{1}{A}\right),$$

or performance gains

$$q_\infty = q_\infty\left(\varepsilon,\alpha\right) = 1 - \varepsilon^2 + \alpha\left(\frac{1-\varepsilon^2}{\ln\left(1/\varepsilon^2\right)} - 1\right)\ at\ 0 < \varepsilon,\alpha < 1$$

shown in Fig. 6.51

In other words, even without slight $(0 < \alpha \ll 1)$ braking $\beta = 2\alpha A$ of the pushed mixture on the auger shell $\bar{r} = 1$ (2), its capacity cannot be increased with an increase in the inner radius $\varepsilon > 0$, similar to the role of Vyshnegradskiy rotational friction in ensuring stable operation of the Watt regulator [14].

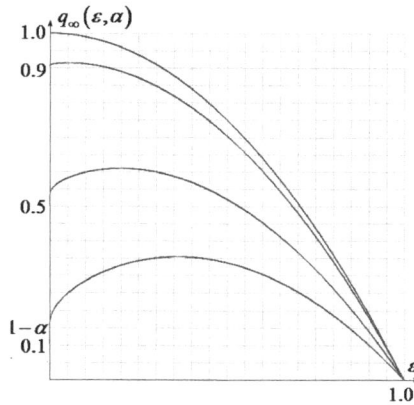

Figure 6.51. Increase in the capacity of the ideal auger from external friction ($0 < \alpha < 1$) with an increase in its inner radius ($0 < \varepsilon < 1$).

The transportation of the medium by the ideal auger with full external braking $\alpha = 1$, occurs as in a real auger in Fig. 6.52. Its performance

$$q_\infty\left(\varepsilon,1\right) = -\varepsilon^2 + \frac{1-\varepsilon^2}{\ln\left(1/\varepsilon^2\right)}, \quad 0 < \varepsilon < 1, \tag{12}$$

is close to the maximum possible values (30%) achieved at a ratio of diameters $\varepsilon = 0.4$ (which exactly corresponds to the parameters of augers of industrial samples of stiff extruders by J.C. Steele & Sons, Inc), i.e., at

$$\frac{dq_\infty\left(\varepsilon,1\right)}{d\varepsilon^2} == -1 - \frac{1}{\ln\left(1/\varepsilon^2\right)} + \frac{1-\varepsilon^2}{\varepsilon^2 \ln^2\left(1/\varepsilon^2\right)} = 0 \ for \ 0 < \varepsilon = \varepsilon_\infty < 1.$$

Thus, the applied formalization of the extrusion briquetting process, based on the concept of the ideal auger, made it possible to obtain analytical dependences that adequately describe the real geometric parameters of the augers of industrial extruders of stiff extrusion, the only manufacturer is still J.C. Steele & Sons, Inc. The obtained ratios are recommended for use in the design process of briquette modifications of the company's extruders.

When considering the processes of extrusion briquetting of plastic masses with a viscosity following a power-law dependence on dissipation, in order to obtain integral characteristics of

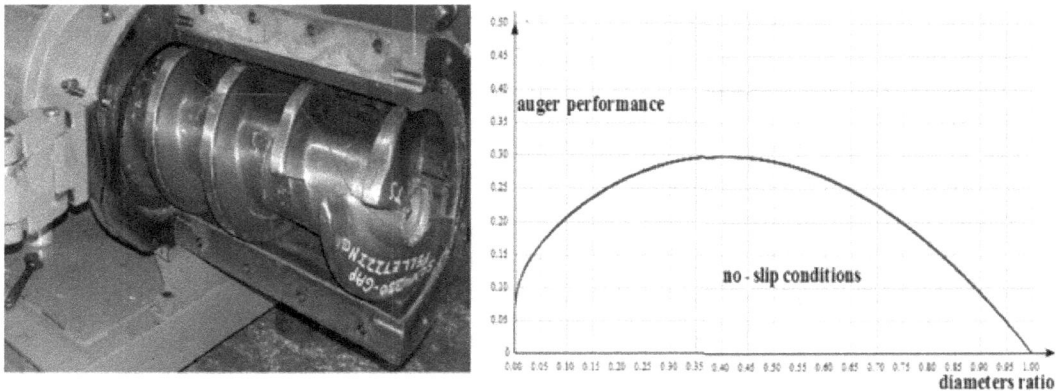

Figure 6.52. External view of the auger of an industrial extruder manufactured by J.C. Steele & Sons, Inc, (left) and the capacity calculated by formula (3) at full external braking.

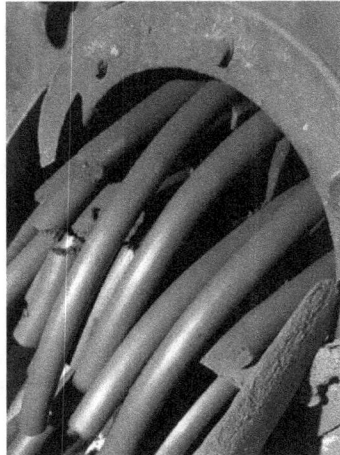

Figure 6.53. Raw brex exiting the extruder.

the extruder performance, it is legitimate to use the model of a stationary flow of a Newtonian incompressible fluid with constant viscosity.

In zone 5, the brex are squeezed out of the holes in the die (Fig. 6.53), which completes the process of their formation.

References

[1] Extrusion in Ceramics. Frank Händle (Ed.). Springer. 2007. 413 p.
[2] Moroz, I.I. 2011. Technology of Structural Ceramics. Moscow: Ecolit, 384 p. (in Russian).
[3] Ruzhinskiy, S. et al. 2006. All about Foam Concrete. 2nd ed., Improved and Expanded. - St. Petersburg, OOO Stroy Beton (Stroy Beton, LLC), 630 p. (in Russian).
[4] Batchelor, G.K. 1967. An Introduction to Fluid Dynamics. Cambridge: Cambridge Univ. Press, 615 p.
[5] Loytsyanskiy, L.G. 1987. Mechanics of Liquids and Gases. Moscow: Science, 840 p. (in Russian).
[6] Abramovich, G.N. 1991. Applied Gas Dynamics. Part 1. Moscow: Science, 600 p.
[7] Bingham, E.C. 1922. Fluidity and Plasticity. New York, London: McCraw-Hill Book Company, Inc., 439 p.
[8] Ishlinskiy, A.Y. and Ivlev, D.D. 2001. Mathematical Theory of Plasticity. Moscow: Publishing House of Physical and Mathematical and Technical Literature, 704 p. (in Russian).
[9] Laenger, K.-F., Laenger, F. and Geiger, K. 2016. Wall slip of ceramic extrusion bodies. Part 2. Process Engineering 93(4-5): 1–6.
[10] Belotserkovskiy, O.M., Betelin, V.B., Borisevich, V.D., Denisenko, V.V., Kozlov, S.A., Eriklintsev, I.V., Konyukhov, A.V., Oparin, A.M. and Troshkin, O.V. 2011. Toward theory of counterflow in rotating viscous heat-conducting gas. Journal of Computational Mathematics and Mathematical Physics 51(2): 222–236 (in Russian).
[11] Troshkin, O.V. 2016. Elements of Mathematical Hydrodynamics and Hydrodynamic Stability. ISBN-978-3-659-93972-3.
[12] Ladyzhenskaya, O.A. 1969. The Mathematical Theory of Viscous Incompressible Flow. 2014 Reprint of 1963 Edition – NY: Gordon and Breach, 224 p.
[13] Joseph, D. 1981. Stability of Fluid Motions. Moscow: World, 638 p.
[14] Pontryagin, L.S. 1974. Ordinary Differential Equations. Moscow: Nauka, 331 p. (In Russian).

6.3.3 Numerical simulation of extruder operation

A technique for modeling the main auger of an industrial extruder should be considered. The extruded material is represented as a set of spherical particles. The description of the movement of the extruded mass is carried out using the method of discrete elements.

To construct mathematical models describing the behavior of granular, free-flowing materials, a continuous or discrete (corpuscular) approach is used. The continual approach assumes the continuity and structurelessness of the medium, the behavior of which is described by the laws of classical mechanics, usually presented in the form of differential equations. As a result, the behavior of individual particles is ignored. The resulting constitutive equations are solved numerically, for example, using the finite element method. The main problem of this approach, as applied to the modeling of loose bodies, is the construction of correct models of the behavior of the material of the medium. Appropriate stress-strain laws for a material may often not exist or may be overly complex. In addition, the macroscopic behavior of loose bodies is most often very strongly dependent on the behavior of the system at the level of individual particles, which cannot be taken into account in the framework of the continuum approach.

In contrast to the continuous one, the discrete approach is based on modeling each individual particle, with the presentation of a bulk material as an idealized set of particles. The macroscopic behavior of the system is the result of the interaction of individual particles, which makes the discrete approach most suitable for studying the phenomenon occurring on the scale of the particle diameter and for modeling the volumetric behavior of the medium. One of the implementations of the discrete approach is the Discrete Element Method (DEM), presented in a number of engineering analysis software packages, such as SIMULIA Abaqus, eDEM, Rocky, etc.

The general concept of DEM is to represent a free-flowing body as a set of non-deformable particles of a simple topology (for example, a spherical shape) with a given mass. Particles in the course of movement can contact each other and surrounding objects. The mathematical model includes two types of constitutive relations describing the motion of particles and their contacting (Fig. 6.54). Using algorithms for finding a contact and the corresponding formulations of contact models, the forces acting on the particles are determined. The particle's acceleration, velocity and position are then calculated using Newton's laws of motion and numerical integration.

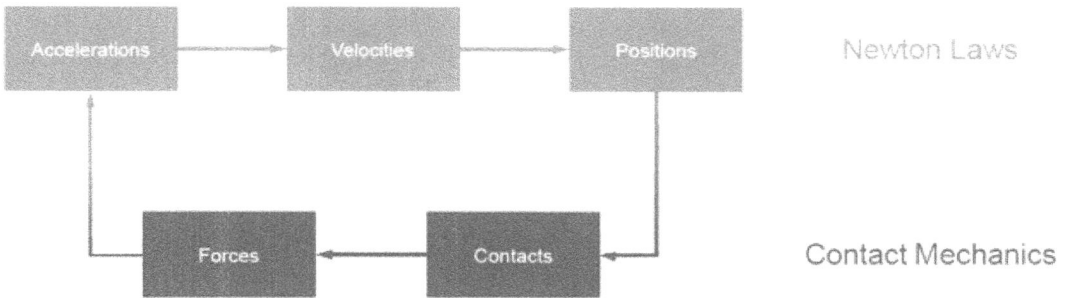

Figure 6.54. General computational scheme of DEM.

Each particle has six degrees of freedom. To calculate the accelerations, Newton's second law is used (Fig. 6.55). The translational motion of the particle:

$$m\frac{dv}{dt} = F_g + F_c + F_{nc},$$
(1)

where v - particle velocity, m - particle mass, t - time, F_g - gravity, F_c - contact loads acting on the particle, F_{nc} - other external loads.

Particle rotation:

$$I\frac{d\omega}{dt} = M,$$
(2)

where I is the moment of inertia of the particle, ω is the angular velocity, M is the moment of external forces.

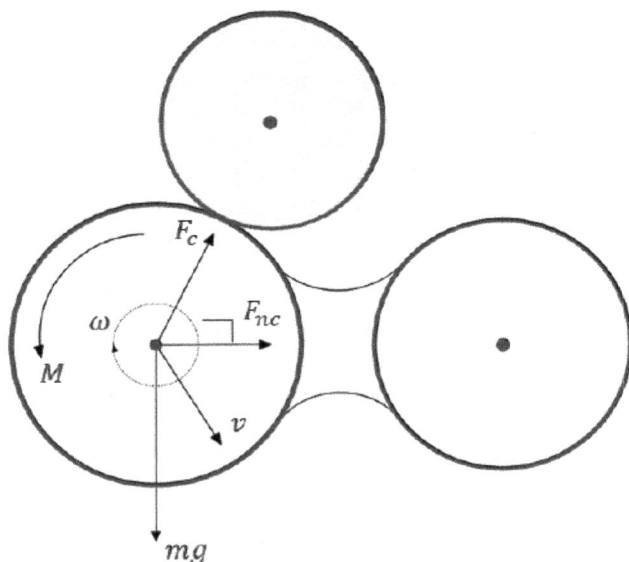

Figure 6.55. Forces acting on the center of mass of the particle.

Accelerations are numerically integrated over a time step to update the particle velocity and position:

$$x(t + \Delta t) = x(t) + v(t)\Delta t$$
$$v(t + \Delta t) = v(t) + a(t)\Delta t,$$

(3)

where v(t) is the speed of the particle, x(t) is the current position of the particle, $a(t)$ is the acceleration of the particle, Δt is the size of the explicit time step (Fig. 6.56).

The choice of the time step Δt is very important in DEM modeling. It should be chosen small enough for two main reasons: to prevent large self-intersections of particles in one integration step, which leads to non physical values of contact forces, and to avoid the undesirable effect of Rayleigh surface waves. The motion is influenced not only by contact interactions with direct "neighbors", but also by the propagation of disturbances from particles located at a great distance (Figs. 6.57 and 6.58). The small size of the time increment prevents the propagation of Rayleigh waves beyond the boundaries of neighboring particles.

The choice of an appropriate time step is often based on data on the propagation velocity of Rayleigh waves. For most problems, $10^{-6} \leq \Delta t \leq 10^{-4}$, which is an order of magnitude larger than the integration step size used in the continuous CFD approach.

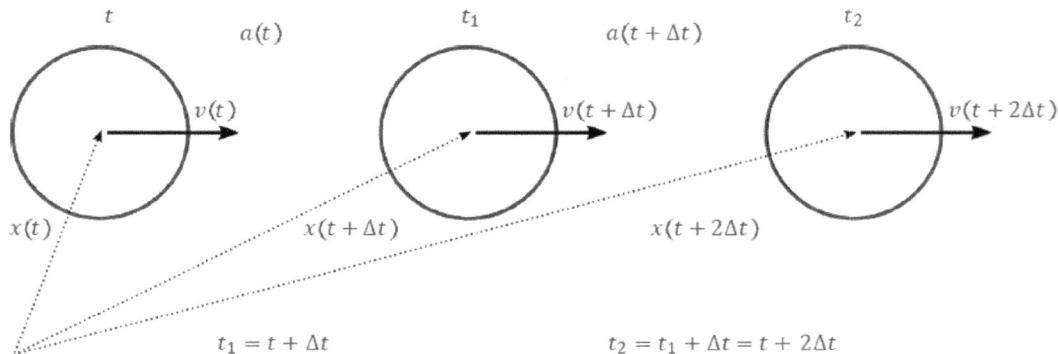

Figure 6.56. Integration scheme.

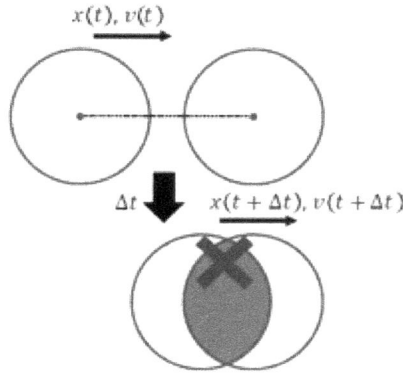

$x(t), v(t)$

Δt $x(t + \Delta t), v(t + \Delta t)$

Figure 6.57. Self-intersection of particles.

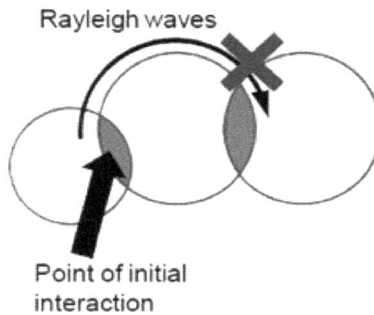

Rayleigh waves

Point of initial
interaction

Figure 6.58. Rayleigh waves.

$$T_R = \frac{\pi R\sqrt{\rho/G}}{0.1631v + 0.8766}, \qquad (4)$$

where T_R is the time of passage of the Rayleigh wave, R is the radius of the particles, ρ is the density, G is the shear modulus of the particles, v is the Poisson's ratio of the particles.

Since the particles in DEM are non-deformable, the mechanical characteristics of the medium are completely determined through the models of contact interactions, which is analogous to the models of materials in the classical finite element analysis. The rigidity characteristics of the medium are realized through the mutual penetration of particles on contact. Contact models determine the relationship between the amount of particle penetration and the resulting force. User-defined properties of the medium (e.g., Young's modulus, Poisson's ratio) are used to calculate contact stiffnesses (Fig. 6.59).

During each time increment, the distance between all particles in the system is checked. For spherical particles, the contact is initialized if the distance between two centers of the spheres is less than the sum of their radii. This process requires significant computational resources and, together with the calculations of contact forces, takes 70–80% of the total volume of calculations. A number of software packages (for example, eDEM) discretize the computational domain into three-dimensional cells, which simplifies contact search algorithms and reduces simulation time. The size of the cells is selected based on the used radii of the particles and their speed of movement (usually 3–5 times less than the radius of the particle) As soon as the domain has been discretized, the cells containing the particles are marked as active and the presence of contacts is checked in them. If they are found, the contact forces acting on the particle are calculated. Next, a new position is calculated and the adjacent cells are activated (Fig. 6.60).

The contact models available at the moment are analytical and applicable, to a greater extent, for spherical particles. The most widespread are the Hertz model, which describes the contacting of

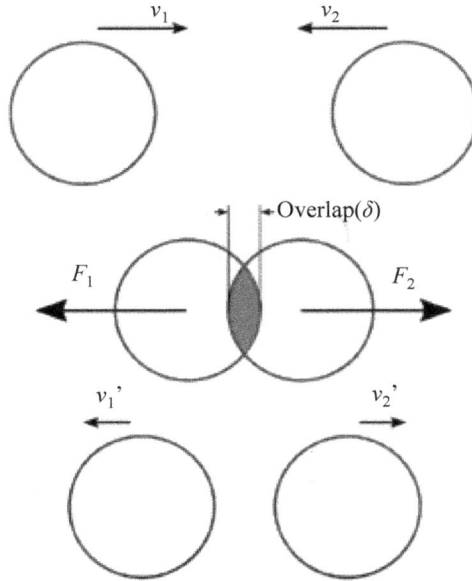

Figure 6.59. Particle contacting scheme.

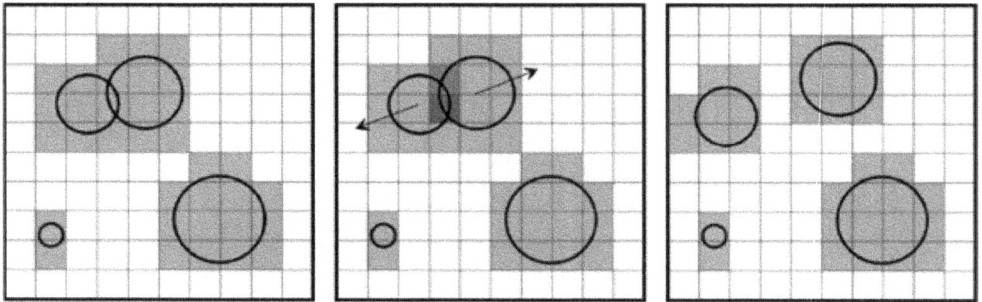

Figure 6.60. Algorithm for searching for contact interactions with cell activation.

two spheres, and Johnson-Kendall-Roberts (JKR) which complements the Hertz model by taking into account the adhesion of particles.

Hertz model. Figure 6.61 shows a schematic representation of the contact stiffness and damping between two particles. Rigidity K_n acts along the normal to the contact area, C_n - determines the damping of particles in this direction.

Hertz's formula for the case of contacting two spherical particles is as follows:

$$F = \frac{4}{3} E^* \sqrt{R} \sqrt{\delta^3},$$ (5)

where F is the contact force, δ is the mutual intersection of particles, R is the reduced radius of the particles:

$$R = \frac{R_1 R_2}{R_1 + R_2},$$

E^* - reduced Young's modulus of particles:

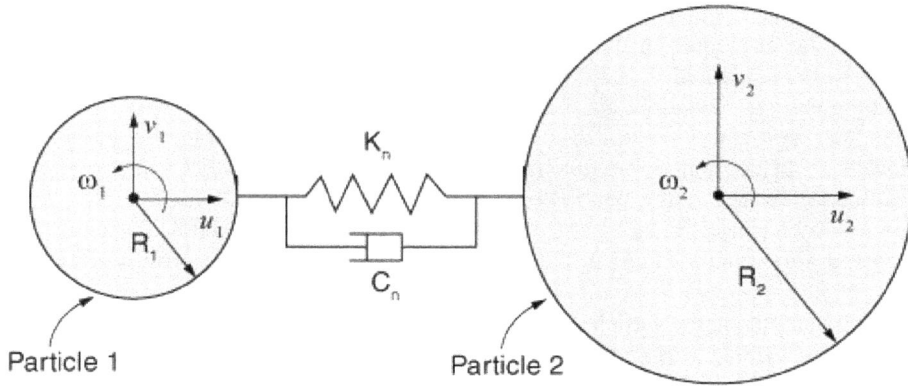

Figure 6.61. Model of contact interactions.

$$\frac{1}{E^*} = \frac{1-v_1^2}{E_1} + \frac{1-v_2^2}{E_2},$$

E_1, v_1 and E_2, v_2 – Young's modulus and Poisson's ratio of each of the particles; R_1 and R_2 – particle radii.

From (5) it follows that the value of the contact stiffness Kn (dimension H/mm) is equal to:

$$K_n = \frac{dF}{d\delta} = 2E^* \sqrt{R}\sqrt{\delta}, \tag{6}$$

The dependence $K_n(\delta)$ has a nonlinear character, however, in practice this feature is often ignored by determining K_{max} - the limiting value of the contact stiffness above which the value of the contact force increases linearly. The K_max value is calculated by substituting in (6) the value $\delta_{max} = 0.005 \dots 0.01R$.

Johnson-Kendall-Roberts (JKR) model. It is an extension of the Hertz model for the case of contacting particles with adhesive bonds. The ratio for the magnitude of the contact force is as follows:

$$F = \frac{4E^* a^3}{3R} - \sqrt{8\pi GE^* a^3} \tag{7}$$

where a – contact area, G – free surface energy (Gibbs energy), which determines the strength of adhesion.

For two spherical surfaces, the dependence $\delta(a)$:

$$\delta = \frac{a^2}{R} - \sqrt{\frac{2\pi Ga}{E^*}} \tag{8}$$

From expressions (7), (8) it follows that at $G = 0$ the JKR model is reduced to a non-adhesive Hertz contact (5).

The adhesion bond between the particles (the second term in expression (5)) is activated immediately after the first contact of the particles. The condition for breaking the connection is to achieve a gap between the particles equal to:

$$\delta_{separation} = -\frac{3}{4} \sqrt[3]{\frac{\pi^2 \Gamma^2 R}{E^2}} \tag{9}$$

The extruded material is represented as a set of spherical particles. The description of the movement of the extruded mass is carried out using the method of discrete elements. To construct mathematical models describing the behavior of granular, free-flowing materials, a continuous or discrete (corpuscular) approach is used. The continuous approach assumes the continuity and structurelessness of the medium, the behavior of which is described by the laws of classical mechanics, usually presented in the form of differential equations. As a result, the behavior of individual particles is ignored.

The computational domain consists of a geometric model of the lower part of the extruder body, the main screw and the extruded mass (Fig. 6.62). The extruder body and main screw were approximated by surface non-deformable finite elements. Each of the bodies was assigned a control point that determines their movement in the computational domain.

The mass to be extruded is a set of 210,000 DEM particles of a spherical shape with a radius of 3.2 mm and a total mass of about 72 kg. The adopted particle characteristics are presented in Table 6.2.

At the initial moment, the particles are located in the upper part of the housing above the surface of the screw. Their filling was carried out by applying gravity. The movements of the auger and housing are limited at this stage.

After filling, the auger rotation is set at an angular speed of 40 rpm. Friction of the auger elements in the clamping area is ignored (Figs. 6.63–6.64).

Hertz and Johnson-Kendall-Roberts (JKR) models are adopted as contact models describing the interaction of particles. The first model describes the behavior of the extruded mass without taking into account the adhesive bonds, the second—takes the adhesion. For JKR, the following level of surface energy (Gibbs energy) $G = 50$ J/m^2 is adopted, which determines the adhesion force.

Below is an illustration of the movement of the extruded mass when the screw rotates with and without taking into account adhesive bonds (Figs. 6.65–6.69).

The entire process can be roughly divided into two stages. Initially, the extruded mass is transferred to the die. Further, the compaction of particles is observed before leaving it. In the model without taking into account adhesion, the material leaves the die for 9 seconds of the process, in the case of adhesion—for 13 seconds (Fig. 6.69).

In the adhesion model, the stagnant zone (Fig. 6.70) is the area of decrease in the cross-section of the extruder body—to reduce it, smoothing of the flow path is necessary.

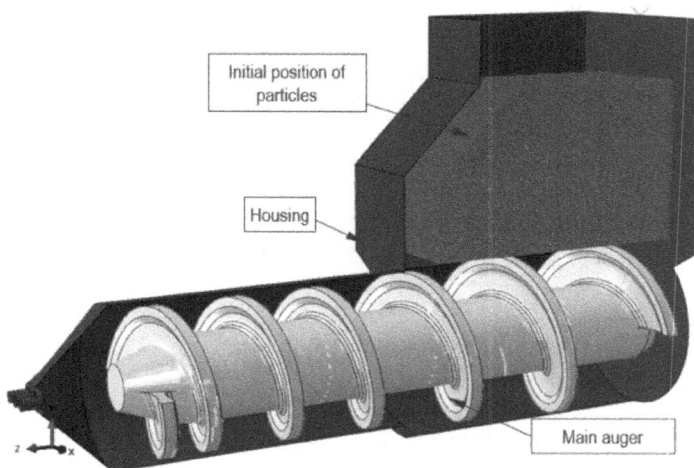

Figure 6.62. Statement of the problem with the initial position of the particles.

Table 6.2. Particles and Material characteristics.

Density, kg/m³	Young's modulus, MPa	Poisson's ratio	Radius, mm	Number of particles	Total weight, kg
2500	1000	0.25	3.2	210 000	72

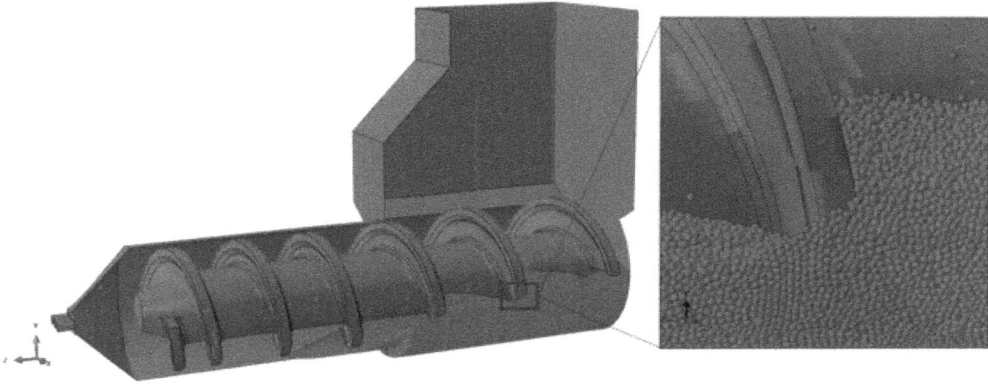

Figure 6.63. Position of particles after backfill.

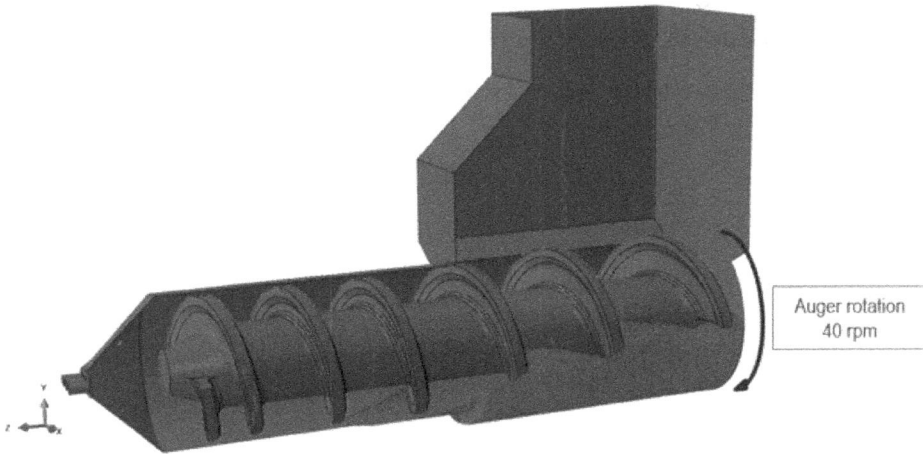

Auger rotation
40 rpm

Figure 6.64. Setting the auger rotation.

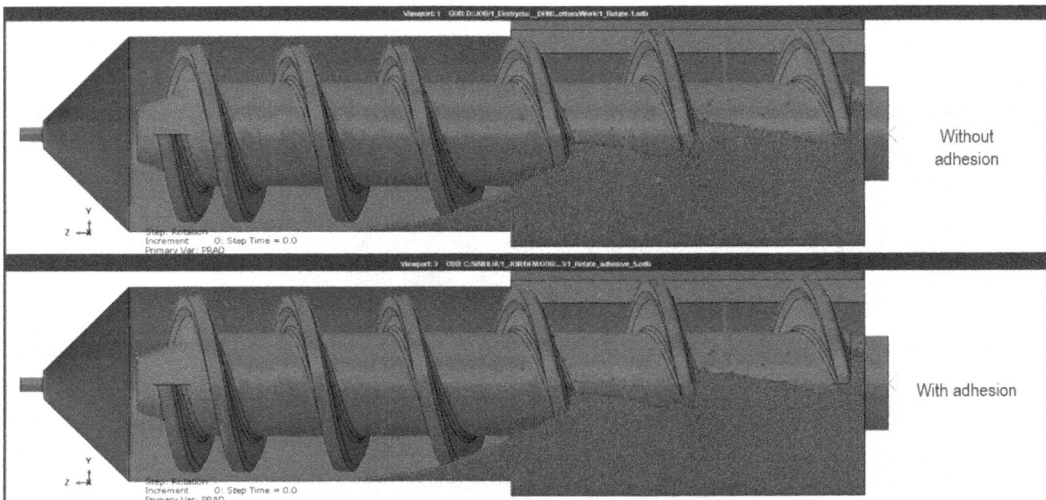

Without adhesion

With adhesion

Figure 6.65. Movement of the extruded mass: process time 0 second.

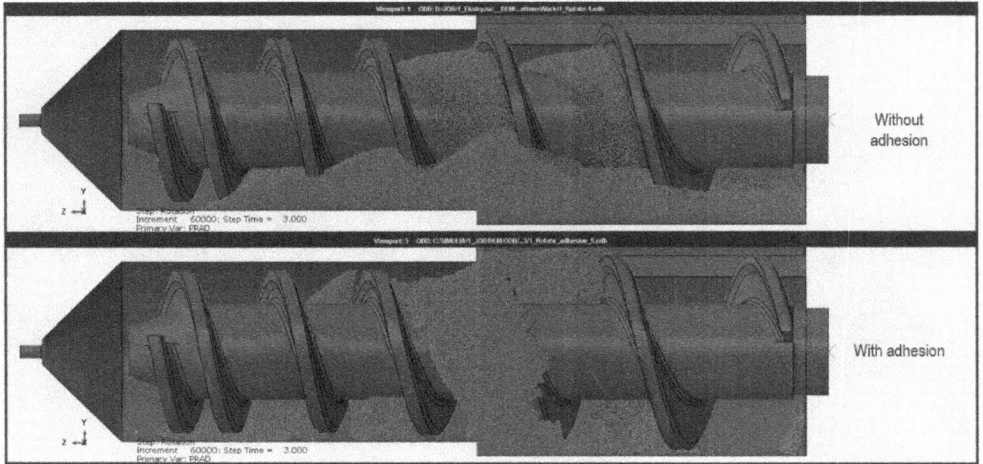

Figure 6.66. Movement of the extruded mass: process time 3 seconds.

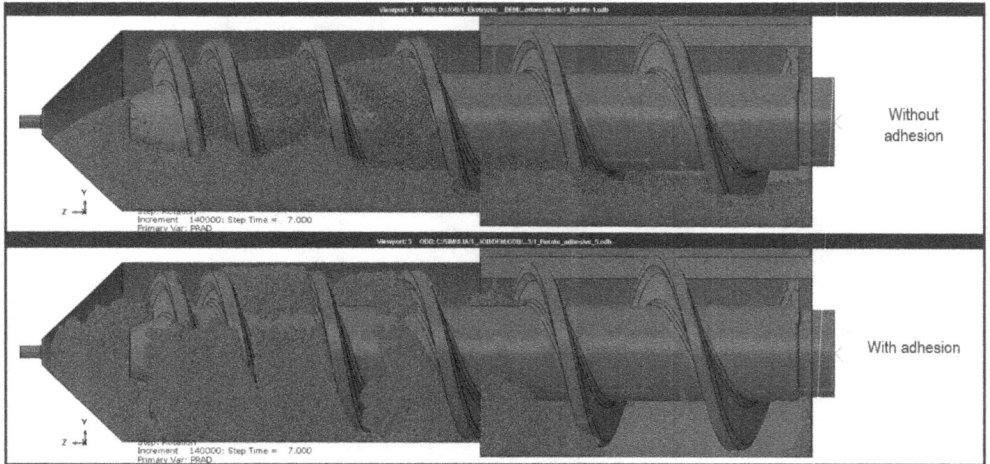

Figure 6.67. Movement of the extruded mass: process time 7 seconds.

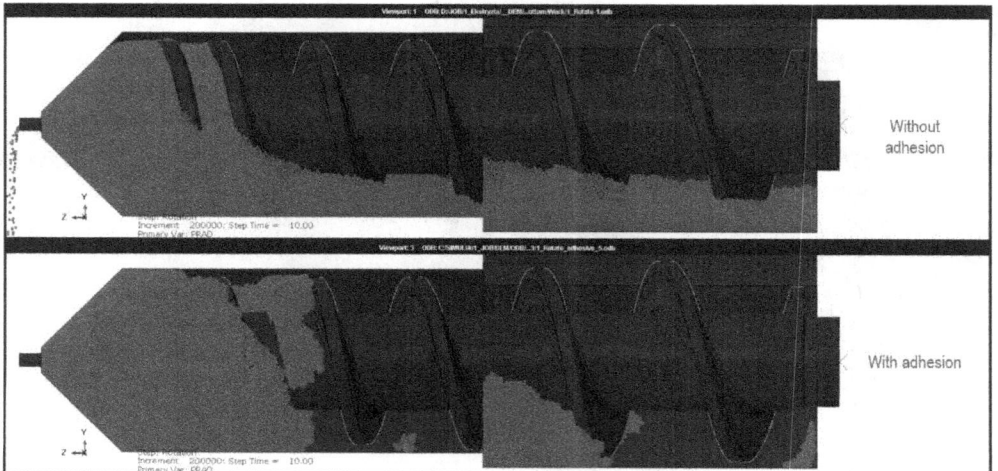

Figure 6.68. Movement of the extruded mass: process time 10 seconds.

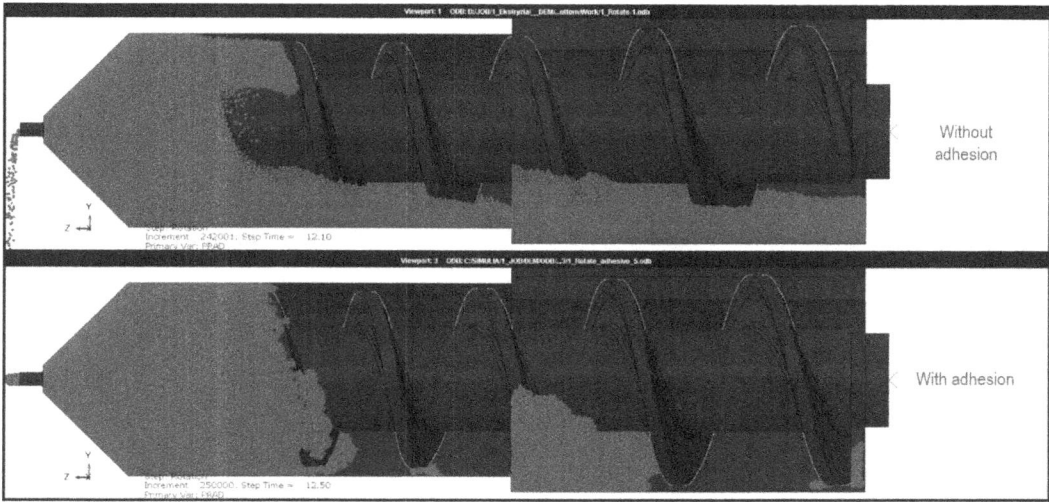

Figure 6.69. Movement of the extruded mass: process time 13 seconds.

Below is the distribution of the speeds of the part when moving (Fig. 6.71). The level of characteristic values of flow velocities is in the range of 0.5–1 m/s.

Figure 6.72 shows the graphs of changes in the time on the auger axis during the extrusion process for the two considered settings. With non-adhesive behavior of the material, the averaged value of the torque at the stage of moving the working mass is 1000 N/m, which corresponds to 4.2 kW. In this case, a torque increase in the moment when the material approaches the die is not observed—with a small relative size, the particles freely leave the die holes. The presence of an adhesive bond increases the averaged value of the torque in the quasi-stationary section up to 2000 N/m (8.4 kW). As the die is reached and the material is compacted, a significant increase in torque is observed up to 6500 N/m (27.2 kW), followed by a steady state regime.

On the basis of the SIMULIA Abaqus software complex, a digital twin of the extruder was created, which made it possible to assess the modes of its operation by varying the following input

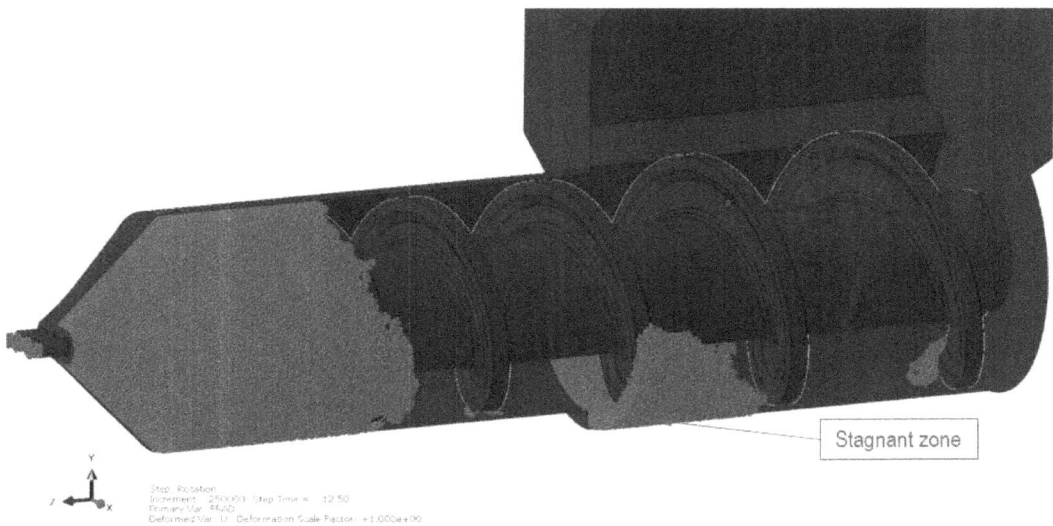

Figure 6.70. Stagnant zones.

V, Magnitude

```
+1.096e+03
+1.005e+03
+9.142e+02
+8.232e+02
+7.323e+02
+6.413e+02
+5.503e+02
+4.593e+02
+3.683e+02
+2.773e+02
+1.863e+02
+9.534e+01
+4.354e+00
```

Y

Z ◄━━━● X

Step: Rotation
Increment   94001: Step Time =   4.700
Primary Var: V, Magnitude
Deformed Var: U   Deformation Scale Factor: +1.000e+00

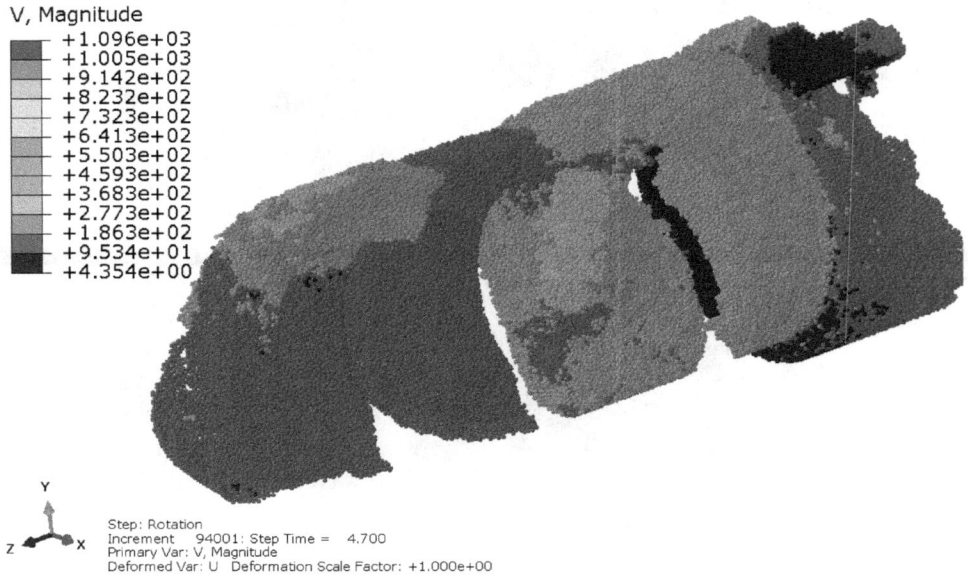

Figure 6.71. Velocity distribution, mm/sec.

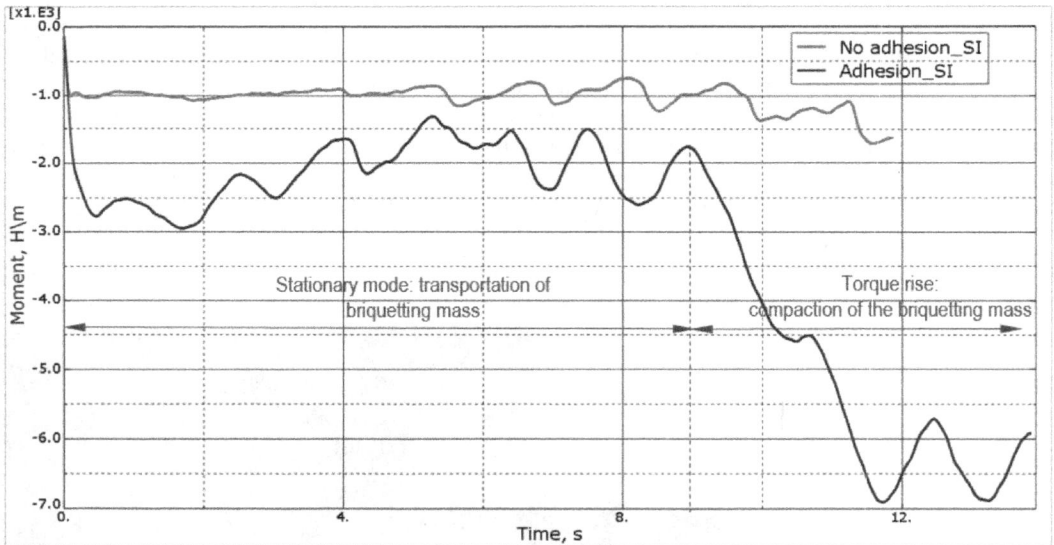

Figure 6.72. Changing the torque on the auger during the process.

parameters: material mass flow rate, material stiffness, shaft rotation speed, blade shape, geometric characteristics of the flowing parts of the body.

Structurally, the extruder consists of three main elements: a fixed body, a rotating screw and a pug mill (Fig. 6.73). The geometric models of each of the elements were built according to the illustrative materials of the manufacturing company, observing the basic geometric proportions. At the same time, the actual dimensions of the structural elements may differ from those taken in the calculation.

On the basis of the geometric model, a finite element model of the extruder was built, consisting of absolutely rigid surface elements. The body is fixed to all degrees of freedom through the control

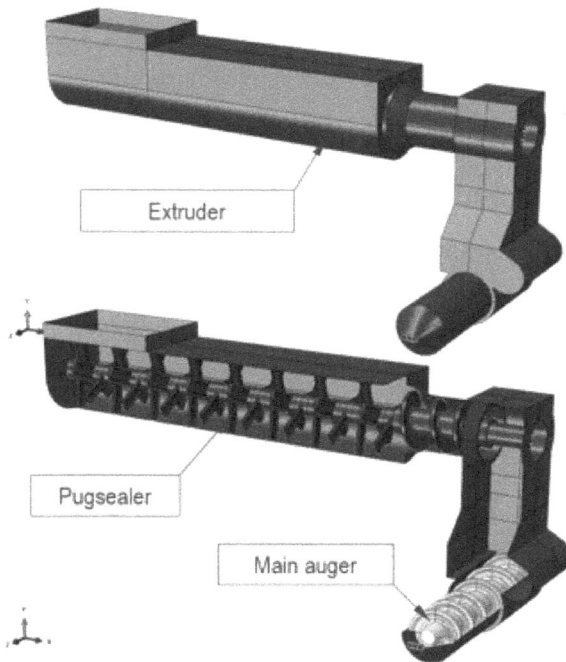

Figure 6.73. Geometric model of the extruder.

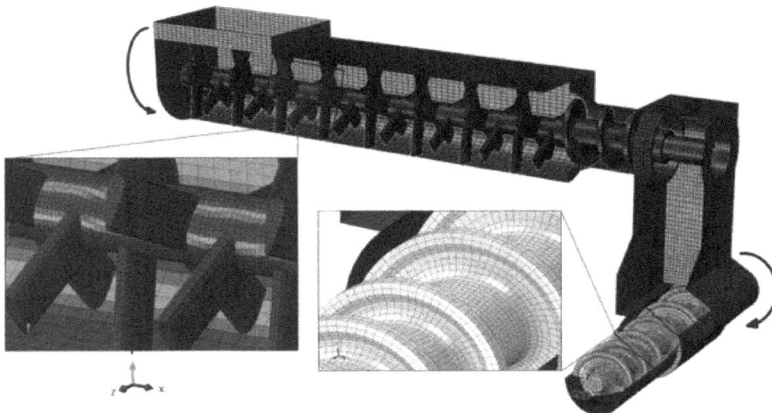

Figure 6.74. Finite element model of the extruder and boundary conditions.

point. The elements of the pug mill and the main auger are set to rotate about their longitudinal axes (Fig. 6.74).

The flow of material entering the extruder was represented as a set of elementary spheres with a diameter of 12 mm and a density of 2500 kg/m³. Spheres-particles were modeled using the discrete element method, which reduces the description of body motion to the motion of a material point located at the center of mass. Contact interaction is determined between particles and extruder surfaces using the Hertz model.

Particles were fed into the computational domain through specialized inflators—particle generators, which allow varying the mass flow rate, radius and initial velocity. After leaving the inflator, the particles fall into the field of gravity, which was determined through the acceleration of gravity of 9.8 m/s². To reduce the simulation time and reach the stationary regime of the material "flow", two inflators were used (Fig. 6.75). An auxiliary inflator was introduced to fill the volume of the area where the pug mill was located and was active for 1.5 seconds. After this time, the generation of particles was carried out only through the main inflator—the actual place of supply of the mixture.

The operation of the auxiliary and main inflators is illustrated in Figs. 6.76–6.77.

The following shows the change in the position of the particles during the operation of the extruder (Figs. 6.78–6.85):

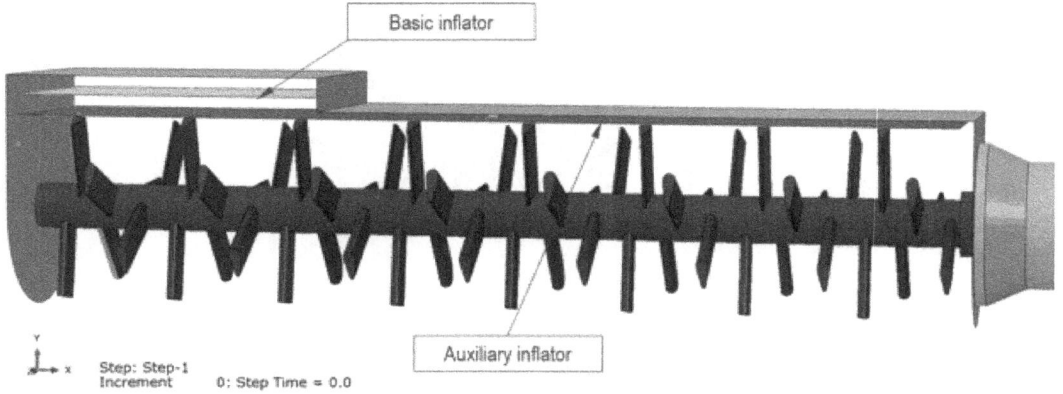

Figure 6.75. Location of particle generators.

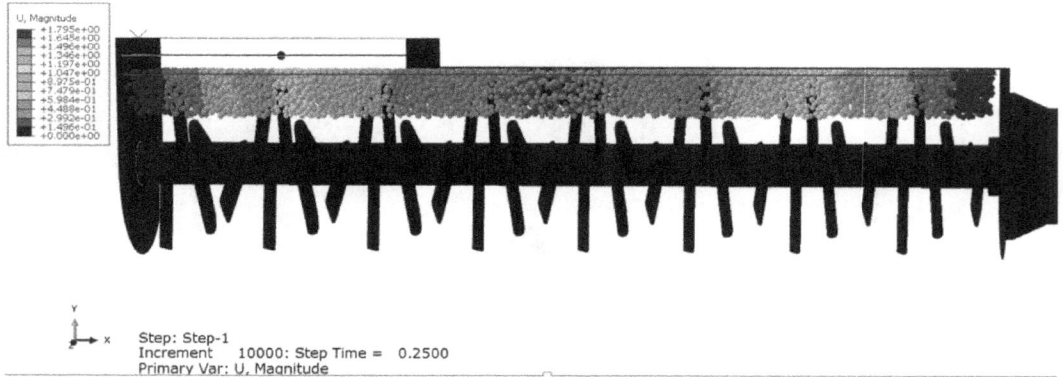

Figure 6.76. Operation of the auxiliary inflator.

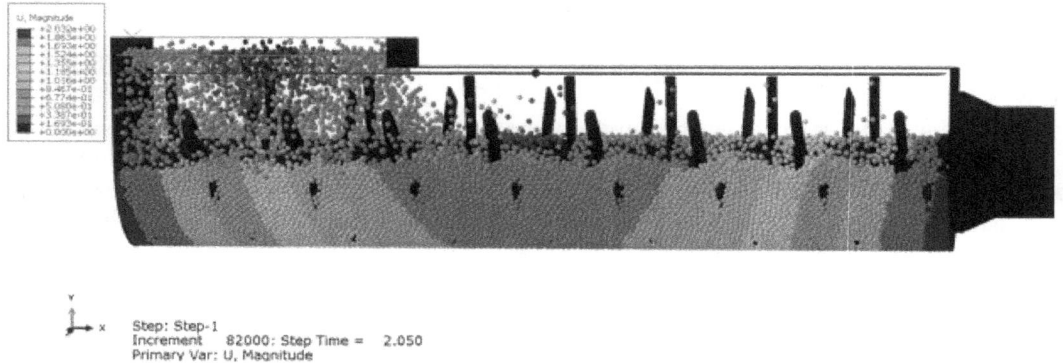

Figure 6.77. Operation of the main inflator.

The task of modeling the operation of the extruder is closely related to the task of preparing the mixture for agglomeration, in particular, the of mixing the briquetted material with a binder. This

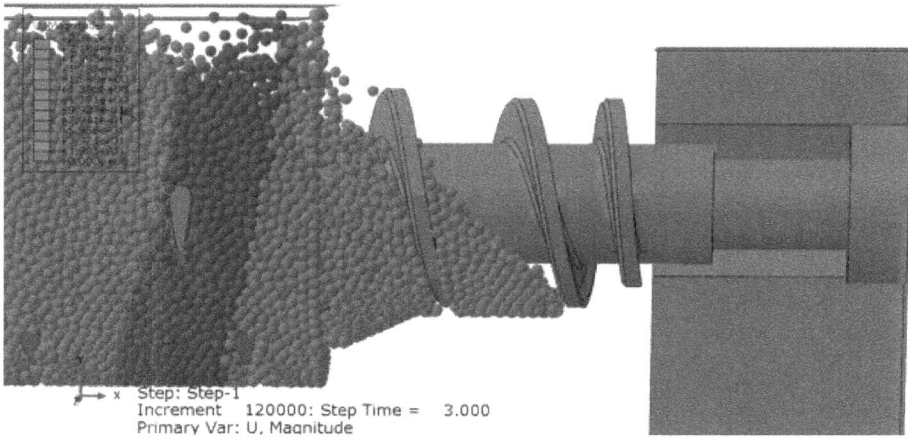

Step: Step-1
Increment   120000: Step Time =   3.000
Primary Var: U, Magnitude

Figure 6.78. Particle movement along the pugsealer. Time 3 seconds.

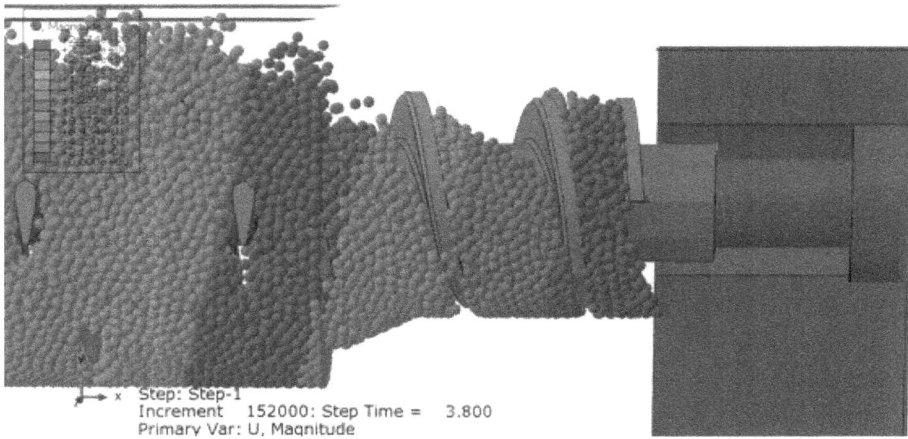

Step: Step-1
Increment   152000: Step Time =   3.800
Primary Var: U, Magnitude

Figure 6.79. Particle movement along the pugsealer. Time 3.8 seconds.

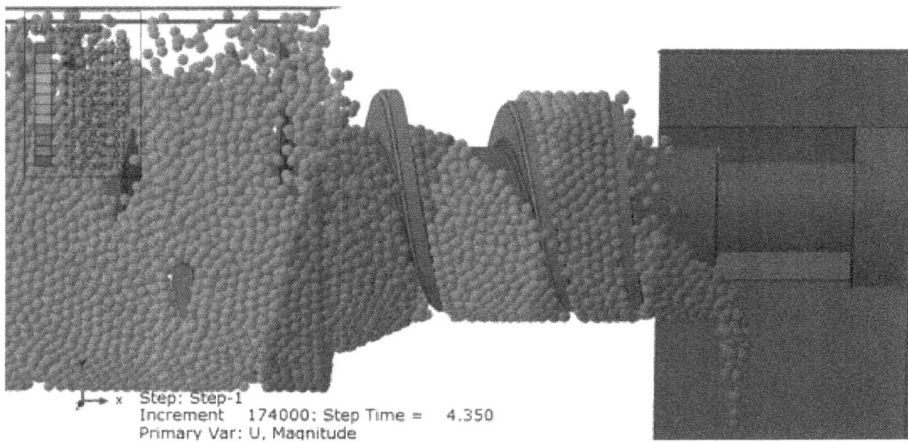

Step: Step-1
Increment   174000: Step Time =   4.350
Primary Var: U, Magnitude

Figure 6.80. Particle movement along the pugsealer. Time 4.35 seconds.

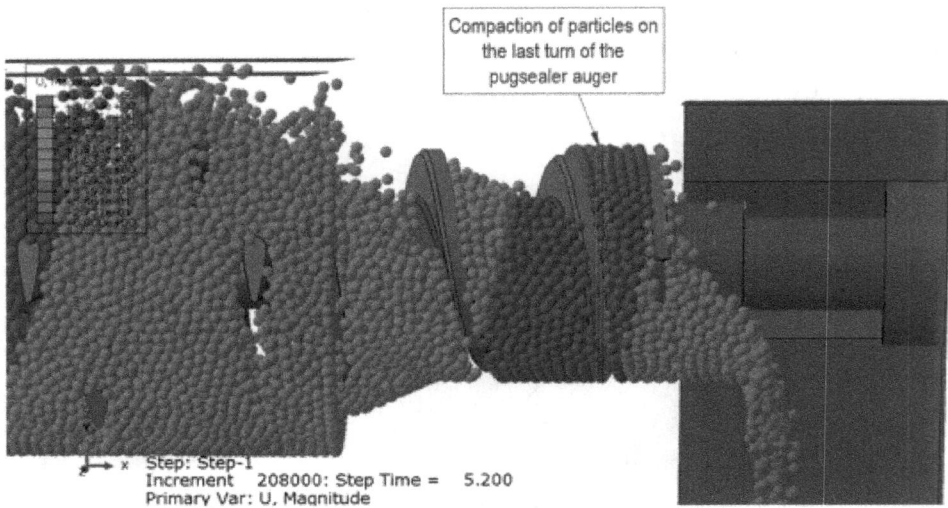

Figure 6.81. Particle movement along the pugsealer. Time 5.2 seconds.

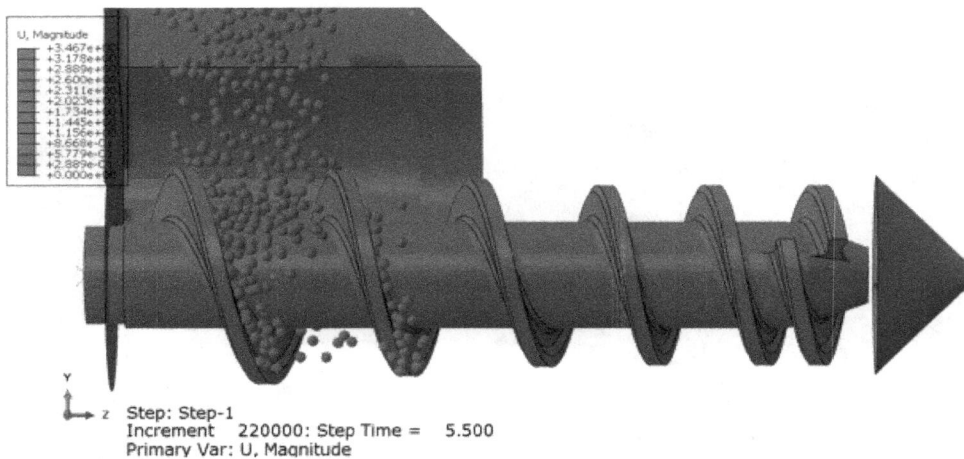

Figure 6.82. Particle movement into the extruder's chamber.

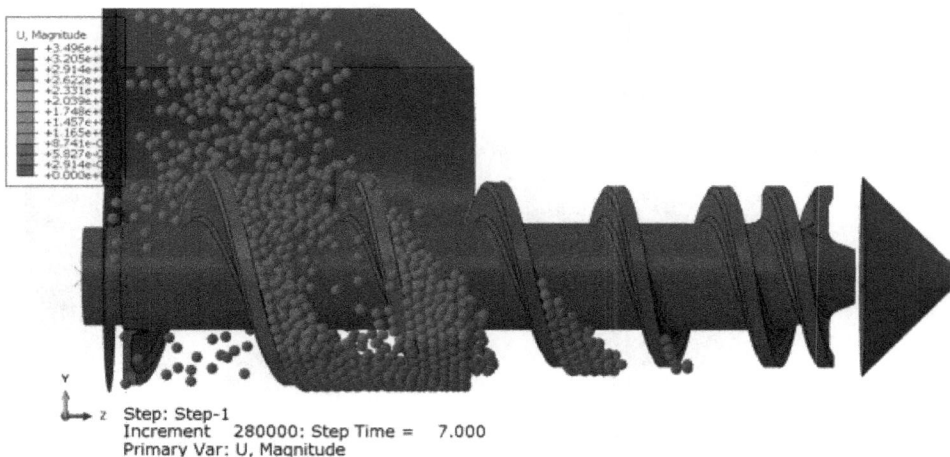

Figure 6.83. Particle movement in the extruder. Time 7 seconds.

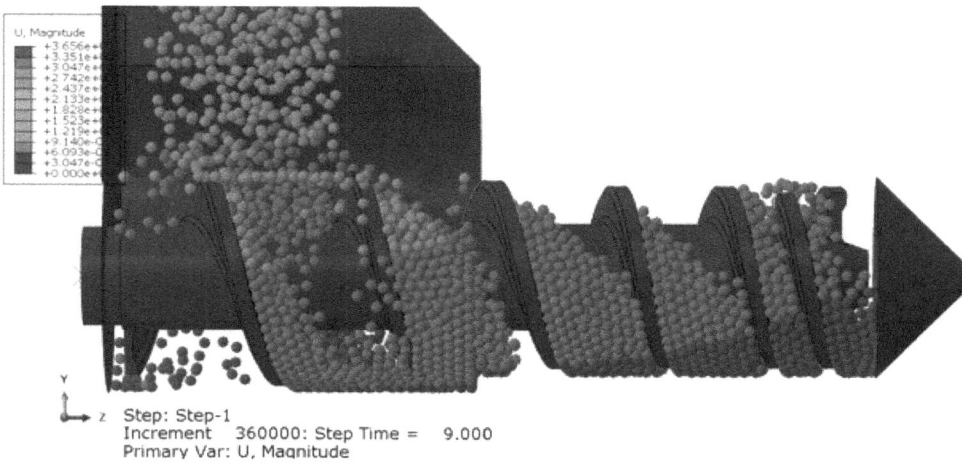

Step: Step-1
Increment 360000: Step Time = 9.000
Primary Var: U, Magnitude

Figure 6.84. Particle movement in the extruder. Time 9 seconds.

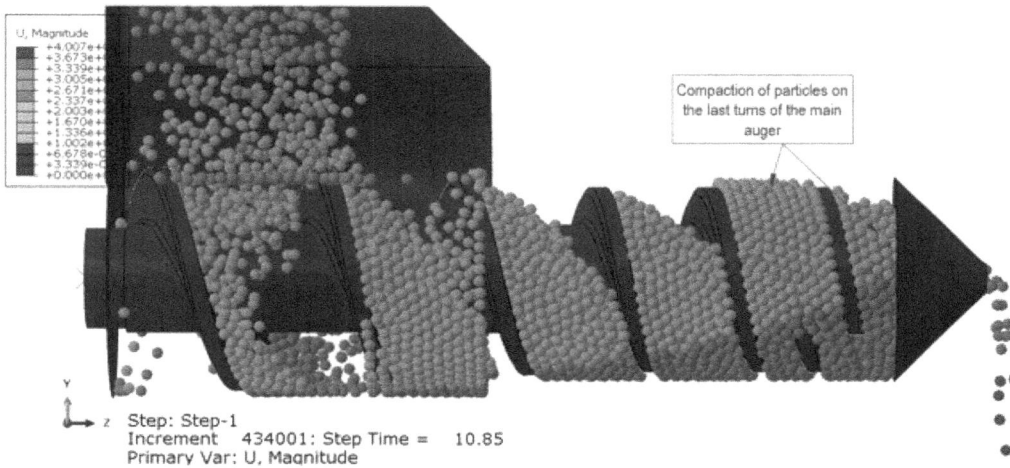

Compaction of particles on the last turns of the main auger

Step: Step-1
Increment 434001: Step Time = 10.85
Primary Var: U, Magnitude

Figure 6.85. Particle movement in the extruder. Time 10.85 seconds.

becomes especially important in the case of using small amounts of binder (1% or less).

The modeling of the process of mixing iron ore concentrate with a binder on the basis of a laboratory mixer of the Eirich company were carried out. Two kinds of mixture were investigated: 90% iron ore concentrate with a particle size of 3 mm + 10% cement with a particle size of 1.5 mm; 99% iron ore concentrate with a particle size of 3 mm + 1% polymer binder with a size of 1.5 mm. If the used proportion of the polymer binder is less than 1, it is possible to recommend its preliminary mixing with a certain volume of briquetted material in a proportion of 10–90 (10% of the binder and 90% of the base material).

Material properties and design parameters of the problem are given in the Table 6.3.

The particles of the mixture were modeled using the discrete element method, which reduces the description of body motion to the motion of a material point located at the center of mass. Adhesion was determined between the particle spheres and the mixer surface based on the *Johnson-Kendall-Roberts* (JKR) model. This model is an extension of the Hertz model for the case of contacting particles with adhesive bonds. The relation for the magnitude of the contact force is given above (7). The dependence δ (a) is given by formula (8). The condition for breaking the bond is determined by the size of the gap between the particles described by formula (8).

Table 6.3. Material properties and design parameters of the problem.

	Density, kg/m³	Particles size, mm	Number of particles in the sample
Iron ore concentrate	4000	3	15 318
Cement	3000	1,5	16 329
Polymeric binder	1000	1,5	4 910

Below are the results of modeling the position of the particles during the operation of the mixer (Fig. 6.86–6.92):

6.3.4 Basic equipment for briquetting by stiff vacuum extrusion

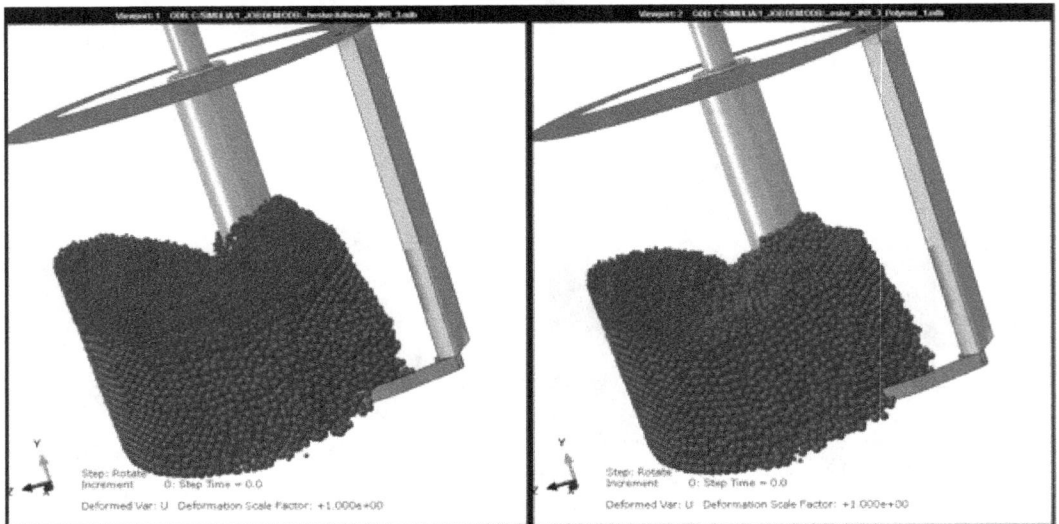

Figure 6.86. Position of the particles after filling the binder. On the left is cement; on the right - polymer binder.

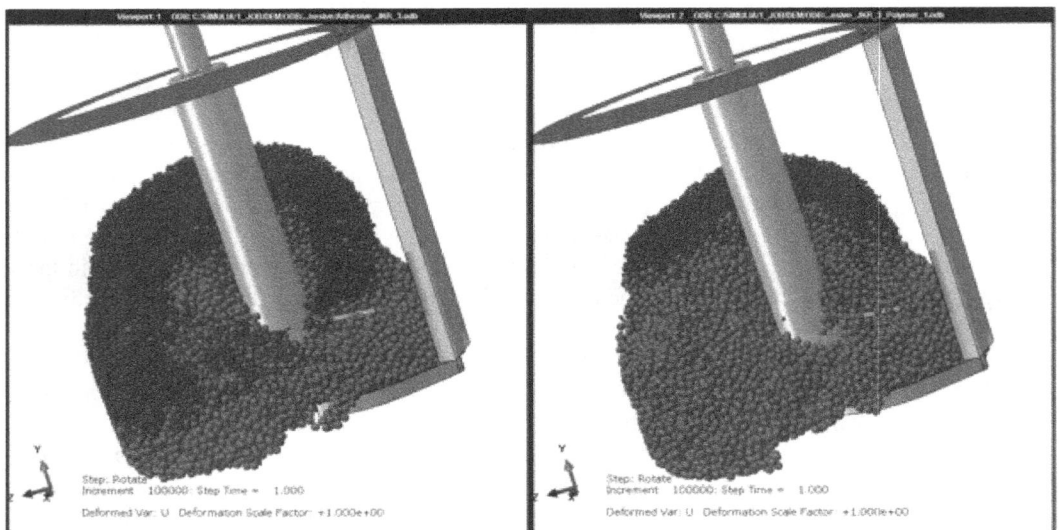

Figure 6.87. Particle position: 1 second rotation.

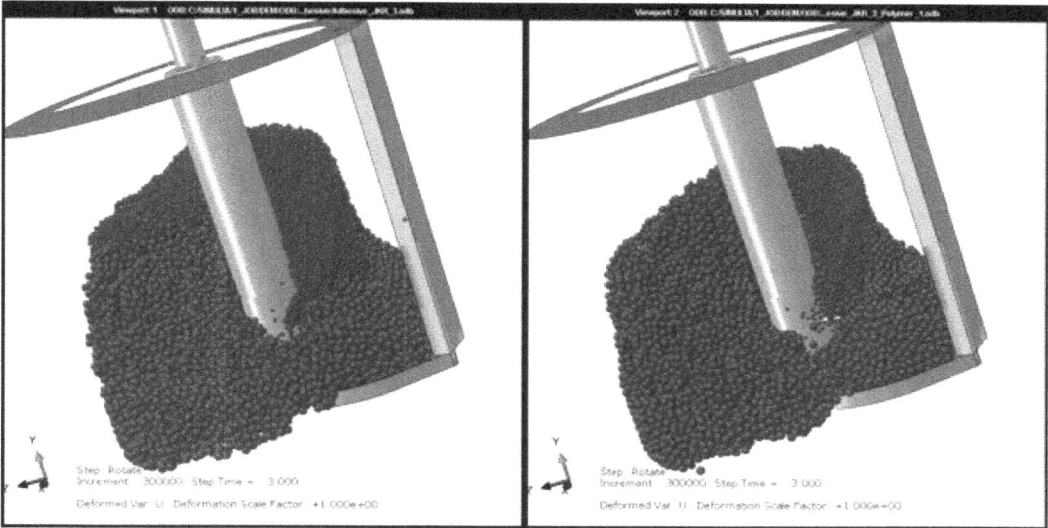

Figure 6.88. Particle position: 3 seconds rotation.

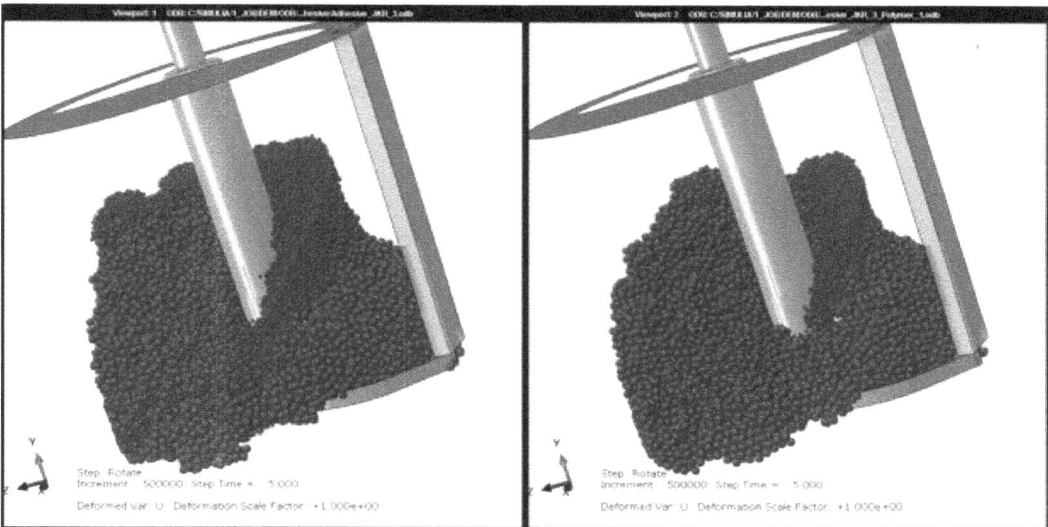

Figure 6.89. Particle position: 5 seconds rotation.

The first successful attempts to de-air the mixture before feeding it into the extruder were made in the United States in 1920, and since 1932 this procedure has been used in Europe. The first experience of stiff extrusion also took place in the United States at the end of the 19th century and is associated with the company "J.C. Steele & Sons, Inc." founded in 1889. However for the first time on an industrial scale, stiff extrusion was implemented in England in 1960, and then in Germany in 1967 by Haendle, which today is part of the J.C. Steele & Sons, Inc. group of companies. It was Haendle who combined the creation of vacuum in the working chamber and the stiffness of the mixture during extrusion. J.C. Steele & Sons, Inc. is considered the creator of modern stiff extrusion technology and the world's largest supplier of stiff extrusion extruders.

As noted above, shear stress plays a significant role in stiff extrusion. This is the effect that the charge briquetted by this method is subjected to at all stages of agglomeration, starting with feeding and mixing in the so-called J.C. Steele & Sons, Inc. uniform feeding feeders (Fig. 6.93). The

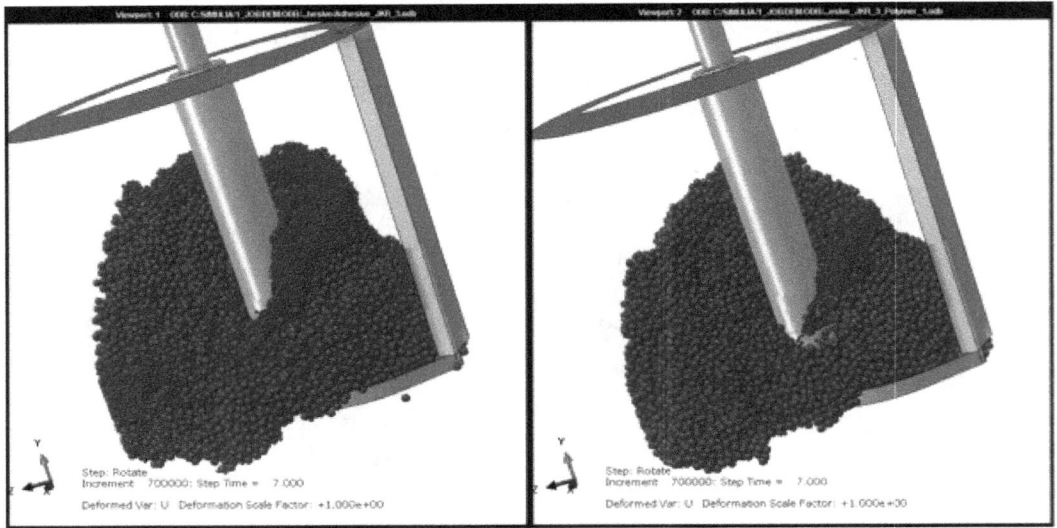

Figure 6.90. Particle position: 7 seconds rotation.

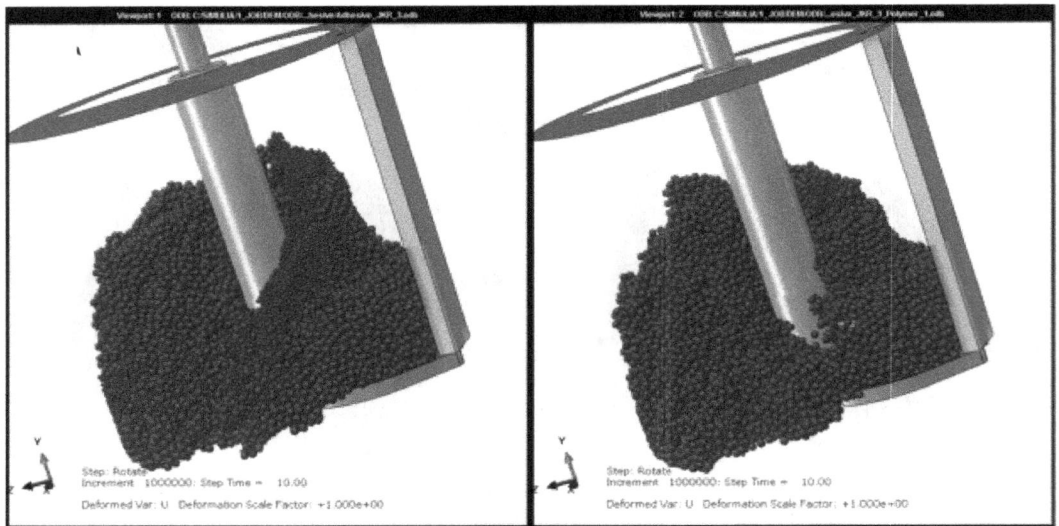

Figure 6.91. Particle position: 10 seconds rotation.

working bodies in them are solid and split augers (Fig. 6.94). The company manufactures feeders with two, four and eight screws. Solid screws only supply the charge, and split screws, in addition to feeding, carry out loosening, crushing and crushing. The capacity of uniform feeding feeders is from 30 t/h to 200 t/h. Single-shaft and twin-shaft pug mills are used for mixing and moistening the batch. The clay melting knives are made of 28RS alloy (28% chromium). The productivity: from 10 to 150 t/h.

J.C. Steele & Sons, Inc. serially manufactures several types of extruders, differing in productivity. The smallest design productivity of the extruders of this company is 5 tons per hour

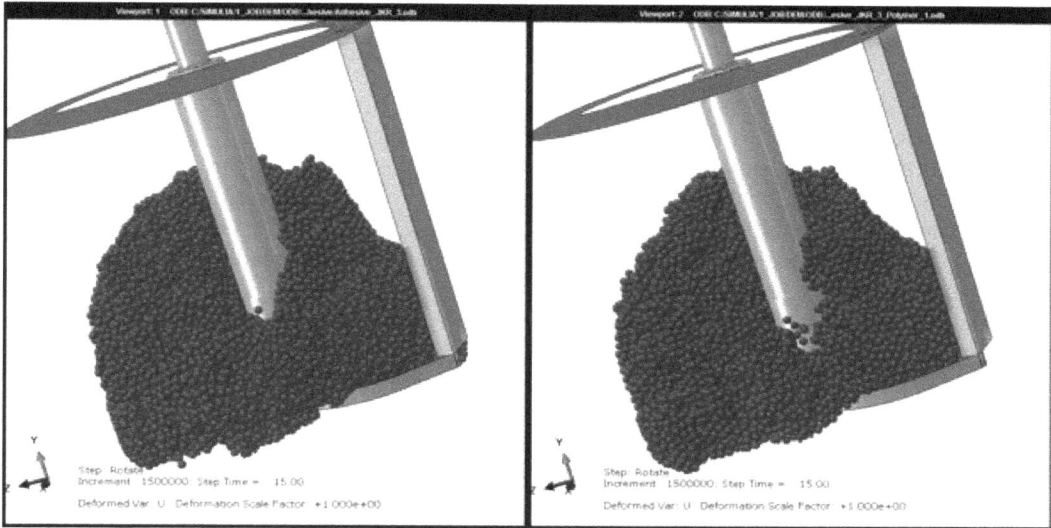

Figure 6.92. Particle position: 15 seconds rotation.

Figure 6.93. J.C. Steele & Sons, Inc. even feeder.

Figure 6.94. J.C. Steele & Sons, Inc. even feeder shafts.

(extruder HD-10, Fig. 6.95). It is clear that the real productivity is determined by the density of the briquetted mass.

In the range of capacities from 20 to 50 tons per hour, the company supplies Steele extruders of the 25th and 45th series (Figs. 6.96–6.97).

The Steele 75 extruder can produce up to 54 tons of briquettes per hour (Fig. 6.98). The most efficient extruder of the company is the Steele 90 extruder, designed for production of 80–90 tons per hour (Fig. 6.99). The 45, 75 and 90 series extruders can be equipped with a hydraulic die changer that completes this procedure almost instantly. The 120 AEX Stiff Extruder and Pug Sealer combination has the same ease of operation and durability as 90 Series extruder, however it differs by 50% more output per hour (Fig. 6.100).

All units and parts of the machines are designed to withstand significant overloads, especially bearings and powerful shafts made of one-piece forging. All elements of the extruder body, the

Figure 6.95. J.C. Steele & Sons, Inc. HD-10 Extruder with a capacity of 5 tons of briquettes per hour.

Figure 6.96. J.C. Steele & Sons, Inc. Steele 25 Extruder capacity up to 20 tons of briquettes per hour.

Figure 6.97. J.C. Steele & Sons, Inc. Steele 45 Extruder capacity up to 47 tons of briquettes per hour.

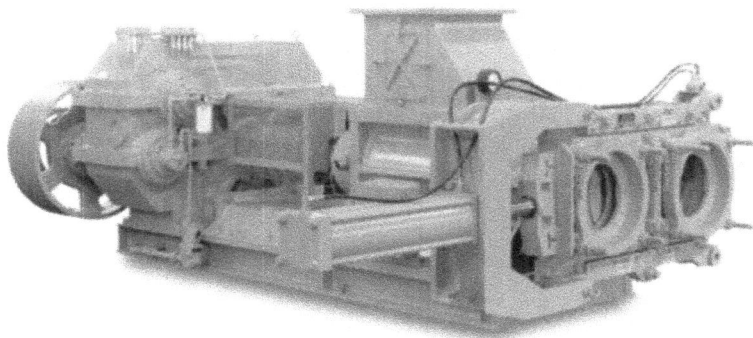

Figure 6.98. J.C. Steele & Sons, Inc. Steele 75 Extruder capacity up to 54 tons of briquettes per hour with hydraulic die changer.

Figure 6.99. J.C. Steele & Sons, Inc. Steele 90 Extruder with a capacity of up to 80–90 tons of briquettes per hour with a hydraulic die changer.

vacuum chamber and the seal assembly of the pug mill are cast from ductile ductile iron. In the design of the extruder and the vacuum chamber of the pug mill, welding is not used. All parts in contact with the processed metal are made of 28PC alloy (28% chromium and other alloying additives). Extruders and pug mills are designed for a long service life (30 years or more) and are easy to maintain and operate. By 2021, the company, together with the member of the group

Figure 6.100. J.C. Steele & Sons, Inc. Steele 120 Extruder capacity up to 120 tons of briquettes per hour.

"J.C. Steele & Sons, Inc." supplied equipment for dozens of briquette factories around the world. Briquettes are used in the charge of blast furnaces and ore smelting furnaces. The annual production of briquettes by the company's extruders exceeds 2 million tons.

CHAPTER 7
Industrial Briquetting in Iron Making

The reasons why roller briquetting is not widely used in blast-furnace production are due to the need to ensure low moisture content of the components of the briquetted charge, which, in turn, limits the applicability of Portland cement and other hydrated binding materials as a binder. For the hardening of the cement stone in such a briquette, there would not be enough water. Organic binders are not known to ensure the integrity of briquettes at temperatures in the cohesion zone.

The search for ways to produce durable briquettes with the required hot-strength level for blast furnace on the basis of roller briquetting has led to renewed interest in thermal briquetting. The companies Köppern and Sahut Conreur own technology for the production of briquettes from a heated mixture based on roller presses of a special design. Full-scale industrial tests carried out in 2011 in Japan at blast furnace No. 3 by Kobe Steel showed the possibility of achieving a positive effect from the use of roller briquettes made from a mixture of preheated coal and iron ore concentrate heated to 600°C [1]. The manifestation of the binding properties of the plasticized coal during heating made it possible to achieve the required hot strength. However, the method of thermal briquetting was not widely used for the production of blast furnace briquettes, since it did not provide a competitive advantage over sinter and pellets. In terms of thermal briquetting, the obvious advantage of cold briquetting, consisting in the absence of heat treatment and undesirable environmental impact is lost.

Another development of the same company, in which the use of roller briquetting is possible, is based on the production of a new agglomerated product, called "ferro coke", obtained by coking the coal charge with the addition of iron ore or flue dust [2]. In contrast to thermal briquetting, the agglomeration of poor ore and coke breeze is carried out using a special binder based on naphthalene-type solvents used in coal mining. The obtained blast furnace briquettes are additionally carbonized at a temperature of 1000°C for 6 hours. Studies have shown the possibility of achieving high metallurgical properties of ferro coke and formed the basis of the project to build a test plant for its production worth US$ 100 million, carried out by JFE Steel with Kobe Steel and Nippon Steel & Sumitomo Metal [3]. The pilot plant will produce 300 tons of ferro coke daily. The main objective of the project is to test the effectiveness of using this new agglomerate product to reduce carbon dioxide emissions and the possibility of achieving the expected increase in the efficiency of blast furnaces by 10%. It is hoped to commercialize ferro coke by 2022.

A concept based on a cold-bonded ferrocoke was proposed in [4]. To reduce coke consumption in blast furnace smelting, the authors recommend using so-called Reactive Coke Agglomerate (RCA) on a cement binder. The essence of this concept is to create a special RCA structure in which high homogeneity and orderliness of the mutual arrangement of iron ore and coke concentrate particles is achieved, ensuring their maximum proximity to each other, which leads to high reactivity and, as a result, to a decrease in coke consumption. Attempts to use roller and other presses for the production of such agglomerated products without their further heat treatment do not allow obtaining a mechanically strong briquette due to the high carbon content recommended by the developers of this technology – 20% of the briquette mass. In this case, the proportion of the cement binder required to provide the necessary strength may exceed values that justify the economic attraction of cold briquetting. Commercialization of this technology was carried out by the method of obtaining cold bonded pellets using specially designed heat and moisture treatment, which significantly accelerated the curing of the pellets to achieve the required mechanical strength [5].

Thus, it is clear that the prospects for using roller briquetting for the production of thermo-briquetted blast-furnace briquettes and ferro-coke are currently awaiting their determination.

In this regard, the metallurgical properties of blast furnace briquettes obtained by the method of vibropressing and stiff extrusion will be considered further. First, the mechanism of preserving the strength of briquettes when they are heated in a reducing atmosphere are of interest.

In Russia the first vibropressing briquetting factory was commissioned at the JSC Tulachermet in 2003 [6]. The capacity of the line enabled a monthly production of up to 8000 tons of briquettes. Two types of briquettes were used as components of the BF charge: the so-called "washing" briquettes and iron and carbon briquettes. It is known that in some cases, technologists were faced with "cluttering" of the hearth of the blast furnace due to the deterioration of the coke's ability to filter when filling the voids between its particles with pieces of slowly moving smelting products. This can result in combustion of air tuyeres, decrease of hearth heating and other phenomena, which reduce the melting performance. One of the most effective means of combating this phenomenon is "flushing" the hearth using liquid slag containing FeO or MnO. Typically, the special sinter made using mill scale serves as the "washing" material. Mill scale briquettes can also be used for such purposes.

The value for the design compressive strength of Tulachermet briquettes was supposed to be at least 6.0 MPa. After drying its values were – 3.83 MPa, after heat-moisture treatment – 6.9 MPa. In total during the course of the existence of the line more than 52 thousand tons of briquettes were manufactured (about 50 thousand tons of briquettes containing iron and carbon, and 2700 tons of washing briquettes). During the unloading of the briquettes from the bunker their bridging was noted [7]. The maximum shares of the briquettes in the charge of the blast furnaces were as follows: BF No. 1 – 32 kg/t of hot metal, BF No. 2 – 56 kg/t of hot metal. With the application of briquettes, flue dust the emission decreased by 4 kg/t. Factor analysis showed that with the application of iron and carbon containing briquettes the specific consumption of dry skip coke decreased by 14.4 kg/t of hot metal, which corresponded to the coke's replacement in the body of the briquettes with coke breeze by a ratio of 1 to 1. Pig iron production increased by 37 tons/day, which is associated with an improvement in the gas permeability of the charge column due to the improved granular composition of the charge. This helped to increase the blast volume and, as a result, to intensify melting. After several years of operations, the briquetting factory was closed as a result of changes in economic conditions affecting the availability of suitable carbonaceous materials for briquetting.

A series of campaigns on the use of vibropressed briquettes of varying compositions on a cement binder in the charge of blast furnace with a volume 1000 m^3 was carried out in 2003–2004 by the Novolipetsk Steel Company (JSC NLMK) [8–10].

The results of experimental melts confirmed the effective application of iron and carbon containing briquettes as a new blast furnace charge component. Due to the presence of carbon such briquettes are self-reducing and their utilization leads to a proportional reduction of coke consumption. The main results of the campaigns are as follows: vibropressing can provide the required values for the mechanical strength of briquettes made of natural and anthropogenic oxide materials with a share of the cement binder of not less than 8–10% by weight of a briquette; the value of the compressive strength for most of the briquettes was not less than 30 kgF/cm^2 and ensured their integrity during overloads and transportation with a less than 5–7% fines generation (–10 mm).

In 2010 the Kosaya Gora Iron works vibropressing briquetting factory (located in Russia) was commissioned for the production of 120000 tons of briquettes from a mixture of iron ore concentrates and fines, and flue dust with a Portland cement binder (not less than 10% by weight of the briquette). The exposure time of the briquettes in a heat treatment chamber (for heat-moisture treatment) was 36 hours. The density of briquettes ranged from 2.0–5.0 g/cm^3. Their compressive strength: not less than 3.5 MPa, and the moisture content was up to 9%. The share of briquettes in the ore part of the blast furnace load is 100 kg per ton of hot metal. Bridging was noted during loading [11]. In total the factory has produced over 300 thousand tons of briquettes. There is almost no information available in the open domain concerning a systematic analysis of the results of the operation of this factory. The factory has practically been out of operation since October 2015.

During March 2012, in Finland at the Raahe Works Company a vibropressing briquetting line for the production of briquettes for blast furnaces was commissioned after the closing of the sinter plant [12]. Briquetted charge consisted of: mixture of pellets fines (71000 tons), coke dust (12000 tons), premix of flue dust and steel scrap (59000 tons), mill scale (59000 tons), briquettes fines (34000 tons) and scrap (15000 tons), slag cement (17000 tons) and rapid cement (19000 tons). The total amount of briquettes recycled via blast furnace process is 283000 tons per year. The briquette size is 60 × 60 mm, weight is 475 g. After the 48 hours of strengthening the cold strength of briquettes according to ISO 4696 was 74% (the share of pieces larger than 6.3 mm). Coke rate was decreased by 6% by feeding 120–130 kg of briquettes per 1 ton of hot metal into the charge of the BF due to the fact that the briquettes contained carbon. It is seen that vibropressing led to the formation of large quantities of briquettes fines (34000 tons per year). It was also impossible to agglomerate efficiently BOF sludge (60000 tons per year; moisture content 33–39%) and BF sludge (30000 tons per year; moisture content 20–25%) by vibropressing.

The results of full-scale testing and of the practical implementation of vibropressing briquetting technology confirm the possibility of achieving the required level in terms of the metallurgical properties of the agglomerated products. However, this technology creates some significant technological limitations, which are difficult to overcome or leads to a large increase in the cost of a briquette.

In April 2011, an industrial briquetting line was commissioned at the Suraj PL plant in Rourkela (India), producing brex using stiff vacuum extrusion technology for their use in a blast furnace. The management of the company decided to implement a project for the production of brex from a mixture of converter sludge, blast furnace dust, Portland cement and bentonite. The main objective of the project was to increase the economic efficiency of the small blast furnace by replacing the purchased ore with cheap brex.

In the period from 2011 to 2018, several more briquetting plants were built, producing brex for blast furnace. The world's largest factory for the production of blast furnace brex was built in 2018, Lipetsk (Russia) at the site of NLMK. Its capacity is 700 thousand tons of brex per year.

Next, the metallurgical properties of blast furnace briquettes obtained by the methods of vibropressing and stiff extrusion will be considered. First in the mechanism of preserving the strength of briquettes when they are heated in a reducing atmosphere is of interest.

References

[1] Kasai, A., Toyota, H., Nozawa, K. and Kitayama, S. 2011. ISIJ Int. 51: 1333.
[2] Yamamoto, T., Sato, T., Fujimoto, H., Anyashiki, T., Fukada, K., Sato, M., Takeda, K. and Ariyama, T. 2011. Tetsu-to-Hagané 97: 501.
[3] Electronic resource. https://asia.nikkei.com/Business/JFE-Steel-catalyst-helps-cut-blast-furnace-emissions.
[4] Higuchi Kenichi. 2013. Development of new burden for blast furnace operation with low carbon consumption. IEAGHG/IETS Iron&Steel Industry CCUS&Process Integration Workshop. 6th November 2013.
[5] Higuchi, K., Yokoyama, H., Sato, H., Chiba, M. and Nomura, S. 2017. Tetsu-to-Hagané 103: 407.
[6] Titov, V.V., Murat, S.G. and Kiselev, N.I. 2007. Ecology and Industry 1: 16–21 (in Russian).
[7] Electronic resource. http://briket.ru/metallurg6.shtml (in Russian).
[8] Kurunov, I.F. and Kanaeva, O.G. 2005. Briquetting—a new stage in the development of raw material agglomeration technology for blast furnaces. Bulletin of Scientific, Technical and Economic Information "Ferrous Metallurgy" 5: 27–32 (in Russian).
[9] Kurunov, I.F., Shcheglov, E.M., Kononov, A.I., Bolshakova, O.G. and others. 2007. Research of metallurgical properties of briquettes from technogenic and natural raw materials and assessment of the effectiveness of their

use in blast-furnace smelting. Part 1. Bulletin of Scientific, Technical and Economic Information "Ferrous Metallurgy" 12: 39–48 (in Russian).

[10] Kurunov, I.F., Shcheglov, E.M., Kononov, A.I., Bolshakova, O.G. and others. 2008. Research of metallurgical properties of briquettes from technogenic and natural raw materials and assessment of the effectiveness of their use in blast-furnace smelting. Part 1. Bulletin of Scientific, Technical and Economic Information "Ferrous Metallurgy" 1: 8–16 (in Russian).

[11] Electronic resource. http://www.kmz-tula.ru/articles-20100728.html (in Russian).

[12] Timo Paananen and Erkki Pisilä. 2015. Improved Raw Material Efficiency in Hot Metal Production; Düsseldorf, 15–19 June 2015, METEC-ESTAD, pp. 1–8.

7.1 Vibropressed Blast Furnace Briquettes

Vibrocompression for the production of blast-furnace briquettes is most widespread in Sweden. The study of the metallurgical properties of such briquettes was first carried out at the Lulea University of Technology [1]. The metallurgical properties of briquettes of two different compositions (93.5% of pellet fines, 6.5% of Portland cement and 62.7% of pellet fines, 30.8% of sludge and dust, 6.5% of Portland cement) produced by the method of vibropressing were investigated. One feature of these briquettes is a tendency to swell considerably during the reduction process using carbon monoxide at 950°C, which is linked with a similar abnormal swelling of the indurated pellets [2–4]. The mechanism of the catastrophic swelling of iron ore pellets is associated with the growth of iron whiskers during wustite reduction, which is enhanced in the presence of CaO [5]. The degree of swelling decreased when adding hydrogen to the reducing gas. It was also noted that the tendency to swell in the presence of cement is reduced when its content in the body of a briquette exceeds 10% (mass). The largest degree of briquette swelling took place with a cement share in the mass of the briquette ranging from 4–6%. It was noted that at high temperatures the converted phases of cement, olivine and wustite react with the formation of $(Ca, Mg, Fe)_2SiO_4$ (dicalcium silicate, forsterite and fayalite). The properties of this phase are determined by the relationship between the values content of pellets fines and cement. The formation of the silicate melt at a lower temperature is also possible (1150°C). These briquettes were also tested in the experimental blast furnace. It is claimed that unlike testing briquettes in laboratory conditions, their reduction in the experimental blast furnace was not accompanied by catastrophic swelling. Reduced briquettes, after extraction from the experimental blast furnace, had a significant wustite core even with fairly profound lowering in the furnace. As an explanation, the authors indicated the following possible reasons: a large volume of briquettes, creation of a slag system $CaO-SiO_2-MgO-Al_2O_3-FeO$ and coating of the wustite grains in a molten slag.

The results obtained allowed to determine the effect of the particle size and phase composition of briquetted materials, their moisture content and the proportion of Portland cement binder on the metallurgical properties of briquettes. However, this did not lead to an understanding of the mechanism for maintaining the strength of such briquettes when heated in a reducing atmosphere.

The authors of [6] approached the study of the metallurgical properties of briquettes produced industrially at the SSAB factory in Raahe came the closest to identifying the mechanism of hot strength of vibropressed briquettes. To study the behavior of vibropressed briquettes when heated in a reducing atmosphere, a blast furnace simulator was used. Briquette components—mill scale (41.2%), scrap materials (23.4%), top dust (9.5%), and aspiration dust from casting and dosing sections (4.8%), coke dust (6.4%) screening of pellets (2.8%) and other anthropogenic materials. Three different combinations were used as a binder: 10% Portland cement; 8% of Portland cement + 3% of crushed blast furnace slag; 8% Portland cement + 6% crushed granulated slag. The briquettes were made on the industrial line of vibropressing; however, for testing in the simulator of the blast furnace, the briquette samples were cut into bars with a triangle-shaped section. However, it is impossible to obtain authentic test results, since in this case the specific surface of the briquette fragment is higher than that of the original briquette, and therefore, the recovery rate is different. The following results were obtained: (1) briquettes showed significantly faster recovery compared to pellets in the temperature range of 780–1100°C, which can be explained by the presence of carbon in their composition and higher porosity, the growth of which, according to the authors, is due to the effects of phase hematite-magnetite transition; (2) swelling of briquettes was observed (25–50%) during the transition wustite-iron at 900–1000°C; (3) the addition of crushed granulated slag contributed to the swelling of the briquettes; (4) the briquettes did not collapse, and the formation of dicalcium ferrite played a role in maintaining strength at high temperatures. In an earlier work on the mechanical strength of the industrial briquettes under consideration [7], it was argued that the replacement of part of Portland cement with slag Portland cement in combination with fly ash leads to an increase in the mechanical strength of the briquette.

The results of the above mentioned blast furnaces campaign with vibropressed briquettes in the ore portion of the charge, conducted at NLMK in 2003, contributed to understanding the mechanism of preserving the strength of briquettes in a blast furnace.

At the first stage, to evaluate the efficiency of the technological scheme for processing iron-zinc-containing sludge, a batch (2500 tons) of briquettes (65% converter sludge, 20% coke breeze and 15% Portland cement) was smelted. Briquette consumption ranged from 50–70 kg/t of hot metal in the first 5 days to 190 kg/t of hot metal in the last 24 hours and averaged 121 kg/t of hot metal. The efficiency of using briquettes was evaluated by comparing the results of the furnace operation in the base period, which included 15 days of operation of the furnace before the melting of briquettes and 7 days after their melting. When melting briquettes, the degree of coke substitution by fine-grained coke was 0.96 kg/kg.

At the second stage, a batch of 2475 tons of briquettes made of a mixture of iron ore concentrate, coke breeze and Portland cement was smelted, with a gradual increase in the share of briquettes in the charge (122, 198, 303 kg/ton of hot metal). The results of the heat confirmed that such briquettes are a complete self-reducing component of the blast furnace charge, the use of which ensures a reduction in coke consumption in the blast furnace smelting, proportional to their consumption. The share of such a component in the blast furnace charge is only marginally limited to a decrease in the furnace productivity due to a decrease in the iron content in the charge and can reach 50% or more.

At the third stage, a batch (2560 tons) of vibropressed briquettes from a mixture of blast-furnace sludge (59%), mill scale (20%), coke breeze (10%) and cement (11%) were smelted in a blast furnace of 2000 m^3. When unloading briquettes from bunkers, their increased bridging was observed. The average consumption of briquettes for the period amounted to 62 kg/t of hot metal, with fluctuations in days from 36 kg/t of iron to 81 kg/t of hot metal. The efficiency of smelting of coke briquettes was evaluated by comparing the melting indices in the experimental and basic periods of the furnace operation. A slight decrease in the productivity of the blast furnace during the melting of briquettes is mainly caused by a decrease in the iron content in the charge, as well as the negative effect of increased basicity and viscosity of the slags formed from the oxides of the gangue of briquettes. There were also lower values of the carbon of coke substitution by coke breeze of briquettes compared to that obtained when melting briquettes from converter sludge (0.96 kg/kg).

The results of the investigation of the metallurgical properties of the laboratory briquettes made by vibropressing, as well as of briquettes used in full-scale testing are described in [8]. In particular, the behavior of carbon-free briquettes made of magnetite concentrate (91.2%) and Portland cement (8.8%) and BOF sludge (91.9%) and Portland cement (9.1%) whilst undergoing heating in the reducing atmosphere has been studied. Since the cement binder maintains the strength of the briquette up to temperatures of 750–900°C, the formation of new solid briquette microstructures was attributed to physical-chemical processes that occur in the body when heating the briquettes in a reducing atmosphere. Such processes are primarily a reduction of iron oxides in the solid-state reaction between wustite, oxides of cement stone and gangue components of a briquette with the formation of calcium-iron-magnesium silicates.

In briquettes made of magnetite concentrate and converter sludge a zonal microstructure with a surface layer of metallic iron has been optically determined. This fact is due to the relatively high density of briquettes, their considerable size and the lack of a solid reductant in their composition.

In the briquettes made of iron ore magnetite concentrate a magnetite to wustite reduction process is clearly observed throughout the volume of the sample, on the surface there is the layer of metallic iron with a thickness of 3–5 mm, forming a superficial shell (skeleton) of a briquette. In deeper layers (up to a distance of 20–25 mm from the surface of the briquettes) only small zones can be seen with particles of metallic iron on the edges of the wustite grains. In the center of the briquette the metallic iron is missing and the entire iron containing phase is represented only by wustite and ferrous olivine. Depending on temperature and duration of a briquette presence in a reducing atmosphere in different parts of the briquette iron-silicate phase was either in a plastic or a near-plastic state and was filling the space between grains of ore in the form of olivine phase.

Due to the large content of SiO_2 in the concentrate (6.3%), the presence of impurities Al_2O_3, MgO in the mineral phases of cement, as well as the extended surface of contact between the concentrate (particle size 70–120 μm) and cement stone particles, a large portion of the active wustite—a product of magnetite reduction—reacts with silica and forms a system Fe_2SiO_4-Ca_2SiO_4 comprising of an eutectic with a crystallization temperature of 1120°C (phase diagram CaO-FeO-SiO_2 in Fig. 7.1). Iron-silicate phase has a low reducibility, which together with the diffusive difficulties and large size of a briquette explains the lack of metal iron in its central part. Conservation of the form of a briquette made of magnetite concentrate as it undergoes heating in a reducing atmosphere is provided by the formation of a surface metal iron frame with a thickness of 3–5 mm, which ensures the strength of a briquette at the stage of dehydration of calcium hydrosilicates in the cement stone. The destruction of the briquette when it is heated to 1150°C is also hampered by the formation of a matrix of iron-calcium olivine in the entire volume of the body of the briquette [9–10].

It is known that the system CaO-FeO-SiO_2 has a eutectic (25% CaO • SiO_2 and 75% FeO • SiO_2) with a crystallization temperature of 1030°C. There are also fields with a crystallization temperature close to 1150°C.

Due to the limited time in the furnace at a temperature of 1150°C and above, the iron-silicate phase does not reach a liquidus temperature and remained in a plastic condition without violating the integrity of the briquette in the already established frame of metallic iron with a melting point above 1500°C.

As the initial materials for the experiments, industrial vibropressed briquettes were used, which included in different proportions ore of borehole hydraulic mining, Portland cement, as well as coke and two types of dust from dry gas cleaning of a blast furnace.

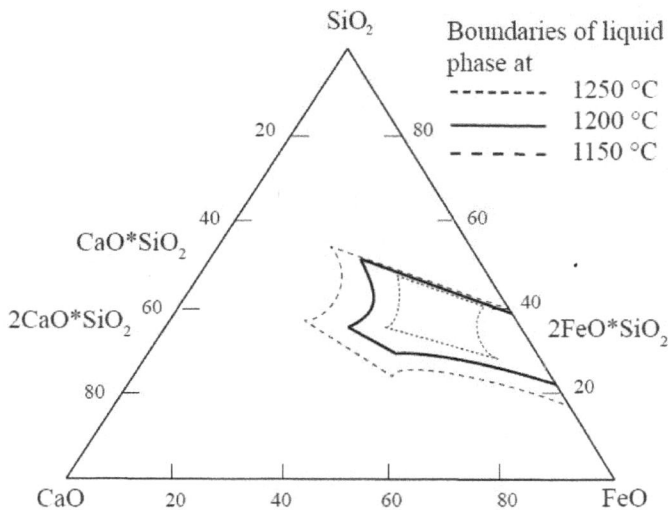

Figure 7.1. Phase diagram of CaO-FeO-SiO_2.

Table 7.1. Chemical composition of some wastes of metallurgical production, mass. %.

Material	Fe$_{tot}$	Mn	C	CaO	SiO$_2$	Al$_2$O$_3$	P	S
Sinter plant dust	41.0	0.23	5.9	13.2	9.1	1.8	0.05	0.51
Flue dust	24.8–39.2	0.24–1.41	20.6–42.5	4.1–9.7	6.4–40.2	1.4–2.8	0.05–0.07	0.2–0.52
BOF dust	60.42	1.2	1.4	6.25	2.0	0.62	0.01	2.01
EAF dust	36.9	3.56	4.26	7.18	3.6	0.78	0.14	0.63
Mill scale	71.83	0.48	0.25	0.28	1.48	0.9	0.07	0.1
BF sludge	0.28	0.65	-	41.8	34.6	14.8	-	1.48
Pellets fines	60.2–64.8	0.03–0.08	-	0.6–2.8	3.0–4.3	0.4–0.8	-	0.01–0.1

The chemical composition of some wastes of metallurgical production used as components of briquettes is shown in Table 7.1.

The briquette is formed in multi-seat molds on wooden or metal technological pallets by vibrocompression, i.e., simultaneous exposure of the molding mixture to vibration and pressing. Depending on the customer's requirements, a briquette of any configuration can be produced with dimensions from 20 × 20 × 20 mm to 1000 × 1500 × 1500 mm. In one cycle (no more than 30 seconds), from 0.05 to 2.25 m^3 of briquettes can be produced.

Vibration is an effective means of mechanized distribution, placement and consolidation of concrete mixes. The main advantage of this forming method is that in the process of vibration, the viscosity of the molding mixture sharply decreases. Under the influence of vibration, friction and adhesion between the particles in the mixture is significantly reduced, as a result of this, mixing of the particles and the compaction of the mixture is facilitated. The moldable mixture as a whole, transforms from stiff and inactive to a highly mobile flowable mass that quickly fills the mold. Following the laws of hydrostatics, the liquefied mixture, when vibrating, exerts hydrostatic pressure on the walls of the mold, while it carefully fills forms even with complex outlines. As a result of a significant decrease in the forces of internal adhesion and friction, the molding mixture is compacted under the influence of gravity. Large particles, sliding mutually, fit very compactly, the voids between them are filled with a binder.

Plasticizing additives, enveloping the binder, give it additional fluidity under the influence of vibration, the molding mixture loses its mobility, and being compacted, acquires even greater structural strength than before vibration. This property of colloidal systems is called thixotropy.

As already noted, it is due to thixotropy that cement and its analogs are practically the only binder suitable for vibropressing briquetting.

The disadvantages of Portland cement are that the sulfur content in the amount of 0.4–1.2%. Portland cement is a complex material obtained by roasting and joint grinding of clay and limestone and containing oxides: CaO 62–67%; SiO$_2$ 20–23%; Al$_2$O$_3$ 4–8%; Fe$_2$O$_3$ 1–4%; MgO 0.5–5%; SO$_3$ 1–3%; K$_2$O and Na$_2$O 0.5–1%. The behavior of Portland cement at high (over 1000°C) temperatures requires additional study. The presence of oxides such as CaO, MgO in the composition of the cement gives rise to the assumption that sulfur will remain in the slag part, and will not pass into the metal melt. In addition, depending on the hardening time of Portland cement (and this process proceeds intensively for 28 days, and then develops slowly), various crystalline hydrates are formed.

In addition, the existence of such a variety of cements such as alumina cements should be noted. Possessing all the physical and mechanical properties inherent in Portland cements, alumina cements have significant differences in chemical composition. The content of basic oxides in alumina cement: CaO 35–40%; SiO$_2$ 4–8%; Al$_2$O$_3$ 35–44%; FeO 4–10.5%; MgO 0.5–5%; SO$_3$ 0.01–0.32%; K$_2$O and Na$_2$O 0.1–1.2%. The use of alumina cement as a binder will allow to limit the amount of sulfur in the briquette. However, it should be kept in mind that alumina cement is a scarce material and its price is five times higher than the price of Portland cement.

Laboratory studies and industrial experiments have shown that the metallurgical properties of briquettes from man-made waste cement-bonded meet the requirements of blast-furnace smelting when used both as lump iron-containing or iron-carbon-containing raw materials, and as a material for washing the hearth of blast furnaces. At the same time, the mechanism of preserving the strength and shape of briquettes when they are heated in a reducing atmosphere is insufficiently studied; the existing explanations for the high hot strength of briquettes from metal siftings cannot be attributed to briquettes made of oxide materials of different genesis.

In order to determine the mechanism of preserving the strength of briquettes from oxide iron-containing materials when they are heated under reducing conditions and to study the physicochemical transformations that occur with the components of the briquette in the working space of the blast furnace, studies were carried out.

The experiments were carried out in laboratory conditions with briquettes $70 \times 70 \times 70$ mm in size of four component compositions presented in Tables 7.2 and 7.3. Portland cement grade M500 of the following chemical composition, %: CaO - 63.7; SiO_2 - 20.6; MgO - 3.7; Al_2O_3 - 6.5; Fe_2O_3 - 1.8; Fe_{tot} - 1.26; SO_3 - 1.5. Briquette No. 4 was made as a reference sample for comparing the results of the physicochemical interaction of the components of the cement stone with oxides of scale, iron ore concentrate and converter sludge.

All briquettes were heated to 1150°C at a rate of 500°C/h in a tubular furnace (inner diameter 100 mm) in a stream of hydrogen (flow rate 200 L/h), followed by cooling to room temperature by blowing nitrogen in the furnace.

The common component for all briquettes was Portland cement of grade 500, the main mineral components of which are alite Ca_3SiO_5 (65.5%) and belite Ca_2SiO_4 (18%). In small amounts, cement contains more complex calcium silicates, which also contain Al_2O_3, Fe_2O_3 and MgO. As the composition of Portland cement consists of mainly minerals tri- and dicalcium silicates, the predominant components of the cement stone are solid solutions of calcium hydrosilicates, which are formed when cement is wetted with water.

Studies have shown that scale briquettes have high cold (82–100 kgF/cm²) and hot strength, retaining their shape after high-temperature heating to 1150–1200°C in a reducing atmosphere. However, the cement stone retains the integrity of the briquette only until the moment of dehydration of calcium hydrosilicates.

The process of dehydration of cement stone is accompanied by the destruction of the crystal lattices of hydrates with the formation of active free oxides of calcium and silicon. Removal of hygroscopic moisture from cement stone occurs when heated to a temperature of 300°C. The process of removing hydrated moisture begins at a higher temperature. The endothermic peak on the heating

Table 7.2. Component composition of briquettes, % mass.

Briquette #	Mill scale	Magnetite concentrate	BOF sludge	Quartz sand	PC	H_2O (+100%)
1	93.4	-	-	-	6.6	9.1
2	-	91.2	-	-	8.8	7.2
3	-	-	91.0	-	9.0	14.8
4	-	-	-	89.3	10.7	9.5

Table 7.3. Chemical composition of briquettes, % wt.

#	Fe_{tot}	FeO	Fe_2O_3	SiO_2	Al_2O_3	CaO	MgO	C	S	B
1	69.07	60.62	31.31	2.14	0.59	4.38	0.35	0.05	0.05	2.05
2	60.60	25.56	58.16	7.58	0.75	5.84	0.68	0	0.08	0.77
3	54.24	60.88	9.84	3.58	0.74	17.56	0.84	2.00	0.05	4.91
4	0	0	0.19	91.50	.070	6.82	0.40	0	0.06	0.075

curve at a temperature of 498°C and a sharp weight loss at this temperature indicates the active decomposition of hydrosilicates. The process of removing chemically bound water from a cement stone, accompanied by a decrease in the mass of the test sample, is observed up to temperatures of 700–750°C. In the absence of a chemical mechanism for hardening a briquette when it is heated in a reducing atmosphere, the briquette's strength is lost as a result of the destruction of the crystal lattices of calcium hydrosilicates during the dehydration of the cement stone. This phenomenon took place in a briquette based on quartz sand, which, after heat treatment, almost completely disintegrated under insignificant mechanical impact (50–100 g).

Optical analysis of thin sections of briquette No. 4 after its heat treatment revealed new formations of wollastonite crystals ($CaO \cdot SiO_2$) in the glass phase at the phase boundary "cement stone - quartz particle". Quartz grains in these areas are notably corroded, which indicates the diffusion interaction in the solid phase of the constituent minerals of this system (Fig. 7.2). In this case, the highly basic silicates of the cement stone are saturated with silica and the system is displaced into the area of existence of wollastonite. According to the results of phase analysis, no other processes in the cement stone during heating of briquette No. 4 in a reducing atmosphere were revealed.

Briquettes of compositions No. 1–3 withstood high-temperature heating up to 1150°C without destruction and retained high strength. Since the cement stone strengthens the briquette only up to a temperature of 700–750°C, the formation of a new strong microstructure of the briquettes is due to the physicochemical processes occurring in the body of the briquettes when it is heated in a reducing atmosphere. These processes are, first, the reduction of iron oxides and reactions in the solid phase between wustite, cement stone oxides and waste rock of iron-containing components of briquettes.

In briquettes No. 1–3 of scale, magnetite concentrate and converter sludge, after reduction, a zonal microstructure is optically diagnosed, as evidenced by the presence of a surface layer of metallized iron. This fact is due to the rather high density of briquettes, their significant size and the absence of a solid reducing agent in their composition.

In briquette No. 2 (from iron ore magnetite concentrate), the process of reduction of magnetite to wustite in the entire volume of the sample is clearly witnessed. However, on the surface of the sample in a layer 3–5 mm thick, almost complete reduction of wustite to metallic iron occurs, which at the same time retains the shape of wustite granules (Fig. 7.3). In deeper layers, at a distance of up to 20–25 mm from the briquette surface, small zones of metallic iron are observed along the boundaries of wustite grains (Fig. 7.4). In the center of the briquette, metallic iron is absent and the entire iron-bearing phase is represented only by wustite and ferruginous olivines.

Depending on the temperature and time conditions of heat treatment in different parts of the briquette, the iron silicate phase was either in a plastic (in the surface layers) or in a molten state, filling the space between the grains of ore components in the form of a glass phase (in the central part of the briquette) (Figs. 7.4–7.6).

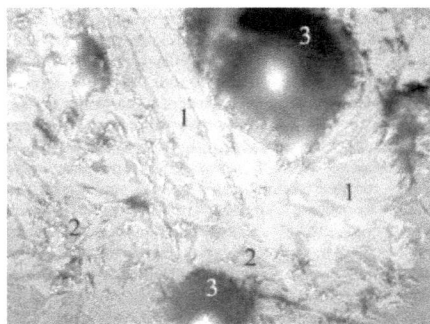

Figure 7.2. Wollastonite (1), glass phase (2), pore (3) at the boundary of cement and quartz. Reflected light, magnification × 500.

Figure 7.3. Surface microstructure of a briquette from magnetite concentrate (up to 5 mm): metal grains (1), wustite (2), olivine phase (3). Reflected light, magnification × 500.

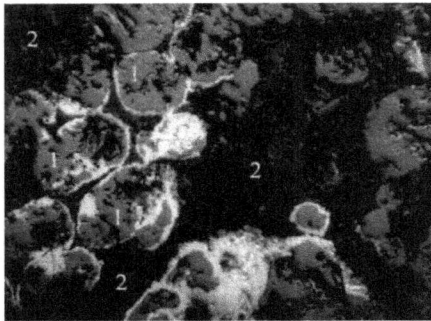

Figure 7.4. Microstructure of the surface layer of a briquette from magnetite concentrate (up to 20–25 mm): wustite grains metallized at the edges (1), olivine phase (2).

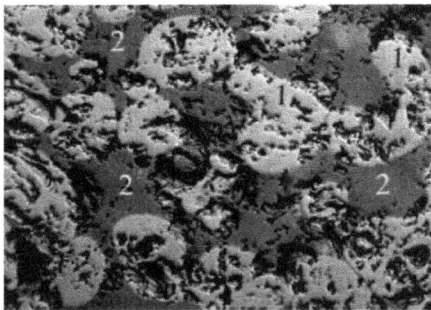

Figure 7.5. Microstructure of the central part of a briquette sample from magnetite concentrate: wustite (1), olivine phase (2). Reflected light, magnification × 500.

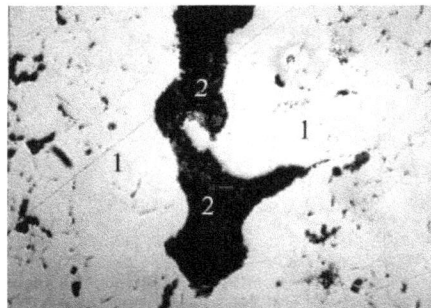

Figure 7.6. Surface microstructure of a reduced scale briquette sample: metallic iron (1), olivine phase (2). Reflected light, magnification × 1000.

Due to the significant content of SiO_2 in the concentrate (6.3%), the presence of impurities Al_2O_3, MgO in the composition of the mineral phases of cement, as well as due to the developed contact surface of the concentrate particles (particle size 70–120 μm) and cement stone, most of the active wustite is a product reduction of magnetite—reacts with silica and forms solid solutions in the Ca_2SiO_4-Fe_2SiO_4 system, which has a melting point of 1120°C, which corresponds to 80% Fe_2SiO_4. This explains the presence in a large part of the body of the briquette of a dense structure formed from the olivine melt. Thus, preservation of the shape of briquette No. 2 from a magnetite concentrate during reduction heating is ensured by the formation of a surface frame made of metallic iron of sufficient thickness (3–5 mm), which provides the strength of briquettes at the stage of dehydration of calcium hydrosilicates of cement stone. The subsequent formation of iron-calcium olivines in the entire volume does not lead to deformation of the briquette when it is heated to 1150°C.

The described mechanism for the formation of the microstructure of a briquette from a magnetite concentrate on a cement bond during its reduction makes it possible to recommend it as a washing material for a blast furnace hearth. The reduction of iron from most of the briquette in the blast furnace will occur only with solid carbon of the coke after the briquette has melted with the formation of ferrous slag in the temperature zone above 1150°C.

The mechanism of formation of the microstructure of briquettes from scale (briquette No. 1) during reduction heating is similar to that described above. As in the concentrate briquette, when heated in a reducing atmosphere, a skeleton of metallic iron was formed in the surface layer of the scale briquette. However, in contrast to the briquette from magnetite concentrate, the metallization of the surface layers of the briquette from scale goes deeper, up to 10–15 mm from the sample surface (Fig. 7.7). The iron-bearing phase of the central zone of the scale briquette is represented mainly by wustite with small zones of metallic iron (Fig. 7.7).

The minimum amounts of nonmetallic components of the charge, especially SiO_2, in the composition of briquette No. 1, as well as a smaller (than concentrate) surface area of contact between scale particles (size 0–5 mm) and cement stone dehydration products (CaO, SiO_2) is the reason for the formation of a reduced sample of only small amounts of silicates of the olivine subgroup. A limited amount of iron-calcium melt contributed to an increase in the iron metallization zone in the briquette.

The system "converter sludge - Portland cement" (briquette No. 3) differs from the charge of the iron-containing briquettes described above. To prepare briquettes from converter sludge, the product of its dehydration and subsequent drying in a rotary kiln was used. This product consists of sufficiently strong spherical granules of 0.5–8 mm in size. Optical analysis of thin sections of a briquette from converter sludge showed the presence of large areas of alite in the structure, the grains of which are located along the boundaries of sludge granules (Fig. 7.8). The shape of the alite grains and the high temperature of its formation indicate that the source of this mineral in the

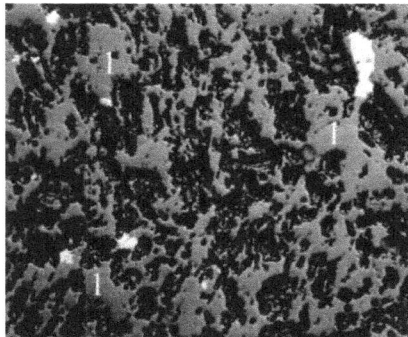

Figure 7.7. Microstructure of the central part of the reduced scale briquette sample: wustite grains (1). Reflected light, magnification × 200.

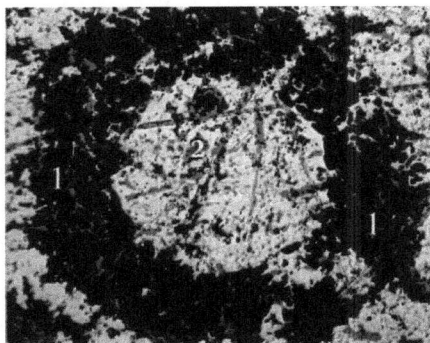

Figure 7.8. Microstructure of a slurry granule with alite (1) located on its surface, and a small amount of olivine phase (2 - gray areas in the metal). Reflected light, magnification × 500.

briquette is converter sludge. High temperatures in the converter and the presence of free lime and silica favor the formation of tricalcium silicate ($3CaO \cdot SiO_2$).

Compared with briquettes No. 1–2, briquette No. 3 in the cold state had a lower crushing strength, but retained its shape and high strength after heat treatment in a reducing atmosphere. As in briquettes No. 1–2, the surface layer of a briquette from converter sludge 6–8 mm thick consists of metallic iron, but unlike the first in briquette No. 3, alite grains are present in the surface layer along with a small amount of olivine phase (Fig. 7.9). Thus, the strength of the briquette and the retention of its shape during heat treatment in a reducing atmosphere, as in the briquettes before, is ensured by the formation of a surface metal frame.

In layers deeper from the surface (10–20 mm), the briquette structure is represented mainly by wustite, metallized along the grain boundaries (Fig. 7.10). The center of the briquette is represented

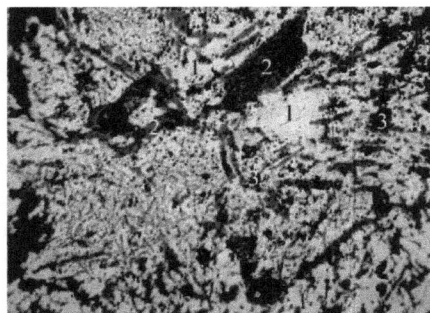

Figure 7.9. Microstructure of the surface layer of the briquette (up to 6–8 mm): metal (1), iron-calcium phase (2), alite (3). Reflected light, magnification × 500.

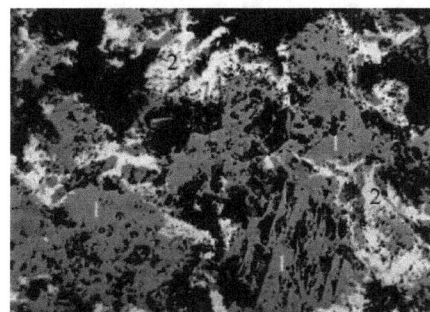

Figure 7.10. Microstructure of the center of the briquette (10–20 mm from the surface of the briquette) from converter sludge: wustite phase (1); metal (2) Reflected light, magnification × 500.

by sintered grains, partially retaining the microstructure of sludge granules and consisting of wustite and alite. There is a small amount of metallic iron along the boundaries of these grains. A higher degree of metallization of briquette No. 3 ensures the content of 2–3% carbon in the converter sludge.

Thus, briquettes on a cement bond, obtained by vibrocompression from dispersed materials containing iron oxides, have a sufficiently high density, which determines the zonal nature of the reduction of iron oxides in the body of the briquette. When briquettes are heated in a reducing atmosphere, they retain their strength up to 700–750°C due to the cement stone, and on further heating, due to the formation of a solid surface metal frame formed during the reduction of iron oxides to iron in the surface layer of the briquette. When heated in a hydrogen atmosphere to 1150°C at a rate of 500°C/h, the thickness of the metal frame in briquettes of iron ore concentrate, converter sludge and scale was 3–5, 6–8, and 10–15 mm, respectively.

As a result of the research, conclusions were drawn about the effect of fine waste on the metallurgical properties of briquettes. The presence of fine waste in the briquette contributes to an increase in the degree of recovery by 10–15%, due to an increase in the contact area during solid-phase reduction, but helps to reduce the softening temperature of the briquette.

The retention of the strength of the composite iron-carbon raw material during reduced conditions under load is explained by the existence of a cement binder up to temperatures of 750–766°C, and in the temperature range 800–1000°C due to the formation of a matrix of fused viscous inclusions of magnetite and reduced iron.

The results obtained, explained the mechanism of hot strength of vibropressed briquettes (No. 2), formed the basis for new series of experimental blast-furnace melts.

In order to clarify the specific consumption of coke and iron-containing materials for the smelting of foundry pig iron using iron-carbon-containing briquettes in the charge in an amount of ~ 200 kg/t, to determine the losses of iron with the blast furnace dust of the dust collector and dry gas cleaning, ladle residues and scrap from the foundry and the balance of the casting department, a blast furnace was carried out.

During the smelting period, a relatively smooth operation of the furnace was ensured.

The amount of charge materials loaded into the furnace was determined according to the readings of an automatic strain-gauge system for controlling weight gain.

The quality characteristics of the raw materials loaded during the smelting period are stable, the ranges of changes in the content of the main element in the raw materials are minimal.

Based on the data obtained on the weight of commercial pig iron (7814.82 tons), as well as on the corresponding weight of the loaded charge materials, Table 7.4 shows the values of the specific consumption of materials per 1 ton of commercial pig iron.

Table 7.4. Consumption of materials for 1 ton of commercial pig iron.

Pellets, kg/t	1352,48
Briquettes, kg/t	196,05
Iron ore, kg/t	97,69
Siderite, kg/t	51,71
Ore total, kg/t	**1697,93**
Mn ore, kg/t	21,51
Wet coke, kg/t	750,79
Coke (dry), kg/t	**717,27**
Coke nut, +10 mm	8,36
Limestone, kg/t	171,30
Dolomite, kg/t	87,76
Total fluxes, kg/t	**259,06**

Table 7.5. Pig iron and slag compositions.

Cast iron composition, %	Si	Mn	S		P	Cr		C	Ti	Fe	
	2,874	0,737	0,015		0,037	0,005		3,902	0,043	92,384	
Slag composition, %	MgO	Al$_2$O$_3$	SiO$_2$	S	CaO	MnO	FeO	TiO$_2$	R$_2$O	0,01	Basicity
	9,07	**8,21**	**38,48**	**1,61**	**40,31**	**0,33**	**0,25**	**0,16**	**1,41**	**Zn**	**1,05**

The specific slag yield was 364.6 kg/t.

Average chemical compositions of pig iron and slag obtained during the balance smelting are presented in the table.

The average value of the temperature of pig iron discharged from the furnace during the period of the balance smelting was ~ 1450°C, the average temperature of the top gas was 242°C.

The iron content in the iron ore part was 59.82%, in the fluxed iron ore part – 54.71%.

Table 7.6 presents the balance of iron with a quantitative indication of losses by various items.

Table 7.6. Iron balance.

PARAMETER	Value	Note
Iron intake, kg/t	1017.16	
Iron consumption in pig iron, kg/t	923.84	
Iron consumption in slag (in the form of FeO), kg/t of pig iron.	0.70	
Losses of iron with slag, kg/t	5.05	1.5% of the mass. in the slag
Loss of iron in the form of pig iron. deposits in buckets (weight increase in 7 days), kg/t	5.92	
Iron losses with ladle residues, kg/t	23.03	80% Fe
Iron losses with scrap from the foundry, kg/t	20.12	80% Fe
Iron losses (scrap from filling machines), kg/t	5.80	85% Fe
Iron losses with emissions from the gas recovery valve, kg/t	0.31	35% Fe
Dust emission in the foundry at the outlet (50% Fe), kg Fe/t	0.25	50% Fe
Iron consumption in blast furnace dust PU + SGO, kg/t	32.14	
Iron content in SGO dust, %	32.59	
The average iron content in the count. dust PU + SGO, %	32.69	
The amount of blast furnace dust PU + SGO, kg/t	90.07	
Including flue dust of the dust collector, kg/t	**52.70**	
The amount of blast furnace dust PU, % of the total output	**58.5%**	
SGO dust output, kg/t	**37.37**	
The amount of dust of the SGO, % of the total output	**41.5%**	
The degree of utilization (extraction) of iron, %	**90.82**	

The degree of utilization (extraction) of iron was 92.384 * 100/1017.16 = 90.82%, where 92.384 is the iron content in cast iron, %; 1017.16 - balance receipt of iron, kg/t of pig iron.

The specific removal of flue dust from the dust collector and dry gas cleaning unit was 53 kg/t and 37 kg/t, respectively (for comparison: the removal of dust from the dust collector and the gas cleaning system in the period preceding the experimental melts was 33 kg/t and 18 kg/t, respectively.).

Thus, the conducted experimental smelting showed that high values of the proportion of briquettes in the ore part of the blast furnace charge are achievable, but due to the insufficiently high hot strength of vibropressed briquettes, the following consequences take place:

1. The control data of the materials loaded into the furnace and the commercial pig iron produced revealed a deviation (decrease) from the planned rate of iron extraction (utilization) of 94.0% for "high" grades of foundry iron. One of the reasons for this deviation is the *increased removal of dust with the blast furnace gas* (dust of the dust collector – 53 kg/t and dust of the gas-cleaning system – 37 kg/t) due to the increased consumption of briquettes up to ~ 200 kg/t.

2. Given that, the modes of effective loading of briquettes in the amount of up to 150 kg/t have been mastered, a further increase in the specific consumption of briquettes requires optimization of loading modes and gas-dynamic parameters.

3. One of the most effective ways to reduce the removal of blast furnace dust is to work with increased pressure under the blast furnace. However, in conditions of the need to maintain the pressure difference in the furnace at a high level, the pressure increase under the top is limited by the power of the available blowing means.

4. The heat balance of the smelting showed that the heat loss with cooling during the smelting of "high" grades of cast iron amounted to 20.9%, which largely exceeds the corresponding typical industry indicators of 14–16% for the smelting of cast iron. In this regard, it should be noted that it is necessary to take measures to build up the scallop in the shoulders and steam in the area of damaged refrigerators due to the installation of copper mini-refrigerators, as well as regular (once every 3–4 months) injection of the injection mass for the armor and refrigerators of the stove, hearth (including tap holes), tuyere zone and boshes to reduce heat loss and ensure gas tightness.

Another series of full-scale testing with significantly smaller amount of vibropressed briquettes in the BF charge (up to 80 kg/t) was made in the blast furnace unit of one of the steel-making enterprises in Western Siberia. Composition of briquettes:

- 75% - iron oxide materials (iron ore concentrate, mill scale, etc.);
- 17% - carbonaceous materials (coke breeze, breeze anthracite, etc.);
- 8–9% - mineral binder (cement).

The chemical composition of iron-containing vibropressed briquettes is presented in Table 7.7.

Table 7.7. Chemical composition of batches of briquettes for experimental blast-furnace smelting, % (mass.).

#	F_{tot}	S	C	CaO/SiO_2
1	47.48	0.12	11.8	1.0
2	49.14	0.12	11.6	1.0
3	46.71	0.12	11.8	1.0
4	46.71	0.12	11.8	1.0
5	46.95	0.12	11.8	1.0
6	47.11	0.08	11.9	1.0
7	48.05	0.11	11.95	1.0
8	47.35	0.12	12.3	0.95
9	47.07	0.12	12.1	1.04
10	48.99	0.12	12.44	1.02
11	48.58	0.07	11.8	0.99
12	46.39	0.08	11.85	1.02
13	45.85	0.09	12.0	1.0
average	47.41	0.11	11.93	1.0

Compressive strength in batches from 2.03 to 8.0 MPa (according to the passport - not less than 5.0 MPa).

The work with briquettes lasted about 19 days. A total of 4947 tons of iron-carbon briquettes were smelted. The increase in the average daily production during the test period amounted to 8 tons.

When removing briquettes from the hopper, they were observed "sticking" into the neck of the hopper (typical for large briquettes, as noted above).

A factorial analysis (Tables 7.8–7.9) that takes into account changes in the iron content in the iron ore part of the charge, the intake of metal additives into the charge, the chemical composition of cast iron and technical analysis of coke, as well as the oxygen content in the blast, the temperature of the hot blast and the pressure on the top, showed that the production of pig iron decreased by 150 tons.

This indicator was influenced by the following factors:

- upset of the furnace during flushing from zinc;

- emergency stopping of the furnace due to lack of electricity;

- uneven delivery of briquette; '

- decrease in the quality of coke during the trial period.

Table 7.8. Analysis of the effect of coke consumption on iron smelting.

Factor	Unit	Factor value			influence on coke	Cokerate Change, %	- reduction ton + increase
		Base period	Testing	Factor change			
Fe in iron-ore part of charge	%	56.04	55.94	−0.1	−1	0.1	25
Consumption of metal additives	Kg/t	29.8	30.3	0.5	−0.03	−0.015	−4
Limestone	Kg/t	0	0	0	0.05	0	0
Lump coal	Kg/t	58.30	50.7	−7.6	−0.8	6.08	388
Coke: ash A	%	11.6	11.7	0.1	1.3	0.13	32
S	%	0.40	0.4	0	3	0	0
strength	%	86.2	85.6	−0.6	−0.6	0.36	88
abrasion	%	10.20	10.8	0.6	2/8	1.68	412
SHUNGIT	Kg/t	28.9	29.8	0.9	−0.6	−0.54	−34
PIG IRON content: Si	%	0.41	0.47	0.06	12	0.72	176
Mn	%	0.66	0.55	−0.11	2	−0.22	−54
P	%	0.09	0.08	−0.01	6	−0.6	−15
S	%	0.029	0.031	0.002	−100	−0.2	−49
BLOW: temperature	°C	1099	1099	0	−0.03	0	0
humidity	g/m'	0	0	0	0.2	0	0
Oxygen content	%	24.98	24.84	−0.14	0.2	−0.028	−7
Natural gas consumption	m³/t	97.4	94.9	−2.5	−0.8	2	128
Top gas pressure	kPa	137	134	−3	−0.02	0.06	15
Crushed electrodes	Kg/t	2.20	3.8	1.6	−0.6	−0.96	−61
SLAG	кг/т	0.0	0	0.0	0.035	0.000	0
Downtime	%	0.62	1.49	0.87	0.5	0.435	107
Damping-down	%	0.44	0.18	−0.26	0.5	−0.13	−32
Actual coke rate	Kg/t	384.1	379.80	-		2.83	
Coke fact	t	24512	24239	-		Total	1115
Coke rate change	Kg/t			17.46		Increase -273	

Table 7.9. Influence of factors on the specific consumption of coke and the productivity of the blast furnace.

Factor		Base period	Testing	Change of factors	Influence on coke rate	Coke rate change	+overspend. -saving kg/t	Coefficients of influence on production	Change in production, %	+overspend. -saving t/day
Fe content in ore part of BF charge.	%	56,04	55,94	-0,1	-1	0,1	1,36	1.7	-0,17	-6
Fe in metallic additives	Kg/t	29,8	30,3	0,5	-0,03	-0,015	-0,20	0,05	0,025	1
Raw limestone consumption	Kg/t			0	0,05	0	0,00	-0,05	0	0
COKE: ash Ac	%	11,6	11,7	0,1	1.3	0,13	1,77	-1.3	-0,13	-5
sulphur S	%	0,4	0,4	0	3	0	0,00	-3	0	0
strength	%	86,2	85,6	-0,6	-0,6	0,36	4,89	0,6	-0,36	-13
abrasion	%	10,2	10,8	0,6	2,8	1,68	22,83	-2,8	-1,68	-59
PIG IRON:										
Si	%	0,41	0,47	0,06	12	0,72	9,78	-12	-0,72	-25
Mn	%	0,66	0,55	-0,11	2	-0,22	-2,99	-2	0,22	8
P	%	0,09	0,08	-0,01	6	-0,06	-0,82	-6	0,06	2
S	%	0,029	0,031	0,002	-100	-0,2	-2,72	100	0,2	7
BLOW:										
temperature	°C	1099	1099	0	-0,03	0	0,00	0,03	0	0
humidity	g/m³			0	0,2	0	0,00	-0,14	0	0
O₂	%.	24,98	24,84	-0,14	0,2	-0,028	-0,38	2,4	-0,336	-12
Natural gas consumption	m³/t	97,4	94,9	-2,5	-0,81	2	2,00	0	0	0
Top gas pressure	kPa	137	134	-3	-0,02	0,06	0,82	0,1	-0,3	-11
Blast furnace discharge schedule.	%	0	0,00	0	-0,05	0	0,00	0,1	0	0
Current downtime	%	0,62	1,49	0,87	0,5	0,435	5,91	-1,5	-1,305	-46
Dumping-down	%	0,44	0,18	-0,26	0,5	-0,13	-1,77	-1	0,26	9
SLAG		0	0	0	0,035	0	0,00		0	0
Shungite crushed stone	Kg/t	28,9	29,8	0,9	-0,6	-0,54	-0,54			
Anthracite	Kg/t	58,30	50,70	-7,6	-0,8	6,08	6,08			
Crushed electrodes	Kg/t	2,2	3,8	1,6	-0,6	-0,96	-0,96			
Metallurgical coke consumption		1359	1378,0	19,0			TOTAL 45,07	TOTAL		-150
Iron production (base)		3537	3545							
Estimated pig iron production		3387								
Actual coke consumption:		384,1	379,8		-4,3					

The results of the experimental heats were as follows:

1. With the introduction of iron-carbon-containing briquettes into the charge, there was no deterioration in blast furnace smelting.

2. With the intake of iron-carbon-containing briquettes into the charge, an increase in the average daily production in the experimental period by 8 tons was observed. A factor-by-factor analysis showed a decrease in pig iron production by 150 tons, which is not associated with the introduction of briquettes into the charge.

3. No significant changes were observed during the removal of blast furnace dust.

4. With the use of iron-carbon-containing briquettes in the amount of 80.0 kg/t of cast iron, a coke saving of 4.3 kg/t of cast iron was obtained.

References

[1] Maneesh Singh and Bo Björkman. 2004. Effect of reduction conditions on the swelling behavior of cement-bonded briquettes. ISIJ International 44(2): 294–303.
[2] Sharma, T. et al. 1991. Effect of porosity on the swelling behavior of iron ore pellets and briquettes. ISIJ Int. 31: 312.
[3] Sharma, T. et al. 1992. Effect of reduction rate on the swelling behavior of iron ore pellets. ISIJ Int. 32: 812.
[4] Sharma, T. et al. 1994. Swelling of iron ore pellets under non-isothermal conditions. ISIJ Int. 32: 960.
[5] Nicolle and Rist, A. 1979. The mechanism of whisker growth in the reduction of wustite. Metall. Trans. B 10: 429.
[6] Kempainen, Iljana, M., Heikkinen, E.-P., Paananen, T., Mattila, O. and Fabritius, T. 2014. ISIJ Int. 54: 1539.
[7] Mäkelä, M., Paananen, T., Heino, J., Kokkonen, T., Huttunen, S., Makkonen, H. and Dahl, O. 2012. ISIJ Int. 52: 1101.
[8] Kurunov, I.F., Malysheva, T.Ya. and Bolshakova, O.G. 2007. Study of the phase composition of iron ore briquettes in order to assess their behavior in a blast furnace. Metallurgist 10: 41–46 (шт Russian).
[9] Kurunov, I.F., Shcheglov, E.M., Kononov, A.I., Bolshakova, O.G. and others. 2007. Research of metallurgical properties of briquettes from technogenic and natural raw materials and assessment of the effectiveness of their use in blast-furnace smelting. Part 1. Bulletin of Scientific, Technical and Economic Information "Ferrous Metallurgy" 12: 39–48 (in Russian).
[10] Kurunov, I.F., Shcheglov, E.M., Kononov, A.I., Bolshakova, O.G. and others. 2008. Research of metallurgical properties of briquettes from technogenic and natural raw materials and assessment of the effectiveness of their use in blast-furnace smelting. Part 1. Bulletin of Scientific, Technical and Economic Information "Ferrous Metallurgy" 1: 8–16 (in Russian).

7.2 Extruded Briquettes (BREX)

7.2.1 Metallurgical properties and optimization of extrusion briquette (BREX) compositions for blast-furnace production

As a possible component of the blast furnace charge, two classes of brex were studied: sludge (based on a mixture of blast-furnace and converter sludge) and ore-coke (a mixture of iron ore concentrate and coke breeze). For the manufacture of the first batch of laboratory brex, sludge and coke breeze were selected, which are generated in the blast furnace and oxygen converter shops of the Novolipetsk Metallurgical Combine OJSC and iron ore concentrate resulting from the enrichment of magnetite ferruginous quartzites at the Stoilensky GOK. The composition of brex is given in Table 7.10.

The brex pilot batches are made on a laboratory computerized extruder complex, simulating the processing of a briquetted mixture in feeders, pug mills and extruders. Brex had a circular cross-section with a diameter of 2.5 mm and a length in the range of 1.5–2.0 diameters. The moisture content in newly formed Brex No. 2 was 12.4%, in Brex No. 4 - 11.9%. The vacuum in the working chamber of the extruder for the production of Brex No. 2 is 20.32 mm Hg, Brex No. 4 - 45.72 mm Hg.

Table 7.10. Components of sludge brex (No. 2) and iron ore concentrate and coke breeze brex (No. 4).

Brex component	Mass content, %	
	Brex No. 2	**Brex No. 4**
Portland cement	9.1	9.0
Coke breeze	-	13.5
Bentonite	-	0.9
Blast furnace sludge	54.5	-
BOF sludge	36.4	-
Iron ore concentrate	-	76.6

To estimate the crush strength of these brex, their cylindrical specimens (diameter 25 mm) 30 mm long were prepared. The load was applied along the vertical axis of the samples. The samples were tested before and after their heating in a reducing atmosphere (50% hydrogen + 50% nitrogen) to 1150°C at a rate of 500°C/hour and subsequent cooling in a nitrogen atmosphere. Five samples of each type were tested. All heat-treated samples retained their shape and size. The chemical composition of the raw and reduced brex is presented in Table 7.11. As can be seen from the test results (Table 7.12), after heating in a reducing atmosphere according to the specified mode, the strength of the samples of sludge brex decreased by 8.2%, while the strength of the ore and coke samples decreased by 14.5%.

The parameter characterizing the relationship of reducibility and strength during reduction is the so-called Reduction Disintegration Index (RDI). This parameter describes the destruction during low-temperature reduction. The mechanical stress resulting from the transformation of the crystal lattice of hematite, leads to strong tensions in certain planes, that leads in the formation of cracks in the brittle matrix.

Table 7.11. Chemical composition of raw and metallized brex.

Brex	Fe_t	Fe_{met}	η_{met}	FeO	CaO	MgO	Al_2O_3	SiO_2	C	Basicity	LOI
No. 2 (Raw)	37.40	-		18.6	15.50	1.26	1.64	7.04	15.7	2.2	19.1
No. 2 (Metallized)	64.5	51.6	0.8	16.8	-			-	9.66		
No. 4 (Raw)	49.20			21.0	6.89	0.58	1.14	8.78	11.6	0.8	12.2
No. 4 (Metallized)	82.6	67.1	0.82	18.9	-			-	7.80		

Table 7.12. Compressive strength of brex.

Sample		Strength after heat treatment, kgF/cm²	Strength before heat treatment, kgF/cm²
No. 2	1	97.7	103.8
	2	124.7	89.4
	3	101.8	112.2
	4	105.0	124.7
	5	118.5	166.3
	Average value	109.5	119.3
No. 4	1	93.5	124.7
	2	95.6	103.8
	3	106.0	110.1
	4	87.3	103.9
	5	93.5	114.3
	Average value	95.2	111.4

Table 7.13. Indicators of hot strength of brex sinters of different composition.

Test material	RDI (+6.3), %
Brex No. 4* (basicity 0,75)	96.5
Brex No. 2* (basicity 1,93)	61.9
Sinter (basicity 1,2)	64
Sinter (basicity 1,4)	60
Sinter (basicity 1,6)	77

* basicity = $(CaO+MgO)/(Al_2O_3+SiO_2)$

The results of tests for hot strength according to ISO 4696, carried out in [1], showed that RDI+6.3 for ore and coke brex is 96.5%, which indicates high hot strength of brex. For sludge brex, the hot strength value is significantly lower (61.9%).

For comparison, the hot strength of sinter of different basicity, obtained from iron ore concentrate (Table 7.13), was simultaneously evaluated according to the same standard.

The hot strength of sludge brex is comparable to the hot strength of sinter with a basicity of 1.2 and 1.4, but inferior to the hot strength of a sinter with a basicity of 1.6. At the same time, the hot strength of the ore and coke brex significantly exceeds the corresponding indicators of all these sinters.

The study of the reduced samples allowed one to establish the mechanism for changing the structure of brex under conditions close to those of the upper horizons of the blast furnace and explain the reasons for the high RDI+6.3 values. The mineralogical study of the structure of raw and reduced brex was carried out on polished thin sections in reflected light.

The structure of raw briquette No. 4 is represented by grains of magnetite, quartz and coke breeze, packed in a binder of Portland cement and a small fraction of magnetite grains (Fig. 7.11). The main carrier of raw briquette strength is cement. The uniform distribution of the particles of materials in the entire volume of the cut slice from the periphery to the center was achieved by good mixing of the charge.

The microstructure of the reduced sample No. 4 (Fig. 7.12) is represented by a system of metallic iron grains and a partially crystallized iron-olivine phase sintered together and inclusions

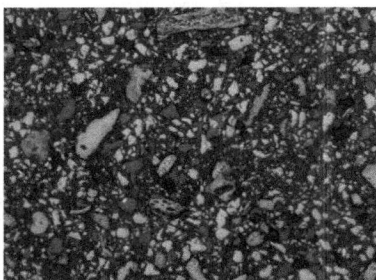

Figure 7.11. Microstructure of the raw brex No. 4; reflective light; 100x magnification.

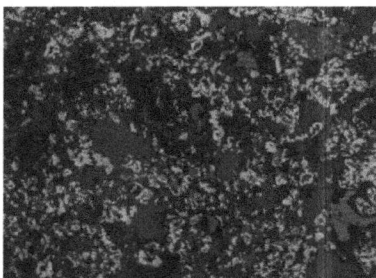

Figure 7.12. Microstructure of the reduced brex No. 4; reflected light; 100x magnification.

of unreacted coke breeze. The strength carrier in the partially reduced brex No. 4 is not only the iron silicate phase, but a framework of metallic iron was also formed from reduced grains of magnetite concentrate.

For comparison, in [1], the metallurgical properties of ore-coke brex based on hematite iron ore from the Belgorod iron-ore district, obtained by the method of hydro mining, were studied. The composition of ore and coke brex on the basis of hydro-ore: hydro-ore - 79%, coke breeze - 15%, Portland cement 5.55%, bentonite - 0.45%. The chemical composition of hydro-ore (%): Fe total - 67.5; SiO_2 - 1.5; Al_2O_3 - 0.3; CaO - 0.2; MgO - 0.3; S - 0.05; P_2O_5 - 0.05. Basicity of brex is 0.56.

A mineralogical study has shown that the bulk of the ore minerals are represented by hematite (Fe_2O_3), less commonly by intergrowths of hematite with magnetite (Fe_3O_4). Iron ore concentrate silicates are most often observed in intergrowths with iron ore minerals. Figure 7.13 shows the microstructure of raw brex.

In order to assess the progress of the iron ore-coke brex reduction process, polished sections of brex samples were analyzed after being heated to temperatures of 900°C, 1100°C, and 1200°C in a reducing atmosphere.

For brex heated to a temperature of 900ºC, the developed process of reducing iron oxides is typical (Fig. 7.14). There is a lot of wustite and magnetite in the central part of brex. Along the periphery, the recovery process is more developed (Fig. 7.15) and the formation of a skeleton is observed, by "merging" the metal grains (Fig. 7.16).

In the course of further heating, at temperatures close to 1000°C, the reduction process is enhanced—the amount of the initial iron bound to hematite is minimized and the iron-containing phases are represented by magnetite, wustite and metallic iron.

Figure 7.17 shows the microstructure of the peripheral part of brex at a temperature of 1100°C, where the formation of a metal frame is clearly visible.

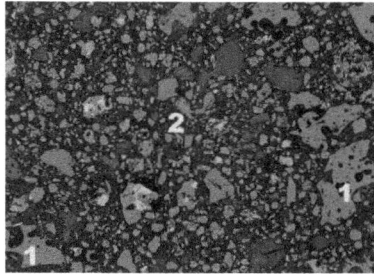

Figure 7.13. Microstructure of the iron ore-coke brex; reflected light; 1 - coke breeze; 2 - grains of Hematite iron ore (white) in cement bond (gray); 50x magnification.

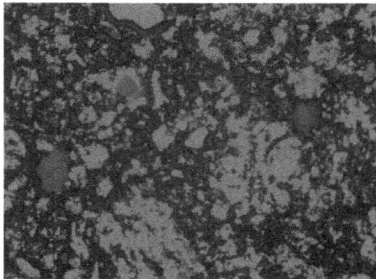

Figure 7.14. Microstructure of the central part of the reduced brex heated to a temperature of 900°C; reflected light; Light gray areas - isolated areas of reduced wustite and magnetite; gray areas - silicate phases; 100x magnification.

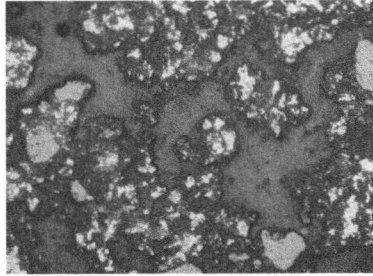

Figure 7.15. Microstructure of the periphery of the reduced brex heated to a temperature of 900°C; reflected light; white is a reduced metal, light gray is wustite; dark gray - silicate phases; 500x magnification.

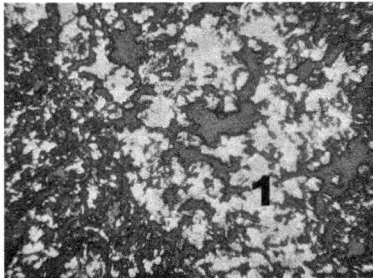

Figure 7.16. Formation of a metal frame at the site of large hematite granules of hydro-ore (1) along the periphery of brex (T = 900°C), reflected light, magnification × 200.

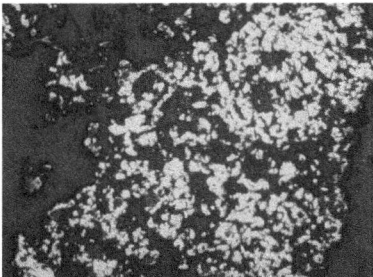

Figure 7.17. Microstructure of the periphery of the reduced brex at 1100°C; reflected light; White areas - metal; gray - transformed mineral phases of the cement; 200x magnification.

A further increase in the reduction time of the brex and temperature up to 1200°C leads to the intensive development of reduction processes. Most of the iron is in metallic form, the rest is preferably wustite. The inclusions of unreacted coke breeze are less than at temperatures below 1200°C, mainly in the center of brex. Unreacted ore quartz grains are also locally observed. Figure 7.18 shows the microstructure of the center of brex at a temperature of 1200°C.

Thus, with an increase in temperature and reduction time of brex, the main strength function is performed by a metal frame formed along the periphery and silicate phases. The strength carrier in brex No. 4 was a matrix of sintered metal grains and an iron silicate phase.

The raw briquette No. 2 consists of blast-furnace and converter sludge, which are anthropogenic fine multi-component products (Fig. 7.19). Unlike briquette No. 4, in the cement bond of this briquette, the particles are more distant from each other.

The structure of reduced briquette No. 2 is represented by scattered fine grains of metallic iron (there is no metal frame made of sintered grains) and the silicate phase (Fig. 7.20).

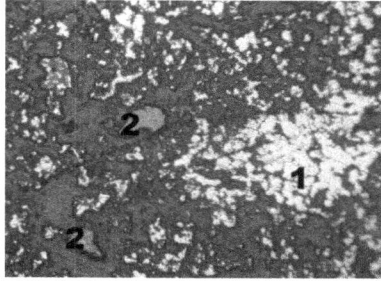

Figure 7.18. Microstructure of the center of reduced brex (at 1200°C); reflected light; 1 - metal; 2 - coke breeze; gray areas - transformed mineral phases of the cement; 200x magnification.

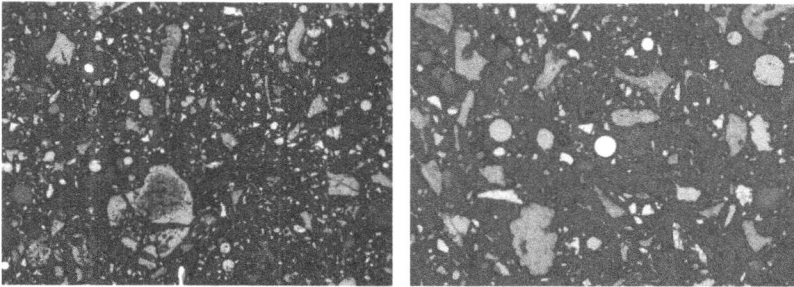

Figure 7.19. Microstructure of raw sludge brex. Reflected light; 100x magnification (left); 200x magnification (right).

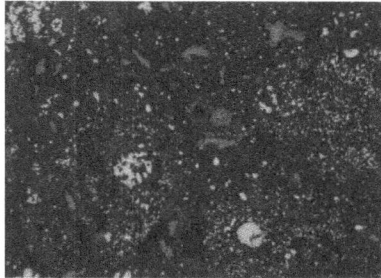

Figure 7.20. Microstructure of the reduced sludge brex. 100x magnification.

Thus, the components of the mechanism of hot strength of brex can be: the formation of a skeleton of sintered metal grains (brex No. 4), the formation of a surface metal shell (brex from hydro-ores), and the formation of a liquid bond (silicate, iron silicate, brex No. 4 and No. 2).

References

[1] Kurunov, I.F., Bizhanov, A.M., Tikhonov, D.N. et al. 2012. Metallurgical properties of brex. Metallurgist 56: 430–437. https://doi.org/10.1007/s11015-012-9593-9.

7.2.2 *Metallurgical properties of industrial brex used as the main component of a blast furnace charge*

The metallurgical properties of brex produced on the above-mentioned industrial briquetting line, which was commissioned in April 2011 at a metallurgical plant owned by Suraj Products Ltd (Rourkela, India) were studied in [1]. The line produces brex for use in small-scale blast furnaces. The layout of Suraj Products Ltd stiff extrusion line is shown in Fig. 7.21.

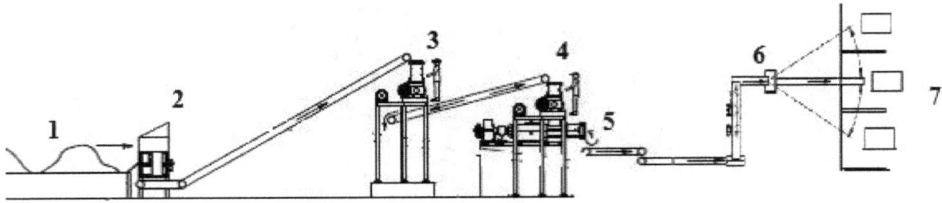

Figure 7.21. Suraj Products Ltd. (India) stiff extrusion line layout: 1- raw material storage; 2 - even feeder with a mixing area for the binder and plasticizer; 3 - primary pug-mill; 4 - pug-sealer; 5 - extruder; 6 - piler; 7 - strengthening unit.

Table 7.14. Chemical composition of the brex and their constituent components.

Brex components shares, %,	Fe_2O_3	FeO	Fe_{tot}	LOI	SiO_2	CaO	MgO	Al_2O_3	TiO_2	K_2O+ Na_2O	C
Iron ore fines 18.9	88.53	-	62.0	4.37	2.71	0.3	-	3.51	0.22	0.2	-
Flue dust 28, 3	53.0	-	37.1	-	6.3	4.9	0.2	5.1	-	-	30.5
LD sludge 47.2	17.9	65.4	63.4.0	-	1.3	8.6	4.9	0.3	-	-	1.5
Portland cement, 4.7	4.2	-	-	1.5	20.5	63.2	2.1	4.5	-	0.6	-
Bentonite 0.9	10.0	-	-	9.4	58.4	0.8	0.3	12.6	-	4.4	-
Brex 100	40.4	30.7	52.1	1.4	4.7	8.3	2.5	2.6	0.1	0.13	9.2

Due to the plasticity of brex, the transportation and storage operations are not accompanied by the formation of fine fractions. In two-shift operation, the briquetting department produces 200 tons of brex per day. In addition to conventional brex, as the main component of the blast furnace charge, washing brex from manganese ore fines are produced at the briquetting site. Brex are currently used as the main and only component of the charge on a small blast furnace with a volume of 50 m³. Before the construction of this line, the blast furnace operated with 100% rich hematite ore using limestone and dolomite as fluxes.

The main components of the brex are LD (Linz-Donau process) sludge, flue dust, iron ore fines, Portland cement and bentonite (Table 7.14).

The main objective of the project was to increase the economic efficiency of blast furnace performance by replacing the purchase of iron ore with cheap brex made of a blast furnace dust and sludge from steelmaking.

The reducibility and metallization of industrially manufactured brex when heated in a reducing atmosphere was investigated in comparison with the recoverability and metallization of iron ore fines.

The following testing methodology was used for this. The test chamber was a hollow corundum tube with an outer diameter of 85 mm and an inner diameter of 75 mm. The brex samples were placed in a graphite crucible placed inside the chamber and covered with coke breeze particles with 1–5 mm in width. The thickness of the coke breeze layer covering the entire crucible was 25 mm. After that, perforated inserts were used to block the crucible from both ends, which provided heat insulation. Water-cooled metal clips were installed on both sides of the corundum tube. The vacuum pump was connected to one of the clips together with an air supply system and a flow meter. Before heating, the internal volume of the corundum tube was evacuated to 0.5 mbar. Then, the heating process started (10°C per minute) and after the temperature reached 800°C, the air supply was switched on at a flow rate of 10 liters per minute until the end of the test. The brex samples were subjected to reducing roasting (in the CO atmosphere) in the 1000–1500°C temperature range at intervals of 100°C. The test was performed at a fixed temperature for the duration of 2 hours. The installation diagram and the sample positioning method are shown in Figs. 5.19 and 5.20 (section 5.2).

The chemical composition of the samples of ore and brex are given in Table 7.15.

Table 7.15. The chemical composition of iron ore and brex, %.

Element	Iron ore	brex
CaO	0.2	12.28
SiO_2	2.21	6.84
Al_2O_3	3.11	2.65
Fe_2O_3	88.53	66.09
MgO	-	1.34
TiO_2	0.22	1.05

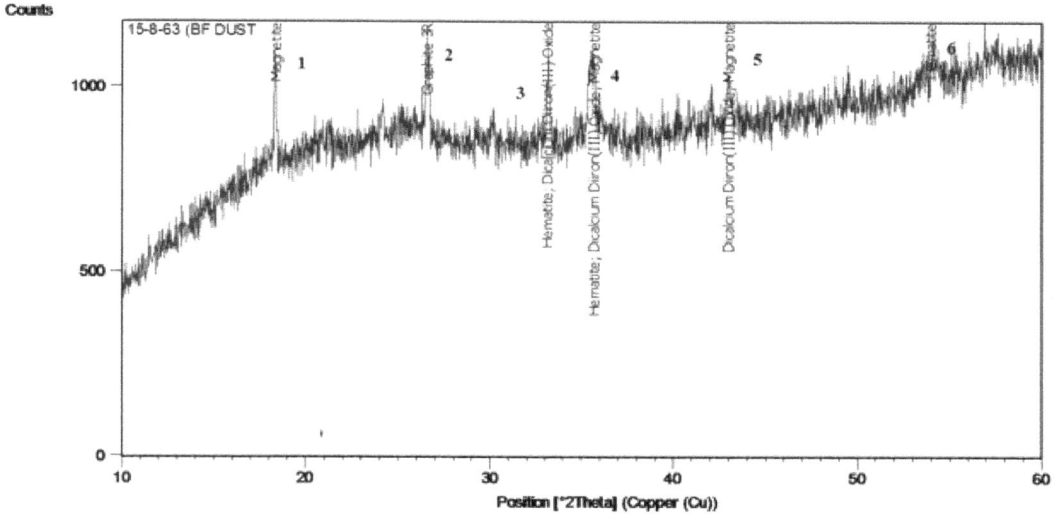

Figure 7.22. Phase composition of flue dust (hematite, magnetite, dicalcium ferrite $2CaO \cdot Fe_2O_3$, graphite).

The phase composition of the components of the iron: iron ore - hematite, goethite, gibbsite, pyroxene, kaolin; top dust (Fig. 7.22)—magnetite, hematite, wustite, graphite, quartz, dicalcium ferrite; Converter sludge—magnetite, wustite, Wollastonite, calcite.

The results of measuring the degree of reduction and metallization of ore samples and brex are given in Table 7.16.

Figures 7.23–7.24 show comparative curves of the degree of reducibility and metallization for samples of iron ore and brex depending on the heating temperature in a reducing atmosphere.

It can be seen that brex are reduced and metallized better than iron ore. This is due to the presence of carbon in the brex in the first place, introduced by blast furnace dust. Closer contact of iron oxide particles with carbon accelerates the reduction of brex compared to iron ore [2].

Table 7.16. Metallization and reducibility of iron ore and brex.

T, °C	Fe_{MET} (ore/brex), %	Fe_{total} (ore/brex), %	Metallization (ore/brex), %	Reducibility (ore/brex), %
1000	2.16/14.20	67.48/60.5	3.20/23.5	35/48
1100	12.69/20.96	69.52/61.7	18.25/33.97	52/62
1200	17.65/30.36	70.25/63.7	25.10/47.6	61/72
1300	30.33/44.74	73.3/65.6	41.30/68.2	65/80
1400	53.75/47.04	80.79/67.46	66.53/69.7	82/88
1500	84.25/79.22	90.4/80.87	93.20/97.95	93/99

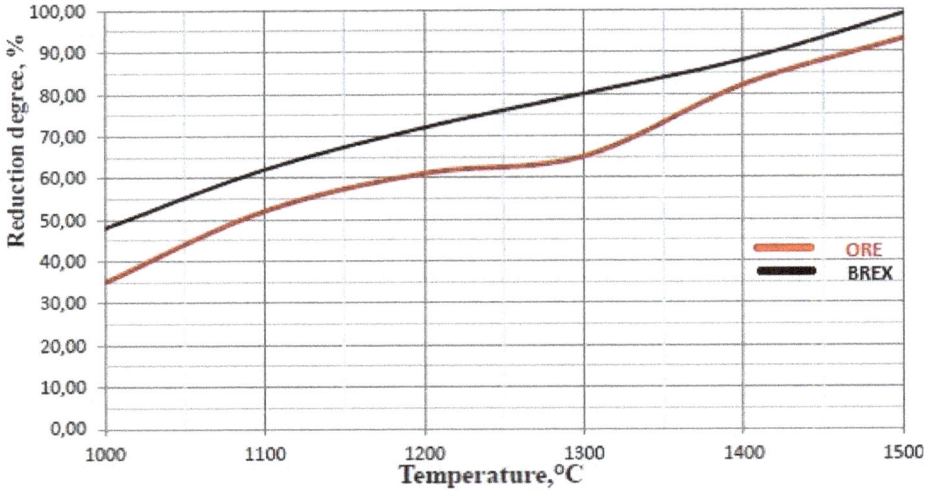

Figure 7.23. Change in the degree of reduction of iron ore and brex when heated.

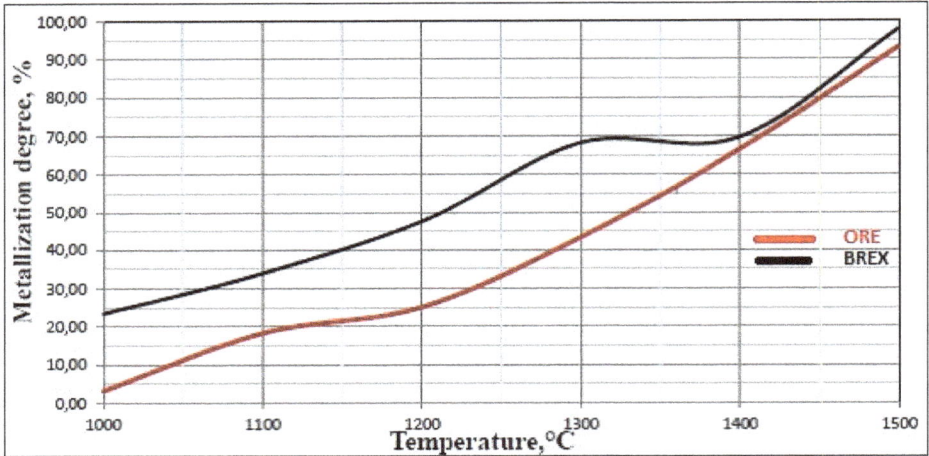

Figure 7.24. Changes in the degree of metallization of iron ore and brex when heated.

The dynamics of changes in the porosity of brex at different temperatures, which was established in [3], also contributes to improving the recoverability of brex. Values of porosity and density of brex and iron ore at different temperatures are given in Table 7.17 and in Fig. 7.25.

At 1000°C, the density value of the brex has a local maximum (2.81 g/cm³). By 1200°C, the transformation of hematite into magnetite is completed. The increase in porosity values in the temperature range of 1100–1200°C is explained by the mechanical stresses in the body of brex, associated with the transition from the hexagonal lattice of hematite to the cubic lattice of magnetite. From 1200°C, there is an increase in the density of brex (from 2.56 g/cm³ to 3 g/cm³ or more) and a decrease in the porosity of brex, which is associated with the formation of an iron silicate melt that fills the pore space and further with the development of the metal structure of brex.

Figure 7.26 shows the microstructure of the sample after heating at 1000°C. The structure of brex contains hematite, magnetite and wustite, fragments of silicate phases and ore.

Monocalcium ferrite ($CaFe_2O_4$), melting incongruently at 1216°C, and dicalcium ferrite are also observed in the structure of brex.

Table 7.17. Porosity and density of brex and iron ore at different heating temperatures.

Brex/Iron ore	Porosity, %	Density (g/cm³)
raw	25.8/18.4	2.60/4.11
500°C	33.66/19.4	2.67/4.04
1000°C	37.36/29.15	2.81/3.63
1100°C	37.28/27.6	2.61/3.84
1200°C	46.87/24.5	2.56/3.92
1300°C	29.71	3.08
1400°C	22.6	3.19

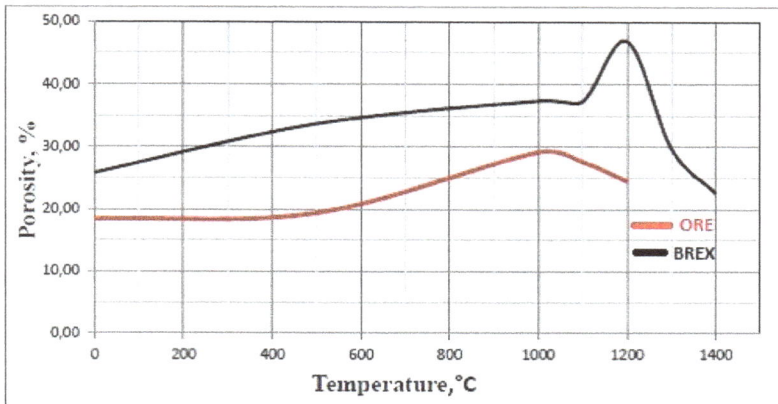

Figure 7.25. Changes in porosity of brex and iron ore when heated to different temperatures.

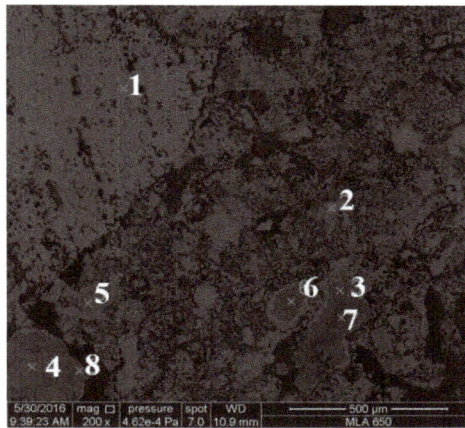

Figure 7.26. Brex structure when heated to 1000°C. 1 - hematite, 2 - magnetite, 3 - gehlenite, 4, 7, 8 - solid solution $CaOAl_2O_3$-$CaOFe_2O_3$, 5 - okermanit-helenite, 6 - gehlenite.

At a temperature of 1100°C (Fig. 7.27), iron oxides are detected that interact with silicates.

At 1100°C, when reduced to wustite, the ferrous iron becomes the melting component and the formation of the iron silicate melt is determined by the area of joint crystallization of olivine and dicalcium silicate.

The structure of the brex at 1200°C is shown in Fig. 7.28.

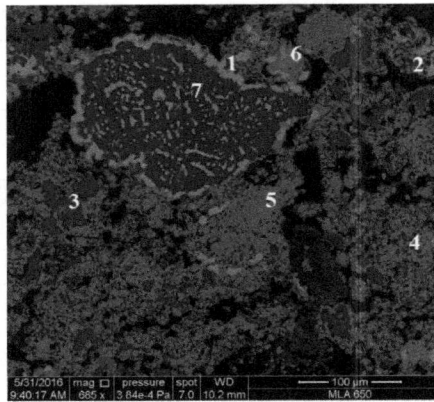

Figure 7.27. Brex after heating at 1100°C. 1, 2 - α-iron, 3 - olivine-dicalcium silicate, 4 - wustit-melilite, 5, 6 - wustite, 7 - wustite-gehlenite.

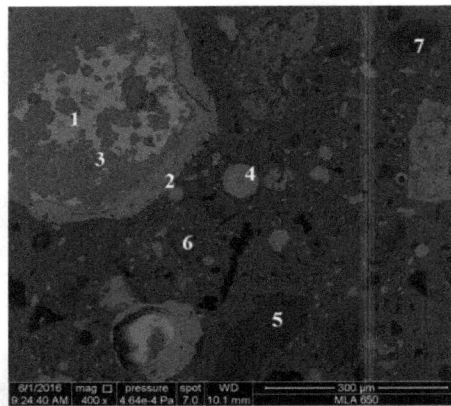

Figure 7.28. Brex after heating at 1200°C. 1 - α-iron, 2 - vustite and melilite, 3, dicalcium silicate, 4 - magnetite, 5, 6 - spinel and dicalcium silicate.

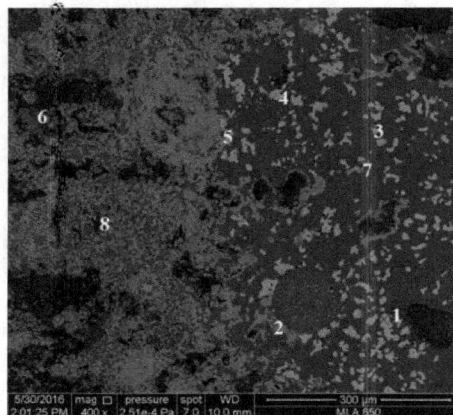

Figure 7.29. Brex after r heating at 1300°C. 1 - α-iron, 2 - wustite, 3 - dicalcium silicate, 4, 7 - okermanite, 6, 8 - gehlenite.

At a temperature of 1300°C (Fig. 7.29), the formation of dicalcium silicate, okermanite and gehlenite is observed. The ore binding consists of metal and wustite grains.

Figure 7.30 shows images of the brex structure at 1400°C. Metal mesh structure. Sintering of metal grains occurs.

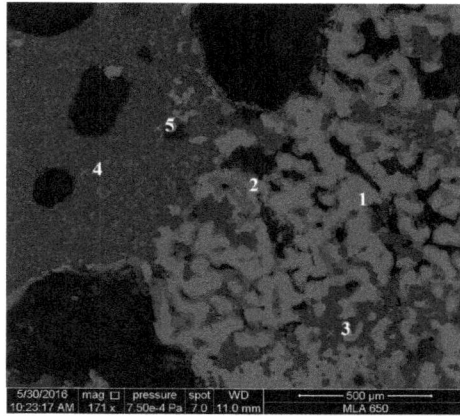

Figure 7.30. Brex after heating at 1400°C. 1 - α-iron, 2, 5 - wustite, 3, 4 - silicates.

Thus, the integrity of industrial brex heated in a reducing atmosphere after the loss of cement stone strength is achieved by the successive action of a number of mechanisms: the formation of the surface metal shell, the formation of the ore bond as a phase with the structure of ferrites and silicates, sintering the grains of the reduced metal.

References

[1] Bizhanov, A., Kurunov, I., Dalmia, Y., Mishra, B. and Mishra, S. 2015. Blast furnace operation with 100% extruded briquettes charge. ISIJ International 55(10): 175–182.

[2] Matsui et al. 2003. ISIJ Int. 43(12): 1904.

[3] Kurunov, I.F., Yashchenko, S.B. and Fursova, L.A. 1986. Theory, technology and equipment for metallurgical production. Calculation of Indicators of Blast-Furnace Smelting on a Computer. Textbook for Course and Diploma Design. Moscow MISIS. 86 p. (in Russian).

7.2.3 Blast furnace operation with 100% brex in charge

The small-scale blast furnace (volume 45 m^3) at Suraj Products Ltd is equipped with a skip hoist with a volume of 0.5 m^3, a double-cone charging device and hydraulic equipment for notch servicing, recuperative stoves and a two-stage dry gas-cleaning system (dust collector + the 7 modules of the bag house filters) and sludge granulation equipment. The furnace has external watering and eight air tuyeres. The hot metal (foundry and pig iron) is immediately formed by the casting machine. The daily production of the blast furnace is 100–130 tons of pig iron.

The extrusion line of the blast furnace unit began operating in May 2011, using 10% of brex in the blast furnace charge. Gradually, the brex content of the charge increased to 80%. From the end of 2013 the blast furnace has been operating using 100% brex charge. Table 7.18 presents the main furnace operating parameters with brex in the charge at Suraj Products Ltd.

When working on the charge with 80% brex and 20% iron ore coke consumption decreased by 150 kg per ton of hot metal (22%) compared with operation on 100% iron ore. The reduction in coke consumption occurred both due to the carbon contained in the brex and due to the removal of limestone and dolomite from the charge. The productivity of the blast furnace that used 80% of brex in the charge decreased mainly due to a 7.2% decrease in iron content in this charge compared with the charge based on iron ore and raw fluxes. The transition to a mono-charge blast furnace operation resulted in a further decrease in coke consumption due to the additional carbon input. The increase in the blast temperature by 100°C also contributed to the decrease in coke consumption. In addition, the use of manganese ore washing brex reduced the viscosity of slags and improved the processing of smelting products. As a result, the furnace productivity was increased due to the improvement of the structure of the charge column, the decrease in the viscosity of the primary slags and the increase

Table 7.18. Furnace operating parameters at Suraj Products Ltd.

Furnace operating parameters	100% ore	80% brex	100% brex
Consumption, kg/t:			
Iron ore	1500	372	-
Brex	-	1425	1960
Limestone	150	-	-
Dolomite	144	-	29
Scrap	132	-	-
Quartzite	-	-	13
Mn ore brex	-	19	75
Coke*	680	530	490
Fe_{tot} in charge, %	57.6	50.4	45.5
Productivity, t/m^3 per day	1.9	1.62	2.0
Blast temperature, °C	925	900	1000
Blast pressure, kg/cm^2	0.5	0.34–0.38	0.38–0.42
[Si], %	1.0–1.8	1.0–1.5	0.8–1.1
[Mn], %	0.2	0.4–0.5	0.7–0.8
[C], %	3.8–4.0	3.75–3.90	3.80–3.95
[S], %	0.050–0.060	0.038–0.050	0.038–0.042
Cast iron temperature, °C	1380–1440	1400–1450	1410–1450
(CaO), %	34.86	33.12	38.0–39.0
(SiO_2), %	31.98	30.23	30.0–32.0
(Al_2O_3), %	23.87	17.98	16.0–18.8
(MgO), %	9.46	9.48	8.0–9.5
(FeO), %	1.01	1.26	0.6–1.15
(MnO), %	0.35	0.75	1.3–1.0

* The size is 15–25 mm

in the overall pressure drop. The reduction in specific heat loss also contributed to a reduction in coke consumption.

The development of new blast furnace smelting technology with the use of brex meant switching to a decreased burden level of blast furnace charge due to the difficulties in the operation of dry gas cleaning system. The gradual increase in the share of brex in the charge led to a significant increase in the humidity of the furnace top gas and a decrease in its temperature. As a result, bag filters began to become clogged with wet dust, and regeneration by their back pressure pulses did not achieve the effect. The reduction in the level of the burden practically did not affect the performance of the blast furnace due to the fact that the reduction of lower iron oxide (FeO) in the brex occurs predominantly in a direct way due to the carbon of coke contained in the brex, and the high reducing potential of the hearth gas and its temperature ensured the reduction of higher iron oxides to wustite.

Despite high specific heat losses due to the small volume of the blast furnace and low blast temperatures, the coke consumption during the operation of the blast furnace in the mono-charge of brex does not exceed 500 kg/t of pig iron. This is achieved because the brex used in the above-mentioned blast furnace are self-reducing due to the presence of fine coke particles of flue dust in close contact with the dispersed particles of ore and steel-smelting sludge containing iron oxides, as well as due to the high basicity of the brex providing obtaining the necessary basicity of blast furnace slag without the use of limestone. To adjust the slag basicity and the content of magnesia in it, additives of quartzite and dolomite are used. As a result, coke consumption for pig iron smelting when operating at 100% brex was reduced by almost 200 kg/t of pig iron as compared with the operation at 100% rich iron ore using limestone and dolomite.

The results of more than 7 years of industrial operation of the stiff vacuum extrusion process line as part of the blast furnace shop showed that brex obtained from natural and anthropogenic dispersed raw materials have optimal and adjustable sizes, controlled chemical composition and high metallurgical properties, are a new type of agglomerated and fluxed component of the blast furnace charge. The work of a small blast furnace on a mono-charge from 100% of brex produced from a mixture of metallurgical wastes and iron ore fines demonstrates low coke consumption (490–500 kg/t of pig iron) at a blast temperature of 1000°C and an iron content in the flux mixture of 45.5%.

7.2.4 *Assessment of prospects for the use of carbon-containing briquettes from iron ore concentrate*

The effectiveness of brex in the charge of large blast furnaces was estimated by the method of mathematical modeling of blast furnace smelting using the DOMNA program developed in [1]. The blast furnace smelting in a furnace with a volume of 4297 m³, operating under the conditions of PJSC NLMK [2], is modeled. Taking into account the basicity of brex, the level of which is influenced by quantities of the cement binder and bentonite (plasticizer), simulation of blast-furnace smelting was carried out for a mixture consisting of three components—sinter, pellets and brex. The share of pellets is determined by the capacity of the pelletizing plant (Stoilensky GOK, 6 million tons of pellets per year). The basicity of a brex made from iron ore concentrate and low caking coal is determined on the basis of the content of cement (6%) and bentonite (1%) in the composition of the brex, taken on the basis of the results of preliminary tests, and also the content of coal in the mixture. At the indicated contents of cement and bentonite, the basicity of brex (B2) is 0.50–0.55. The basicity of the sinter is determined on the basis of the accepted concept of replacing the sinter with brex and their basicity. When replacing 50% of the sinter in the blast furnace charge with brex, the basicity of the sinter should be between 2.8 and 3.2. It should be noted that with such basicity in the structure of the sinter the phases predominate, providing an increase in its strength compared to the sinter with basicity in the range of 1.5–1.7 characteristic of the sinter, produced in NLMK. The coal content in brex was determined by the method developed based on the results of studying the structural composition of brex after heating in a reducing atmosphere to 1400°C, when, after reaching almost complete metallization, unreacted coke breeze particles remained in the brex structure due to its excessive content in the mixture for briquetting.

The consumption of carbon-containing material in the charge for briquetting is calculated based on the stoichiometric ratio of oxygen and iron in brex (O/Fe) by the time they reach the cohesion zone, where the temperature exceeds 1100°C and reduction takes place only with the participation of solid carbon. The content of carbon in coal or other carbon-containing material in fractions of a unit, the total iron content in iron ore material, the degree of oxidation of iron in this material (O/Fe) at the arrival of brex in the cohesion zone and the proportion of mixture cement + bentonite in the mixture for briquetting.

To simulate the blast smelting, the calculated compositions of sinters with a basicity of 1.70 and 3.02 and the composition of brex from Stoilensky GOK concentrate were used, the proportion of carbon in which was calculated (Table 7.19).

The simulation was carried out for the same composition of hot metal and its temperature ([Si] = 0.4%; [C] = 4.8%; $T_{pig\ iron}$ = 1500°C) and for the same reduction efficiency (the degree of approximation to the equilibrium composition of the gas in the wustite reduction zone). The results showed that due to the carbon contained in the brex, the coke consumption for hot metal production is reduced compared to the base case by 10%. When pulverized coal is injected with a flow rate of 160 kg/t of hot metal, coke consumption of 284 kg/t of hot metal is achieved, and when blowing natural gas with a consumption of 125 m³/t of hot metal, coke consumption of 354 kg/t of hot metal is achieved (Table 7.20).

Table 7.19. Chemical composition of raw materials, sinters and brex, %.

Materials	FeO	Fe_2O_3	SiO_2	Al_2O_3	CaO	MgO	C	SO_3
Portland cement	-	4.71	20.64	4.98	63.58	1.15	-	2.55
Bentonite	0.5	4.37	59.25	14.27	2.07	3.62	-	0.14
Caking coal	-	1.2	2.7	1.5	0.4	0.1	68.9	0.36
SGOK pellets	1.51	90.31	7.04	0.32	0.22	0.45	-	0.12
SGOK Iron ore concentrate	29.2	62.3	6.62	0.18	0.26	0.1	-	0.05
Sinter (170)	11.81	66.0	6.70	0.72	11.37	2.51	-	0.05
Sinter (3.02)	10.0	60.8	6.30	0.7	19.0	3.0	-	0.06
Brex	24.83	53.28	7.67	0.71	4.09	0.19	5.5	0.22

Table 7.20. Results of the blast furnace smelting simulation when using a traditional charge and a new charge with one third iron-carbon-containing brex.

BF operation parameters	Basic variant	Variant 1	Variant 2
Sinter consumption B2 = 1.7, kg/t	1109	-	-
Sinter consumption B2 = 3.0, kg/t	-	557	575
SGOK pellets consumption, kg/t	546	557	541
Brex consumption, kg/t	-	557	575
SGOK iron ore consumption, kg/t	-	17	-
Fe content in charge, %	58.2	57.45	57.15
Coke rate, kg/t	391	354	284
Natural gas consumption, nm^3/t	125	125	35
Pulverized coal consumption, kg/t	-	-	160
Blast rate, m^3/min	7483	7568	7340
Blast temperature, °C	1240	1240	1240
O_2 content in blast, %	30.5	30.5	30.5
Blast humidity, g/m^3	10	10	20
Top gas yield, m^3/t	1545	1540	1470
Top gas pressure, kPa	240	240	240
CO, %	24.4	24.9	26.2
CO_2, %	23.2	22.6	23.9
H_2, %	9.7	9.9	8.2
Slag ratio, kg/t	318	314	323
Slag basicity, B2	1.01	1.01	1.02
Capacity, t/day	12465	12624	12708
Capacity, t/m^2·day	92.48	93.66	94.3
Reduction efficiency, %	94.2	94.2	94.2

7.2.5 *Metallurgical properties of blast-furnace BREX based on hematite concentrate*

Metallurgical properties of brex made from hematite iron ore concentrate have been studied, including with addition of coke fines and/or flux, i.e., mechanical or hot strength, reduction capacity, temperature for the start and range of softening.

Concentrate from hematite iron ore may be classified as a fine and specific surface according to Blaine » 2500 cm^2/g. Magnetite and hematite grains have extended shape in cross section, and the

Table 7.21. Chemical composition of the brex components.

Material	Fe$_{tot}$	FeO	Fe^{+2}	Fe$_2$O$_3$	SiO$_2$	Al$_2$O$_3$	CaO	MgO	TiO$_2$	MnO
Concentrate	66.10	3.20	2.48	90.98	1.83	0.10	0.75	0.22	0.004	0.030
Material	S	P	C	P$_2$O$_5$	CO$_2$	Calcin. loss		K$_2$O	Na$_2$O	Fe$_{mag}$
Concentrate	0.0034	0.025	0.40	0.057	1.47	3.05		0.038	0.008	2.8

Table 7.22. Chemical composition of brex.

Material	Fe$_{tot}$	FeO	Fe$_2$O$_3$	SiO$_2$	Al$_2$O$_3$	CaO	MgO	Calcin. loss	K$_2$O	Na$_2$O
Bentonite	-	-	3.15	65	19.50	0.95	2.75	-	0.5	2.4
Limestone	0.25	0.01	0.35	0.75	0.60	67	-	30	-	-
Portland cement	-	4.70	-	21.55	-	65.32	1.45	-	-	-

presence of non-ore minerals (quartz) is noted. The chemical composition is provided in Table 7.21. The chemical composition of other brex components is given in Table 7.22.

Sedimentation analysis was conducted in a HELOS analyzer that is intended for measuring particle size in the range 0.1–3500 pm. Sedimentation analysis results of concentrate are given in Fig. 7.31, where the integral volumetric distribution of particles with size less than that prescribed (Q^3) and distribution density (q^3) are provided. The content of fines in concentrate with size not less than 45 pm is 93.85%, which corresponds to the same index for ore concentrate, i.e., 45 pm more than 90%.

Coke fines particles of the order of 98% have a size less than 2.36 mm (standard screen size No. 8 in the USA).

Experimental brex were prepared by means of a computerized laboratory extruder modeling all the main stages of extrusion pelletizing, including treatment of a mixture in an open pug mill, their removal in a hermetically sealed pug mill connected with the extruder working chamber, and capable of stiff extrusion. The level of vacuum in the extruder working chamber was maintained at 3848 mm Hg. Extrusion was completed by means of a briquetting die with an opening of 19 mm.

Tests were conducted for brex without flux and highly basic material, including the addition of solid fuel. The brex components and chemical composition are provided in Tables 7.23 and 7.24. Pelletized material without flux is used for increasing the iron content of a blast furnace charge, whereas fluxed pelletized materials are mainly used for introducing flux into blast furnaces operating on a charge with low basicity. Moreover, addition of solid fuel to brex reduces coke consumption during smelting in a blast furnace.

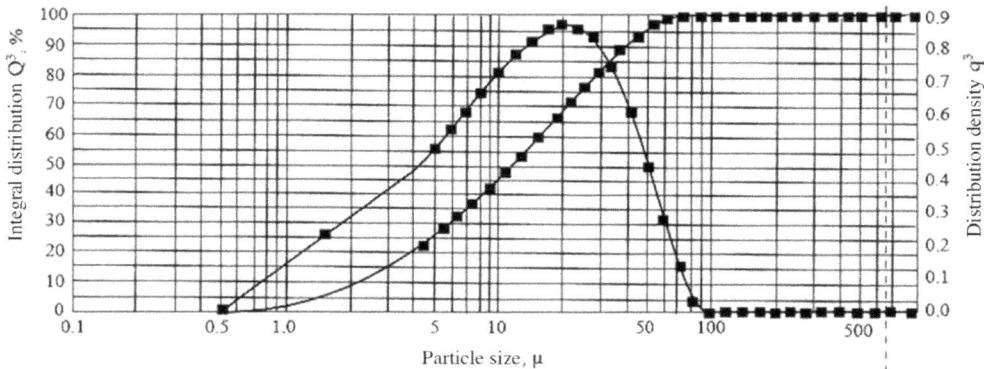

Figure 7.31. Results of hematite iron ore sedimentation analysis.

Table 7.23. Test brex component composition (Fraction, %).

Charge component	Brex 1	Brex 2	Brex 3	Brex 4
Concentrate	90.50	72.70	86.60	69.70
Bentonite	1.00	1.50	1.50	1.50
Limestone	-	17.30	-	16.80
Portland cement	8.50	8.50	8.50	8.50
Coke fines	-	-	3.40	3.50

Table 7.24. Brex chemical composition.

Brex number	Fe_{tot}	FeO	CaO	SiO_2	MgO	Al_2O_3	Fe_{met}	Calcin. Loss	Basicity, B2*
1	59	2.84	5.9	3.74	0.26	0.59	0.6	5.28	1.58
2	48.3	2.17	17.5	3.94	0.43	0.76	0.53	8.93	4.44
3	56	2.5	6.6	4.14	0.42	0.78	0.6	7.7	1.59
4	46.6	2.04	16.3	4.12	0.5	0.88	0.53	11.62	3.96

* B2 is CaO/SiO_2

Raw brex density was measured using a lifting force method (Archimedes principle) on calibrated electron scales. Moisture content was measured by a balance method. Brex were held under atmospheric conditions for two days in a location with controlled environmental temperature in the range 23–26°C.

In order to determine brex mechanical strength a method of triple throwing from height of 2 m on to a steel surface was used. This test reflects most completely the specific action on a briquette of impact loads during transportation from the briquette plant to a blast furnace. Action of compressive and wear loads develops during briquette storage in stack or bunker and during movement to a distant user. During performing compressive strength tests, it should be noted that scrap limits should differ in relation to briquette weight and size since breakage occurs due to a critical amount of structural inhomogeneity that increases with an increase in specimen size. Even for the same material with an identical concentration of structural defects the probability of breakage with an increase in volume increases exponentially. Briquette strength in compression also depends on applied load intensity. It is not entirely correct to use a drum sample for determining impact strength and briquette wear using rejection limits adopted for pellets as a result of fundamental differences between the structures of these pelletized products. Brex have a variable larger mass than pellets and are subjected during falling on to a steel surface in a rotary drum to considerably greater impact action, and their overall amount is less than for pellets with an identical loading weight for a test batch (15 kg). As a result of this, brex mainly fall on to a steel drum surface, whereas pellets fall on to a bed formed by pellets falling earlier that leads to more intensive breakage of brex and to a reduction in yield of class of particles with a size > 5 mm. The features of brex behavior under action of impact and static loads have been analyzed before [3–5].

A test batch of brex with volume of 200 cm³ (500 g for the given mixture) was placed in a hermetically sealed polyethylene package and thrown three times on to the steel plate from a height of 2 m. Then the sample was weighed and the weight fraction of particles with size less than 4.7 mm was determined.

Physicomechanical properties of the test brex (density moisture content, dropping strength) are provided in Table 7.25. It should be noted that almost all test samples (raw materials and strengthened) have good strength indices. An addition of coke fines (3.4%) led to a reduction in strength during dropping tests that is especially typical for raw brex. Testing showed that addition of slaked lime improves brex stability towards breakage, especially for raw materials.

The reduction capacity of brex was determined according to GOST 17212-84 in accordance with which reduction of a sample with carbon monoxide is accomplished under certain temperature

Table 7.25. Brex physical properties and mechanical strength.

Parameter	Brex 1	Brex 2	Brex 3	Brex 4
Brex moisture content, %	10.0	12.6	9.3	12.7
Brex density, g/cm³	3.25	2.77	3.14	2.73
Dropping strength (particle fraction less than 4.75 mm) (raw brex), %	8.4	0.3	23.3	0.2
exposure 24 hours	4.4	1.9	3.4	2.2
exposure 3 days	2.2	2.2	4.4	2.1
exposure 7 days	1.8	1.8	3.1	2.0

Table 7.26. Results of measuring brex reduction capacity by GOST 17212-84.

Parameter	Brex 1	Brex 2	Brex 3	Brex 4
Weight before reduction, g	500	500	500	500
Weight after reduction, g	406.45	418	396	415.25
Reduction capacity, %	66.9	74.74	86.7	65.6
Gas composition, %				
CO	35	35	35	35
N_2	65	65	65	65
Gas consumption, dm³/min				
CO	10	10	10	10
N_2	20	20	20	20
Heating temperature, °C				
first 40 (50) minutes	600	600	600	600
next 175 (16) minutes	1160	1160	1160	1160

conditions (up to 1100°C), and the degree of reduction was determined from results of chemical analysis for the original and reduced samples, or according to weight loss of oxygen during reduction. Test results are provided in Table 7.26.

The data provided point to a relatively high degree of reduction capacity for test samples.

These indices are comparable with the reduction capacity of pellets made from hematite iron ore from the same deposit [6]. The reduction capacity of sinter from hematite concentrate of the same deposit is within the range 37.7–48.7 (with basicity 1.35–1.45). Unfluxed brex with addition of solid fuel (brex 3) demonstrated very good reduction capacity (i.e., 86.7%), that may be explained by acceleration of the reduction process as a result of the immediate vicinity of carbon and iron oxides within the brex body [7]. Fluxed brex 2 also demonstrated good reduction capacity. It is well known that gasification of carbon accelerates in the presence of CaO [8]. However, as shown in [9], there is a threshold value for CaO content affecting the decrease in reduction capacity. The rate of reduction of CaO-containing FeO-agglomerates increased to a range of 2.5 wt.% CaO, and then decreased with an increase in CaO in relation to the intermediate phase of bicalcium ferrite formed. This may explain the marked difference in the effect of CaO on reduction capacity in regions below and above the solubility limit for CaO in FeO in whose presence there is dissolved CaO and $Ca_2Fe_2O_5$ forms correspondingly. The possibility of forming calcium ferrite has been explained before [10] (Fig. 7.32).

Brex 4 demonstrated the least reduction capacity among the test samples. Lowering of reduction capacity in this case is probably connected with a reduction in brex porosity, which could be caused by pore blocking by iron metal formed due to excess reducing agent. Reduction capacity is also affected by the formation of a metal skin at a surface described before that retards the penetration of gaseous reducing agent into the body of a briquette [11].

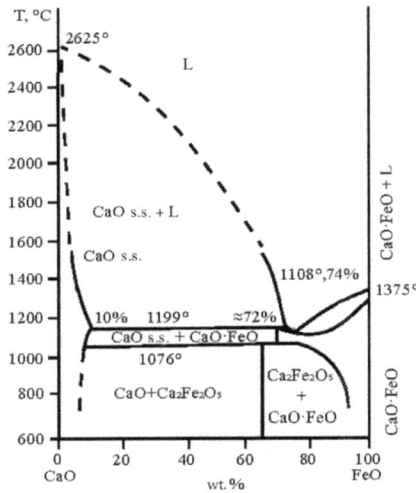

Figure 7.32. Cao-FeO system phase diagram.

Brex hot strength (strength of low-temperature reduction, i.e., Low-Temperature Disintegration LTD) was determined in accordance with ISO 13930. The essence of the ISO 13930 procedure includes treatment of a sample with gaseous reducing agent in a rotating drum (diameter 150 mm, length 540 mm, wall thickness 5–7 mm, with two shelves located diametrically on the inner surface, width 20 mm and thickness 4 mm) with a prescribed temperature regime following test sample screening with respect to particle size classes: + 6.3, – 3.1, – 0.5 mm.

Test performance conditions: drum rotation frequency 10 ± 0.2 min[1]; volumetric consumption of reducing agent 20 dm^3/min; reducing gas composition (20 ± 0.5)% CO, (58 ± 1.0)% N$_2$, (20 ± 0.5)% CO$_2$, (2 ± 0.2)% H$_2$; permissible impurities 0.1% O$_2$, 0.2% H$_2$O; temperature: in the first 45 minutes furnace temperature increases uniformly to 500°C. Three hours after the start of a test the furnace heaters are switched off and separated from the drum. After 10 minutes the drum rotation mechanism is switched on and instead of reducing gas a neutral gas is fed in order to cool a sample. After reducing the temperature to 200°C neutral gas supply is terminated. A sample in the drum is air cooled to room temperature after which it is extracted.

The hot strength index is considered as the ratio of the corresponding class to the overall specimen weight after reduction:

- LTD$_{+6.3}$ is a refinement index, expressed as a percentage with respect to weight, + 6.3 mm of screen fraction (so called strength during milling or hot strength);
- LTD$_{-3.15}$ is milling index, expressed as a percentage with respect to weight, – 3.1 mm of screened fraction (so-called milling index);
- LTD$_{-0.5}$ is refinement index expressed as a percentage with respect to weight, – 0.5 mm screen fraction (so-called wear index after milling).

The results are given in Table 7.27 for testing hot strength after low-temperature reduction. It is seen that the values of LTD+6.3 are very high for unfluxed brex samples 1 and 3, i.e., 62.55 and 60.19% respectively. Previously LTD characteristics have been studied for sinter of hematite ores of the same deposit [12]. The results are provided in Table 7.28 for testing low-temperature reduction of sinter with three different basicity values.

It is seen that values of refinement index (LTD$_{-3.15}$) of brex numbers 1 and 3 are considerably higher than for sinter of the most similar basicity (1.42). Attention is drawn to the high values of brex wear index (LTD$_{-0.5}$) in all basicity ranges. This may be explained from the point of view of the inaccuracy of direct application during brex testing of limits for drum test scrapping used for

Table 7.27. Testing by dynamic method for determining low-temperature reduction-disintegration indices (ISO 13930).

Parameter	Brex 1	Brex 2	Brex 3	Brex 4
Specimen initial weight, g	500	500	500	500
Weight after reduction, g	463.6	497	480.5	457.4
Grain size composition after testing, g				
+6.3 mm	290	263.5	289.2	233
+3.15 mm	43.8	80.1	35.5	90.5
+0.5 mm	48	67.2	60.2	64.1
−0.5 mm	81.8	86.2	95.6	69.8
Refinement indices %:				
$LTD_{+6.3}$	62.55	53.02	60.19	50.94
$LTD_{-3.15}$	28.00	30.87	32.42	29.27
$LTD_{-0.5}$	17.64	17.34	19.90	15.26

Table 7.28. Low-temperature refinement-degradation indices for sinter based on hematite ore.

Low-temperature refinement-degradation parameters	Basicity		
	1.42	1.45	1.35
$LTD_{+6.3}$	55.9	46.5	43.5
$LTD_{-3.15}$	22.5	30.5	31.5
$LTD_{-0.5}$	7.9	10.2	11.1

sinter and pellets mentioned above. Higher impact strengths, which applies to brex exceeding with respect to density and size pellets and sinter, leads to more intense formation of fines. It is evident that wear mainly applies to fragments of broken brex. The results obtained are also in agreement to a sufficient extent with data obtained before for testing roller briquettes according to ISO 4696-1 [13] for which the refinement index $LTD_{-3.15}$ was about 30 for briquettes produced using not less than 2–40 wt.% binder material and with a pressure significantly exceeding that in the extruder working chamber (3.5 MPa).

The cohesion zone is determined for the temperature for the start and end of softening of iron ore materials or the softening temperature range. These indices are some of the important properties of iron ore raw material determining charge gas permeability in the cohesion zone. The possibility of expanding the cohesion zone using several forms of iron ore raw material with different charge softness [14] should be considered, and also the prospect of forming several cohesion zones.

Measurement of the temperature for the start of softening and the softening temperature range for brex was accomplished according to GOST 26517-85 (Fig. 7.33). A rod with a sample is placed in a crucible (with sample heating temperature to 1500°C and above), a plunger descends under action of a load and the chamber is sealed. The control and recording system, supply of inert gas and furnace electric heater are switched on. When the temperature in the chamber reaches 800°C the mechanism for moving the tape of the unit recording rod movement is switched on. Testing in order to determine the temperature for the start of softening is conducted with a heating rate of 10°C/min, the volumetric gas supply rate to the heating chamber is 0.5 dm/min and rod pressure in a sample is 0.1 MPa. With rod movement from the zero position to a value reaching 40% of the original layer height, the recording device, the furnace electric heater, the gas supply system and also the device for monitoring temperature is switched off. Tests are conducted for two samples in parallel.

The results for determining the start of softening and the softening temperature range for brex are provided in Table 7.29.

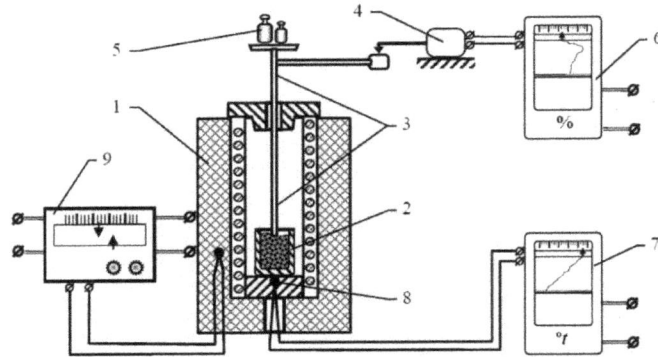

Figure 7.33. Device for determining brex softening temperature and ranges: (1) heating furnace; (2) crucible; (3) shaft with sample; (4) plunger with support bearing; (5) load; (6) device for monitoring and controlling specimen heating temperature; (7) device for automatic recording specimen heating temperature and rod movement; (8) thermocouple; (9) furnace temperature regulator.

Table 7.29. Temperature for the beginning and end of brex softening and disintegration range.

Parameter	Brex 1	Brex 2	Brex 3	Brex 4
Start of softening temperature, °C	860	960	930	880
End of softening temperature, °C	1190	1090	1220	1150
Disintegration range AT, °C	330	130	290	270

It is seen that all samples demonstrate a relatively low temperature for the start of softening compared with sinter and pellets. The highest temperature for the start of softening was recorded for fluxed brex 2, i.e., 960°C, which may be explained by additional strengthening due to lime binder properties. Brex 2 also has a very low softening range (AT = 130°C) compared with that for pellet softening. Unfluxed brex 1 without solid fuel has the lowest temperature for the start of softening, i.e., 860°C. Addition of coke fines to brex reduces the effect of additional strengthening due to adding lime (reduction in temperature for the start of softening to 930°C and expansion of the range to 290°C in brex 3). On the whole it is seen that the temperature for the start of softening correlates with the hot strength value (LTD+).

One reason for early softening of brex may be related with the loss of the cement binder properties at 700–750°C, which leads to a significant change in brex phase composition and a reduction in mechanical strength. However, as demonstrated earlier [15], under blast furnace conditions this does not lead to complete briquette breakage since from cement mineral phases there is formation of new liquid binders that together with partial metallization provide briquette integrity up to complete melting.

In order to demonstrate phase transformation in cement stone during heating in [16] thermal analysis was performed for a sample in a NETZSCH STA 449C instrument. A sample of cement stone was heated at a rate of 20°C/min in an argon atmosphere up to melt formation (T = 1290°C). The cement stone dehydration process is accompanied by breakdown of the hydrate crystal lattice with formation of free calcium and silicon oxides. According the STA results (Fig. 7.34) removal of hygroscopic moisture (8.95 wt.%) from cement stone proceeds on heating to 300°C with a thermal maximum at a point corresponding to 120°C. The endothermic peak on the heating curve at $T = 498$°C and clear weight loss at this temperature points to active breakdown of hydrosilicate. The process of water chemical bond removal from cement stone, accompanied by a reduction in test specimen weight, is observed up to $T = 700$–750°C.

It is well known, as noted in Section 7.1, that in the CaO-SiO_2-FeO system there is a eutectic consisting of a mixture of 25% $CaOSiO_2$ and 75% $FeOSiO_2$ (Fig. 7.1) with crystallization temperature

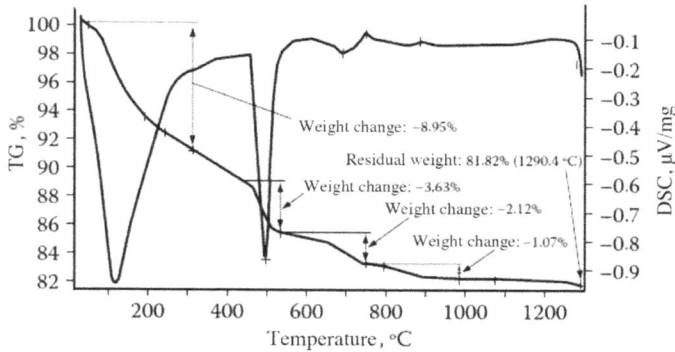

Figure 7.34. Differential curve of thermal effects (DSC) and specimen weight change curve (TG) for cement on heating in Ar atmosphere to 1290°C [16].

at 1030°C [17]. Within the system there is also a region of composition with a crystallization temperature close to 1150°C.

Apparently due to limited exposure time in a furnace at 1150°C and above the iron-silicate phase formed before this within the body of a briquette did not reach the liquidus temperature and remained in a plastic condition, not disrupting briquette integrity for the carcass formed from iron metal with a melting temperature above 1500°C.

In addition, it is well known that solid phase decomposition of ferroakermanite $Ca_2FeSi_2O_2$ (cement clinker component) proceeds at 775°C [18] with formation of iron pseudowollastonite and monticellite. These transformations of phase composition may lead to an increase in briquette porosity and thereby become a reason for earlier softening compared with pellets and sinter during whose product cement is not used.

Conclusions

1. Features of the mineralogical structure of hematite ores make it impossible to achieve satisfactory sinter and pellet quality. Cold pelletizing of hematite ore may be accomplished efficiently by technology of stiff vacuum extrusion SVE that makes it possible to obtain the required briquette metallurgical quality for blast furnace conversion.

2. Ore and ore-coal brex of hematite iron ore concentrates demonstrate strength suitable for blast furnace conversion. Scrapping limits during performance of tests in a drum should be adjusted taking account of the significant difference in levels of mechanical action on brex compared with sinter and pellets.

3. With regard to high-temperature metallurgical properties, brex occupy an intermediate position between pellets and sinter.

4. The reduction capacity of test brex is comparable with that of pellets of iron ore concentrate for ore from the same deposit. Very high values of reduction capacity are typical for ore and coal brex. Addition of flux also increases the reduction capacity compared with ore brex. The reduction capacity of brex is considerably reduced with addition of flux and solid fuel that may be connected with a reduction in brex porosity (due to pore blocking and formation of a metallic skin over a brex surface).

5. Tests for low-temperature reduction-refinement of hematite iron-ore concentrate brex showed strength with refinement ($LTD_{+6.3}$) and grinding index ($LTD_{-3.15}$) at levels suitable for blast furnace smelting. An increase in wear values during grinding ($LTD_{-0.5}$) is connected with a different level of mechanical action on brex compared with sinter and pellets in drum tests.

6. Early brex softening is connected with a reduction in strength due to phase transformation of cement at 700–750°C and solid phase decomposition of ferroakermanite at 775°C. Retention if brex hot strengthen up to moment of melting is connected with formation at 1150°C of liquid binder based on cement mineral phases and with formation of a metal carcass.

7. Early (compared with sinter and pellets) brex softening that may lead to an increase in cohesion zone, should be considered in selecting blast furnace charge composition for combined smelting with sinter and pellets.

8. The results of studies demonstrate that brex based on hematite may be considered as a promising charge component for a blast furnace charge.

9. Brex Nos. 1 and 3 are distinguished by high iron content and quite good metallurgical properties. Brex Nos. 2 and 4 may be used in cases when special smelting regimes are required in blast furnaces using briquettes with high basicity.

References

[1] Kurunov, I.F., Yashchenko, S.B. and Fursova, L.A. 1986. Theory, technology and equipment for metallurgical production. Calculation of Indicators of Blast-Furnace Smelting on a Computer. Textbook for Course and Diploma Design. Moscow MISIS. 86 p.

[2] Kurunov, I.F., Filatov, S.V. and Bizhanov, A.M. 2017. Using mathematical simulation to evaluate the efficiency of iron ore–coal brexes in blast-furnace smelting. Metallurgist 60: 1022–1024. https://doi.org/10.1007/s11015-017-0402-3.

[3] Bizhanov, A.M., Kurunov, I.F., Durov, N.M., Nushtaev, D.V. and Ryzhov, S.A. 2012. Metallurgist 56: 489–493. https://doi.org/ 10.1007/s11015-012-9603-y.

[4] Bizhanov, A.M., Kurunov, I.F., Durov, N.M., Nushtaev, D.V. and Ryzhov, S.A. 2012. Metallurgist 56: 736–741. https://doi.org/ 10.1007/s11015-013-9644-x.

[5] Bizhanov, A.M., Kurunov, I.F., Podgorodetskyi, G.S. and Nushtaev, D.V. 2014. Metallurgist 58: 640–647. https://doi.org/10.1007/s11015-014-9970-7.

[6] Bersenev, I.S., Gorbachev, V.A., Sudai, A.V., Sapozhnikova, T.V. and Zinyagin, G.A. 2013. Study of metallurgical properties of pellets of hematite ore of the Bol'shetroits deposit. Teor. Tekhnol. Metall. Proizvod 3: 3–4.

[7] Matsui, Y., Sawayama, M., Kasai, A., Yamagata, Y. and Noma, F. 2003. ISIJ Int. 43: 1904–1912. https://doi.org/10.2355/isijinternational.43.1904.

[8] 13. Nomura, S., Kitaguchi, H., Yamaguchi, K. and Naito, M. 2007. ISIJ Int. 47: 245–253. https://doi.org/10.2355/isijinternational.47.245.

[9] Kim, W., Lee, Y., Suh, I. and Min, D. 2012. ISIJ Int. 52: 1463–1471. http://dx.doi.org/10.2355/isijinternational.52.1463.

[10] Allen, C. and Snow, R.B. 1955. The orthosilicate-iron oxide portion of the system $CaO-FeO-SiO_2$. J. Amer. Cer. Soc. 38: 264–272. https: //doi.org/10.1111/j.1151-2916.1955.tb14944.x.

[11] Kurunov, I.F., Bizhanov, A.M., Tikhonov, D.N. et al. 2012. Metallurgical properties of brex. Metallurgist 56: 430–437. https://doi.org/10.1007/s11015-012-9593-9.

[12] Bersenev, S., Lopatin, A.S., Belogub, E.V., Chesnokov, Yu.A., Anisimov, N.K. and Maistrenko, N.A. 2017. Metallurgical properties of sinter of oxidized iron-quartzite concentrate. Chern. Met. BNTiEI 3: 48–54.

[13] Lemos, L.R. et al. 2015. Reduction disintegration mechanism of cold briquettes from blast furnace dust and sludge. J. Mater. Res. Technol, 4(3): 278–282. http://dx.doi.org/10.1016Zj. jmrt.2014.12.002.

[14] Tomash, A.A. et al. 1999. Change in porosity of a multicomponent blast furnace charge during softening. Vestn. Priazov. Gos. Tekhn. Univ., Ser. Tekhn. Nauk, PGTU, Mariupol'.

[15] Bizhanov, A., Kurunov, I., Dalmia, Y., Mishra, B. and Mishra, S. 2015. Blast furnace operation with 100% extruded briquettes charge. ISIJ International 55(10): 175–182.

[16] Kurunov, I.F., Malysheva, T.Ya. and Bol'shakova, O.G. 2007. Study of the phase composition of iron ore briquettes with the aim of evaluating their behavior in a blast furnace. Metallurg 10: 41–46.

[17] Levin, E.M., Robbins, C.R. and McMurdie, H.F. 1964. Phase Diagrams for Ceramist, American Ceramic Society, Columbus: OH.

[18] Bowen, N.L., Schrairer, J.F. and Posnjak, E. 1933. The system $CaO-FeO-SiO_2$. Amer. J. Sci. 26: 193–284.

7.2.6 *Metallurgical properties of brex based on hematite iron ore with addition of coke breeze or EAF dust*

Of particular interest is the study of the metallurgical properties of brex based on hematite ore in combination with coke breeze and EAF dust. The properties of the brex of the following compositions were studied:

1. Brex No. 1: Ore 84%, coke breeze 16%, bentonite 0.5% (over 100%), Portland cement 6% (over 100%);
2. Brex No. 2: EAF dust 90%, coke breeze 10%, bentonite 1% (over 100%), Portland cement 6% (over 100%);
3. Brex No. 3: Ore 65%, EAF dust 20%, coke breeze 15%, bentonite 1% (over 100%), Portland cement 6% (over 100%).

For elemental analysis of samples, a set of analytical methods was used: atomic emission with inductively coupled plasma and infrared absorption to determine S and C. Atomic emission analysis of samples was performed on an iCAP 6300 spectrometer (Thermo Electron Corporation, USA, with radial plasma observation). The optimized design of the spectrometer allows simultaneous measurement of any analytical line in the range from 166 to 847 nm with high resolution. The device is equipped with a new solid-state semiconductor detector and a powerful high-performance optical system (Echelle optical system).

The device also uses an improved semiconductor solid-state detector with high sensitivity and resolution—a Charge Injection Device (CID). The spectrometer is controlled by the ITEVA computer program in Russian. When determining by the atomic emission method of analysis with inductively coupled plasma, the samples were introduced into the discharge in the form of solution aerosols. The composition of the solutions should adequately reflect the composition of the initial samples for the elements to be determined and not interfere with the correct measurement of their analytical signals.

Therefore, special attention was paid to the operation of decomposition of solid samples in order to achieve a complete (100%) transfer of the determined elements from the initial sample to the analyzed solution, to prevent contamination of the final solution with the determined elements and to achieve the minimum acidity and salinity of the final solution. Taking into account these requirements, a method for dissolving dust and cinders was developed, including sequential processing of weighed portions of analyzed samples with nitric (diluted 1:1) and perchloric (concentrated) acids, evaporation of solutions to wet salts, dissolution of wet residues with distilled water and dilution of solutions until the total concentration is reached, not exceeding 400 µg/ml.

Sulfur and carbon contents were determined by infrared absorption on a CS-230IH analyzer (LECO, USA). The principle of operation of the analyzer is based on the combustion of samples of materials placed in special ceramic crucibles in an induction furnace and subsequent measurement of the content of C and S in gaseous CO_2 and SO_2 by infrared absorption.

Microscopic studies were carried out on polished thin sections in reflected light. The phase analysis of brex was carried out on a Leica DM IL inverted laboratory microscope with HC high-resolution optics manufactured by Leica Microsystems (Germany).

Brex No. 1. The bulk of ore minerals is represented by hematite (Fe_2O_3), less often intergrowths of hematite with magnetite (Fe_3O_4). Iron ore concentrate silicates are most often observed in intergrowths with iron ore minerals. Coke breeze is evenly distributed throughout the entire volume of the brex (Fig. 7.35).

All components in the original brex (ore, silicate and coke) are separated by cement, and therefore the strength of 200 kgF/cm² is apparently related to the binding effect of the cement itself.

Table 7.30. Results of analysis of the chemical composition of brex.

Element	Content, % mass		
	Brex No. 1	Brex No. 2	Brex No. 3
Al	0,74	0.69	0.66
C	10.0	5.8	9.3
Ca	2.4	7.2	4.5
Cr	0.063	0.21	0.11
Cu	0,0026	0.12	0.043
Fe	44.3	29.7	37.4
K	0,10	0.76	0.65
Mg	0,21	0.99	0.52
Mn	0,015	1.1	0.43
Na	0,43	1.1	0.77
Ni	0.008	0.025	0.010
Pb	0,012	0.64	0.27
S	0.22	0.69	0.43
SiO_2	17.0	-	-
Ti	0,040	0.047	0.046
V	< 0.001	0.007	0.003
Zn	< 0.05	7.4	3.0

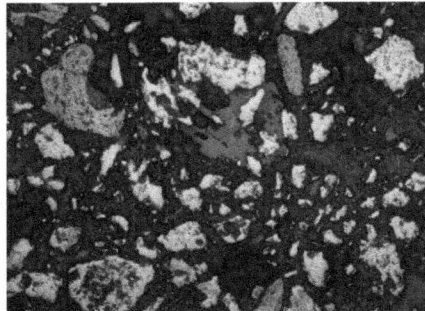

Figure 7.35. Fragments of hematite grains are white, silicates in intergrowths with hematite are gray. X500 (all drawings of the samples were made from the polished surface of the brex and filmed in reflected light; in all the pictures, the brex is based on cement, and the coconut fines stand out against the background of cement in gray-yellow shades).

Brex No. 2. The sample consists of large volumes of the smallest grains of wustite dust strengthened by silicate glass phase (Fig. 7.36).

The optical examination did not reveal minerals of natural origin (hematite, magnetite, silicates) in brex. Separate blocks of wustite-silicate dust in places of accumulation of coke breeze fragments are metallized (Fig. 7.36). Wüstite-silicate compositions in brex are different in phase composition and microstructure. Often, in the total mass of wustite, granular accumulations are observed, which, like blocks, consist of wustite in a glass phase.

An increase in the strength of finished brex was established to 360 kgF/cm², which is apparently due to the presence of dense wustite-silicate metallized compositions in their composition, the contacts between which strengthen the brex as a whole.

Brex No. 3. Fragments of natural hematite and its intergrowths with magnetite and concentrate silicates prevail on the polished surface of the brex. The dust is mainly represented by a globular composition of wustite with a glass phase (Fig. 7.37).

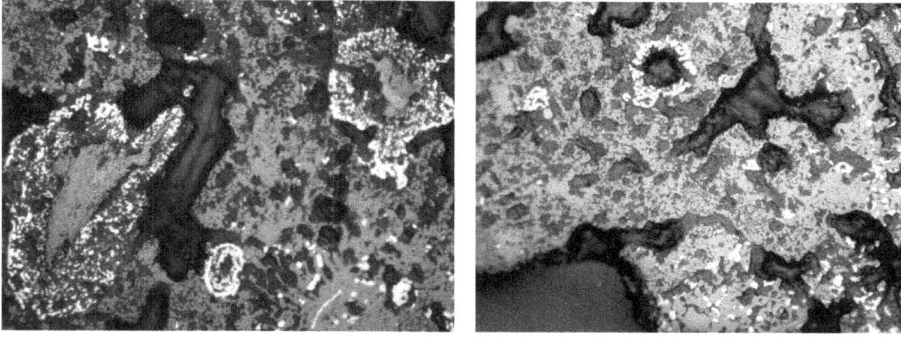

Figure 7.36. Different sections of the brex based on the EAF dust. Masses of wustite dust - gray, hardened by silicate glass phase - dark gray; the metal phase around the pores and debris of coke breeze is white, x500.

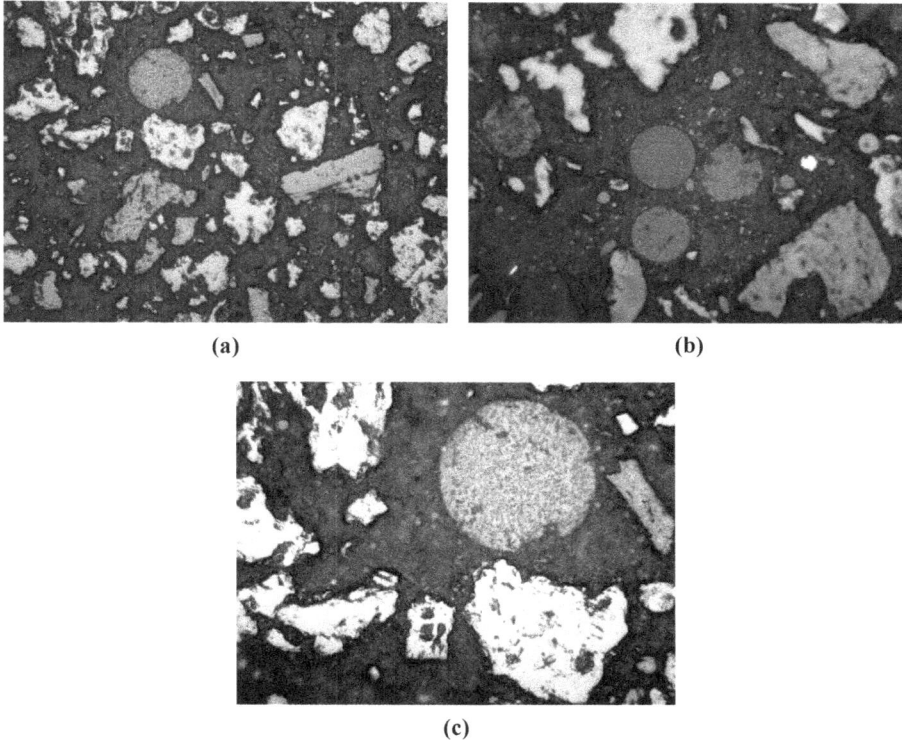

(a)

(b)

(c)

Figure 7.37. Scattered fragments of ore and technogenic components of the brex, similar to those of the brex of compositions No. 1 and No. 2. Hematite - white, wustite globules with glass phase - dark gray; (a) and (b) - magnification 500, (c) - 900.

Compared to brex with iron ore concentrate and EAF dust, a notably larger amount of coke dust is observed in brex from a mixture of their charges.

Structurally, sample No. 3 is similar to No. 1, in the composition of which there are no direct contacts between the phase components, and then cement remains the carrier of the strength of the brex.

Brex were subjected to thermal resistance tests in two versions of holding over a melt heated to 1450°C and 1250°C. Both types of tests were carried out using a Tamman furnace (Fig. 7.38). In the first case, the holding time did not exceed 3–5 minutes. The purpose of this simulation was the thermal stability of the brex to a sharp temperature drop. In the second variant, an attempt was made to simulate the behavior of the brex on the bottom of a metallization furnace, or for a brex based

Figure 7.38. Tamman oven for thermal testing of brex.

on hydro-ore, its behavior under conditions of freezing of the charge in a blast furnace. Duration of keeping the brex at such temperatures is 15–20 minutes. We were primarily interested in the qualitative behavior of the brex at such heating. No corresponding atmospheric modeling has been performed.

In the Tamman furnace (in a graphite crucible), a blast furnace slag melt heated to 1450°C was induced. Slag composition:

MgO	Al_2O_3	SiO_2	S	CaO	MnO	FeO	TiO_2	R_2O
7.29	6.40	39.70	0.87	43.40	0.31	0.22	0.15	1.47

The atmosphere is airy. After complete melting of the slag, the Brex samples were suspended above the melt mirror for 2–5 minutes, and then immersed in the melt. All tested samples withstood sharp heating from room temperature to 1450°C.

Brex No. 1. It was not preliminarily dried. The test duration is 5 minutes. Brex withstood the shock without cracking. After complete melting, the crucible with the brex melted in the slag was removed and drained (Fig. 7.39).

Brex No. 2. It was not dried earlier. The test lasts 9 minutes. Brex withstood the shock without cracking. The melt temperature varied in the range from 1406 to 1422°C. The sample began to crack as it dipped into the oven. It did not sink into the melt and was drowned forcibly. At the 8th minute, intense boiling was observed, which stopped after a minute. The crucible was removed and drained (Fig. 7.40). The same sample was tested by us after preliminary drying at a temperature of 100°C. The melt temperature is 1450°C. Test duration is 5 minutes. The sample is not cracked. Immersed in the melt independently, causing active mixing. Came up in 2 minutes. Complete melting at a temperature of 1490°C.

Figure 7.39. Brex No. 1 after melting.

Figure 7.40. Brex No. 2 after melting.

Figure 7.41. Brex No. 3 after melting.

Figure 7.42. Brex sample in a crucible.

Brex No. 3. It was not preliminarily dried. The test lasts 4 minutes. Brex withstood the shock without cracking. The melt temperature varied in the range from 1490°C to 1499°C. In the melt, it split into two fragments. The crucible was removed and drained (Fig. 7.41).

In the second series of tests, the brex samples were kept above the melt mirror without immersion in it. After 15 minutes of exposure, the samples were removed from the oven (Fig. 7.42).

The behavior of the brex was studied when they were kept for 15 minutes at a temperature of 1250°C. Then the samples were removed and cooled with liquid nitrogen (Fig. 7.43).

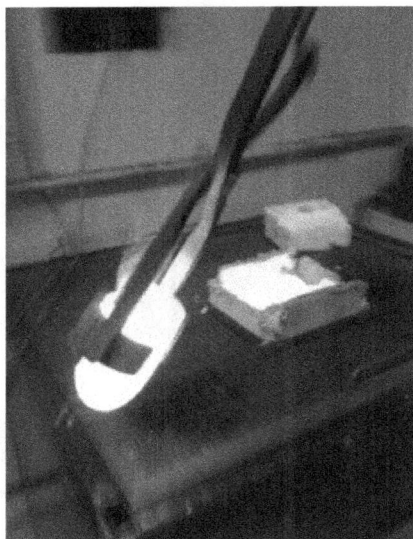

Figure 7.43. Removing the brex from the laboratory oven and cooling with liquid nitrogen.

Figure 7.44. Brex No. 1.

Brex No. 1. Temperature range 1240–1270°C (Fig. 7.44).

Brex No. 2. Temperature range 1250–1265°C. The test lasts 12 minutes. Active gas evolution (Fig. 7.45)

Brex No. 3. Temperature range 1264–1289°C. The test lasts 15 minutes. Intense gas formation (Fig. 7.46).

The reducibility of the brex was determined using the standard method used (GOST 28658-90, ISO 7215-85), which provides for continuous weighing of the sample during its isothermal reduction at 900°C with carbon monoxide. The sample was heated and cooled after reduction in a nitrogen atmosphere. The degree of reduction was calculated from the weight loss of the sample, taking into account the mass fraction and oxidation of iron in the initial samples. The recovery curves are shown in Figs. 7.47–7.49.

Figure 7.45. Brex No. 2. Gassing during brex aging.

Figure 7.46. Brex No. 3. Gassing during brex aging.

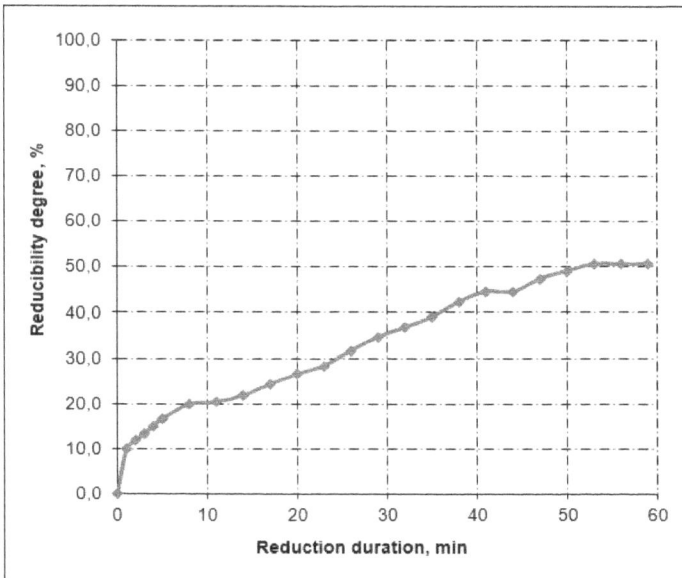

Figure 7.47. Reduction curve brex No. 1.

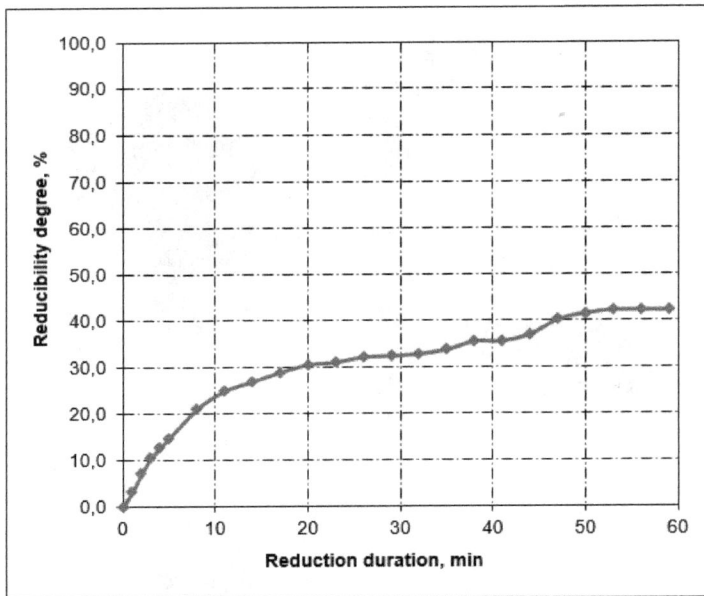

Figure 7.48. Reduction curve of brex No. 2.

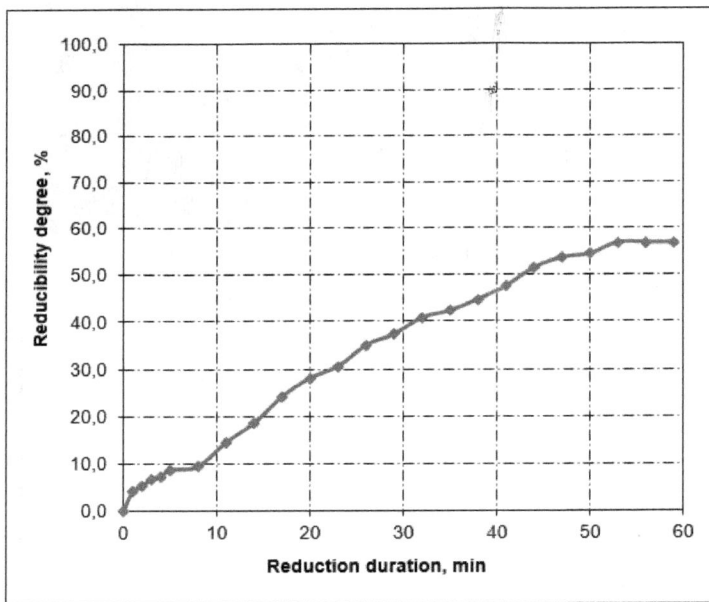

Figure 7.49. Reduction curve of brex No. 3.

The weight loss of the specimens in these tests is not only due to the reduction of iron oxides, but also because of dehydration of cement stone and calcium hydroxide (brex No. 3). For this reason, the calculated value of the degree of recovery in the brex No. 3 was higher.

Hot strength was determined as the strength of the restored samples according to the method for determining the cold strength of briquettes and on the same equipment. The results are shown in Table 7.31.

Table 7.31. Hot strength of brex.

Brex	Compressive strength kgF/cm²
Brex No. 1	53,2
Brex No. 2	47,9
Brex No. 3	X*

* quantity not measured.

Evaluation of the strength of recovered samples of brex from hydro-ore showed that the strength of brex No. 1 dropped by almost half, while brex No. 2–4 times. Such a decrease in strength, significantly exceeding the decrease in the strength of the brex as a result of their heating in a reducing atmosphere to 1150°C, is explained by the decomposition of the cement bond in the brex and their insufficient residence time at a temperature of 750–800°C for the formation of a strengthening matrix from iron-calcium silicates and the resulting metallic iron framework. In the first case, however, the formation of such a matrix in the brex body practically compensated for the loss of strength due to the disintegration of the cement bond. As a result, the assessment of the strength of the restored samples showed only a slight decrease in strength compared to the original samples.

The results of testing the metallurgical properties of brex based on hydro-ore with the addition of EAF dust and coke breeze showed that:

The hot strength of brex meets all the requirements of metallurgical processes, and in some cases even exceeds the hot strength of the sinter.

The composition based on hydro-ore (No. 1 and No. 3) can be recommended for use as a component of the blast furnace charge.

Compositions No. 1 and No. 36 can be recommended as an EAF charge components.

7.2.7 *Metallurgical properties of brex based on magnetite iron ore with addition of coke breeze*

Stiff extrusion briquettes of four compositions based on magnetite iron ore concentrate with the addition of coke breeze and carbon-free ore briquettes with different types of combined binder were investigated as experimental samples. Figures 7.50 and 7.51 show the appearance of a sample of a mixture of iron ore concentrate and coke breeze. Granulometric composition of concentrate particles - 99% with sizes less than 0.6 mm; for coke breeze - 96% of particles with sizes less than 2.36 mm.

Brex compositions and their physical parameters are given in Table 7.32. Table 7.33 shows their chemical composition.

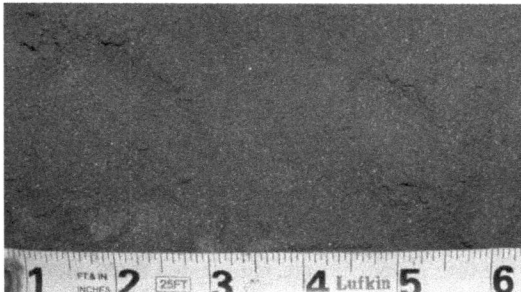

Figure 7.50. Magnetite iron ore concentrate sample.

Figure. 7.51. Coke breeze samples.

Table 7.32. Brex compositions, densities and physical parameters of extrusion.

Sample No.	No. 01	No. 02	No. 03	No. 04
Iron Ore Concentrate	93.0%	94.0%	86.0%	87.0%
Coke Breeze			7.0%	7.0%
Bentonite	1.0%		1.0%	
PCM 500 Cement	6.0%		6.0%	
Hydrated Lime		3.0%		3.0%
Molasses		3.0%		3.0%
Brex Diameter (mm)	19.0	19.0	19.0	19.0
Extrusion WBM	8.5%	6.4%	8.5%	6.8%
Compacted Density, g/cm³	3.44	3.30	3.31	3.28
Cure Temperature °C	22	22	22	22

Table 7.33. Chemical compositions of brex.

Brex No.	Al	Ca	Fe	Mg	Mn	Nb	Ti	V
01	0,74	1,64	60,6	3,29	0,34	0,023	0,51	0,075
02	0,59	1,38	59,9	3,71	0,34	0,020	0,50	0,075
03	0,78	2,42	55,0	3,07	0,32	0,016	0,50	0,069
04	0,63	1,12	56,3	3,60	0,32	0,020	0,47	0,070

Brex No.	K	Na	P	Si/SiO_2	C/CO_2	S/SO_3	Ta
01	0,051	0,046	0,059	1,3/2,78	0,28/1,03	0,24/0,60	< 0,005
02	0,100	0,025	0,059	0,93/1,99	1,5/5,51	0,28/0,70	< 0,005
03	0,051	0,039	0,049	1,51/3,23	6,78/24,8	0,30/0,75	< 0,005
04	0,120	0,028	0,045	1,02/2,18	7,16/26,3	0,27/0,68	< 0,005

In the samples under study, three types of binders (their combination) were used:

- binders of volumetric interaction (cement)
- film interaction binders (molasses, bentonite)
- binders in which the binding properties are manifested after chemical interaction between its components or components and the briquetted material (slaked lime).

Samples No. 01 and No. 03 are made using cement and bentonite as a binder mixture; samples No. 02 and No. 04 are made using molasses and slaked lime as a binder mixture.

When using binders of volumetric (matrix) interaction, bond-bridges are formed throughout the body of the briquette, i.e., particles of material are in a continuous network of binder. The common component of briquettes No. 01 and No. 03 is the presence of Portland cement in their composition.

A film binder (in this case, molasses) is used to form bonds, which are usually created during drying and aging of the agglomerated material, and the effectiveness of this type of binders is determined by the surface area of the wetted particles of the briquetted material.

In the manufacture of the studied brex in two samples No. 02 and No. 04 molasses was used (3% of the charge materials). Molasses is a waste product from sugar beet processing and contains up to 60% of unrecovered carbohydrates. It consists of 20–25% water, about 9% organic nitrogenous compounds (mainly amides), 58–60% of residual carbohydrates, mainly sucrose and raffinose. In addition, molasses contains up to 7–10% ash (minerals). In the absence of an opportunity to investigate the initial charge materials, it can be assumed, according to literature data, that molasses of the above-mentioned composition was used.

When using slaked lime, the strength of the briquettes is determined by its interaction with air carbon dioxide and the simultaneous recrystallization of calcium oxide hydrate. When hardening mineral binders, the setting time is essential—i.e., the period during which there is a progressive increase in strength.

Carbon-containing additives: in addition to the iron-containing material, carbon-containing components may be included in the briquettes, which contribute to the reduction of iron oxides in the body of the briquette and the subsequent carburization of iron. The basic principle of operation of briquettes of this class is the direct reduction of iron oxides with carbon due to the numerous and highly developed surface contacts of these components inside the briquette. In the studied samples No. 03 and No. 04, coke breeze is used as a charge component. Adding a carbon-containing material to the charge for briquetting ensures the production of complex, self-reducing briquettes, the use of which reduces the consumption of the reducing agent in the processes of obtaining the primary metal. Samples No. 01 and No. 02 do not contain coke breeze, so there is no potential to use the carbon potential of these samples, however, the economic effect of their use is achieved by improving the gas dynamics of the process and increasing the mass fraction of iron in the composition of the charge.

To determine the compressive strength, the brex were placed one at a time on the bottom plate of the sample holder of the test press and a constantly increasing load was applied until they broke (Fig. 7.52). The load is registered on the indicator using a strain gauge. Table 7.34 shows the value of the load at the time of the beginning of destruction (cracking) of each brex.

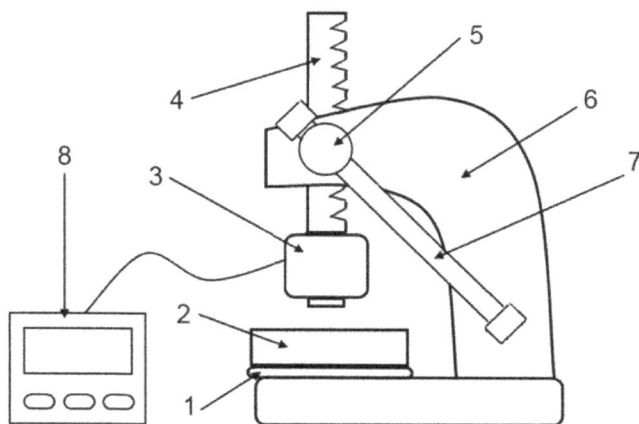

Figure 7.52. Diagram of a lever press. 1 - stand; 2 - bowl; 3 - strain gauge sensor; 4 - stock; 5 - pinion shaft; 6 - body; 7 - lever; 8 - digital indicator.

Table 7.34. Compressive strength of original brex, kg/brex.

Sample No.	The force that one brex can withstand before breaking (cracking)						
01	57						
02	204						
03	128	169.6	127.5	181.4	125.8	196	167
04	50	67	43				

In the study, the best performance was shown by sample No. 03 is partly due to the properties of the cement-bentonite binder and is the result of the formation of coagulation structures in the cement-bentonite-water system, leading to a modification of the properties of the binder.

Sample No. 01 was supposed to demonstrate a ductile-plastic nature of fracture and a decrease in the likelihood of brittle fracture, in which fines are formed (similar to sample No. 03), but after the first load it collapsed.

The use of molasses and lime as a binder in sample No. 04 did not demonstrate high resistance to vertical loading during testing. Although, there is a possibility that this sample will show better results under controlled parameters of preliminary hardening aging.

The strength of the brex for dropping after a day, three days and a week of aging are given in Table 7.35.

For a more detailed study and increasing the significance of the conclusions, it is necessary to take into account the dynamics of changes in the density, porosity and strength of the brex in the process of their hardening aging immediately after production. As shown earlier, the graph of the strength gain of the brex has a pronounced local maximum on the third day, followed by softening during the next day and ending with a further increase in strength. In this case, the strength of the brex immediately before softening is about 88% of the strength of the brex after a week of hardening aging. The nature of the change in open porosity practically repeats the nature of the change in strength, with the exception of the first days of hardening. The decrease in porosity at this time is associated with the swelling of bentonite filling the pore space.

Table 7.35. Drop-test results.

1-Day Drop Test (CPFT* 4.75 mm)				
Brex Wet Basis Moisture	8.4%	4.0%	8.4%	4.8%
3 Drops	10%	17%	11%	32%
5 Drops	16%	26%	18%	38%
7 Drops	23%	32%	25%	47%
3-Day Drop Test (CPFT 4.75 mm)				
Brex Wet Basis Moisture	6.7%	1.7%	6.9%	1.7%
3 Drops	6%	3%	8%	14%
5 Drops	11%	4%	13%	19%
7 Drops	16%	5%	19%	25%
7-Day Drop Test (CPFT 4.75 mm)				
Brex Wet Basis Moisture	1.7%	0.6%	1.7%	1.2%
3 Drops	5%	1%	6%	2%
5 Drops	8%	2%	9%	3%
7 Drops	11%	3%	14%	4%

* CPFT stands for Cumulative Percent Finer Than

For a quantitative assessment of the intensity of the reduction process, the indicator "degree of reduction" is used, which is the ratio of the amount taken away during the reduction of oxygen to all the oxygen associated with iron oxides.

The method for determining reducibility consists in continuous monitoring of the change in sample mass during reduction with hydrogen at a temperature of $800 \pm 10°C$ and the determination of the degree of reduction based on the results of chemical analysis of the original sample and the loss of oxygen mass during reduction.

Sampling and preparation of samples is carried out in accordance with GOST 26136-84. Sample size 10 ... 15 mm. To improve the accuracy of the experiment, duplicated studies are carried out from one sample of iron ore material.

Determination of reducibility is carried out on a continuous weighing installation as shown in the Fig. 7.53.

Hydrogen, carbon monoxide, as well as their mixtures with water vapor, carbon dioxide and neutral gases can be used as a reducing agent. In this work, the reducing gas is a technical grade A 99.99% hydrogen according to GOST 3022-80. Gas consumption 2 l/min.

At the beginning of the experiment, the furnace heated to a temperature of $800 \pm 10°C$ is purged with nitrogen from a cylinder, the flow rate is 2 l/min, these actions are necessary in order to displace the oxygen in the air from the furnace in order to avoid the formation of an explosive mixture when hydrogen is supplied to the furnace. The flow rates of hydrogen and nitrogen are controlled by rheometers.

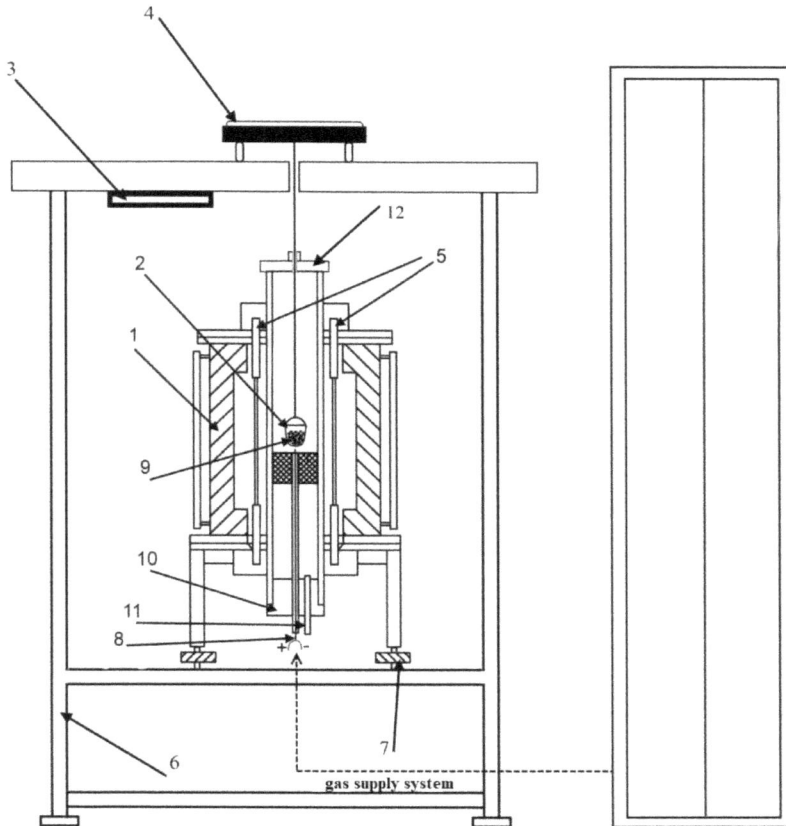

Figure 7.53. Installation for determining the reducibility of iron ore materials. 1 - silite furnace; 2 - basket; 3 - table cooling fan; 4 - electronic scales; 5 - heaters; 6 - stand; 7 - adjusting screws; 8 - thermocouple; 9 - sample; 10 - plug; 11 - fitting; 12 - cover.

The pre-weighed sample in the basket is suspended from an analytical balance (Sartorius LP-1200S) and kept in a nitrogen flow until the balance reading stops fluctuating. The sample basket is in the reaction tube of the oven. The temperature in the working space of the tube is measured with a thermocouple 5 mm below the bottom of the basket. Then hydrogen is fed into the furnace at a flow rate of 1 l/min and nitrogen is shut off. The weight loss of the sample is recorded using an analytical balance: the first 5 minutes every minute, the next 45 minutes every 3 minutes or continuously. After that, the sample is heated from 810 to 1000°C and, after reaching the desired temperature, is kept in a stream of nitrogen (flow rate 2 l/min).

At the end of the experiment, the gas supply is stopped and the basket is removed. The sample to be removed has a temperature of 600 ... 700°C. An important condition for carrying out the experiment is the exact observance of the constancy of temperatures and the flow rate of the reducing gas.

For the reliability of the results, at least two experiments are carried out for each sample, if the difference in the degree of recovery during the experiment does not exceed 5%, then the data are averaged, if the difference is > 5%, then another experiment is carried out and the data are averaged.

To calculate the degree of reduction of a sample, it is necessary to calculate the amount of oxygen bound to iron in a given sample.

$$Fe_{tot} = 0.01 \cdot Fe_{tot} \cdot P$$

$$FeO = 0.01 \cdot FeO \cdot P$$

$$Fe^{2+} = 0.78 \cdot FeO$$

$$O(Fe^{2+}) = FeO - Fe^{2+}$$

$$Fe^{3+} = Fe_{tot} - Fe^{2+}$$

$$O(Fe^{3+}) = 0.429 \cdot Fe^{3+}$$

$$O_{tot} = O(Fe^{2+}) + O(Fe^{3+})$$

where:

Fe_{tot}, FeO – the content of iron and iron oxide in the original sample according to the data of chemical analysis, % of the mass;
Fe_{tot}, FeO – the amount of oxide and iron oxide in the original sample, g;
Fe^{2+}, Fe^{3+} – the amount of ferrous and ferric iron in the original sample, g;
$O(Fe^{2+})$ and $O(Fe^{3+})$ – the amount of oxygen associated with ferrous and ferric iron in the original sample, g;
O_{tot} – the amount of oxygen included in the iron oxides in the original sample, g;
P – sample weight, g;

The degree of reduction is calculated as the ratio of the loss in the mass of the sample to the total mass of oxygen bound to iron in the test sample:

$$RI = \frac{\Delta P}{O_{tot}} \cdot 100\%$$

where:

ΔP – loss of sample weight, g;
RI – reduction rate, %.

Laboratory tests were performed with all samples, but Sample No. 01, No. 02 and No. 04 turned out to be unstable and destroyed after thermal loading at the first stage of the experiment.

Sample No. 03 showed high performance properties, respectively, an experiment to determine the degree of recovery and reactivity was carried out with this sample.

The results are shown in Fig. 7.54 and in the Table 7.36.

Table 7.36. Mass loss of the sample No. 03 under different reduction modes.

Time, min	Weight loss of sample with basket*, g (Hydrogen)	Weight loss of sample, g (Nitrogen)	Weight loss of sample, g (Nitrogen)
0	29,066		
1	28,62		
2	28,457		
3	28,303		
4	28,145		
5	28,013		
8	27,609		
11	27,222		
14	26,844		
17	26,487		
20	26,137		
23	25,812		
26	25,488		
29	25,195		
32	24,906		
35	24,646		
38	24,395		
41	24,159		
44	23,941		
47	23,711		
50	23,544		
53	23,346		
53		23,346	
56		23,301	
59		23,315	
62		23,321	
65		23,322	
68		23,32	
71		23,316	
74		23,302	
77		23,285	
80		23,254	
83		23,245	
83			23,245
86			23,215
89			23,188
92			23,17
95			23,139
98			23,102
101			23,058

* Basket mass 61.76 g

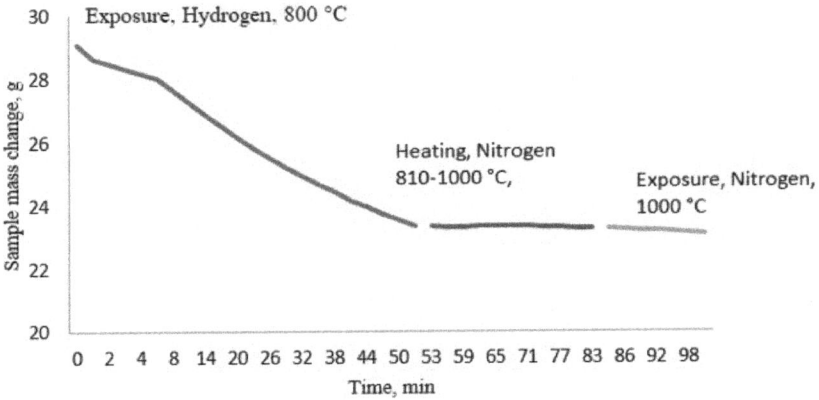

Figure 7.54. Change in weight of sample No. 03 on reduction.

Figure 7.55. Brex sample No. 03 appearance after reduction.

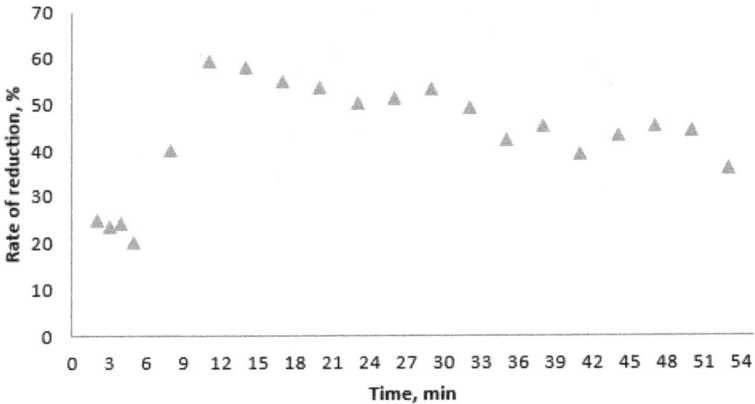

Figure 7.56. Sample No. 03 reduction rate.

The sample mass stabilized after 50 minutes of exposure in hydrogen. Starting from 53 minutes, the process of heating and holding the sample at 1000 degrees, the mass stabilized.

The appearance of the brex sample No. 03 after reduction (no destruction and melting when heated) is shown in the Fig. 7.55.

Brex No. 03 showed good reducibility; at the stage of holding in a stream of hydrogen, there is an active decrease in mass, which indicates an actively proceeding process of reduction by gas. The avalanche-like increase in the rate of reduction of briquettes is explained by the creation of a good contact of oxides with carbon (Fig. 7.56). At this stage, the restoration is accompanied by the formation of a framework that ensures the strength of the sample further during the process.

The brex under study includes a carbon-containing component (coke breeze), which itself is the source of the formation of reducing gases. Accordingly, at the next stage (heating up to 1000 in a neutral atmosphere), the brex are reduced, mainly due to the carbon of the coke. The degree of development of this process depends on the size of the coke particles, the number of contacts of the reducing agent with iron oxides, that is, in brex containing fine materials, the solid-phase reduction reaction proceeds more efficiently.

To assess the development of the processes of reduction of iron oxides with carbon, which is part of the brex, an experiment of isothermal heating of sample No. 03 was conducted, since the briquette, which includes a carbon-containing component, is itself a source of formation of reducing gases.

The experiment is carried out on a continuous weighing installation similar to the installation for determining the reducibility of iron ore materials.

A pre-weighed sample in a basket is suspended from an analytical balance (Sartorius LP-1200S), placed in an oven heated to a predetermined temperature (1000°C) and kept in a nitrogen flow.

Change in mass of sample No. 03 with isothermal heating in an inert atmosphere is shown in Fig. 7.57. The appearance of the brex after isothermal heating in an inert atmosphere is shown in Fig. 7.58.

The loss in mass during the experiment indicates an active course of solid-phase reduction reactions inside the brex.

A characteristic feature of the iron ore concentrate used in the experiments is a high content of magnesium oxide (magnesia, MgO) mass fraction of 5–9%, which is due to the uniqueness of the iron ore of the Kovdor deposit.

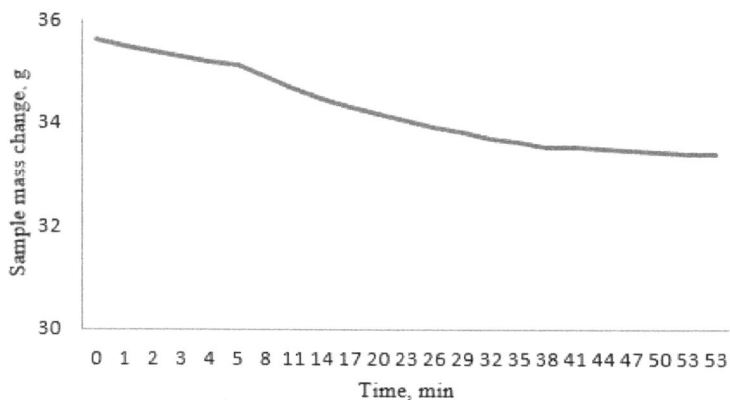

Figure 7.57. Change in mass of sample No. 03 with isothermal heating in an inert atmosphere.

Figure 7.58. Sample briquette No. 03 after isothermal heating in an inert atmosphere.

Brex with the use of such a concentrate in their composition require more careful control of the phase composition, because their composition provides conditions for the formation of solid solutions in the $CaO-MgO-FeO-SiO_2$ system. The formation of a difficult-to-reduce iron-calcium-magnesium silicate melt will occur at a higher temperature, and the reduction of iron from such a melt will be possible only with solid carbon.

Thus, it can be concluded that:

1. Cold strength of briquette No. 03 is not a limiting factor when using these types of briquettes in mining-type metallurgical units (for example, in a blast furnace). This is explained by the properties of the cement-bentonite binder and is the result of the formation of coagulation structures in the cement-bentonite-water system, leading to a modification of the properties of the binder.

2. Sample No. 01 was supposed to demonstrate a ductile-plastic nature of fracture and a decrease in the likelihood of brittle fracture, in which fines are formed (similar to sample No. 03), but after the first load it collapsed.

3. The use of molasses and lime as a binder in sample No. 04 did not show high resistance to stress during testing. Although, there is a possibility that this sample will show better results under controlled parameters of preliminary hardening aging.

4. It was found that the briquette of composition No. 03 is most intensively reduced in the first hour of the experiment at T = 800°C. At this stage, the restoration is accompanied by the formation of a framework that ensures the strength of the sample during the further process.

5. It was found that sample No. 03 is itself a source of reducing gases. At the stage of isothermal heating in a neutral atmosphere, reduction occurs mainly due to the carbon of the coke. The degree of development of this process depends on the size of the carbonaceous material.

6. It was found that when heated in sample No. 03, a metal frame is formed in the surface layers and in the entire volume of the briquette, the formation of a metal frame is the reason for maintaining the shape of the brex during high-temperature heating.

7. It was seen that according to the results of experiments to determine the degree of reducibility of sample No. 03, it can be assumed that when using it as a charge component in pig iron smelting, coke savings of about 40 kg/t of pig iron can be achieved (preliminary assessment, calculation of technical and economic indicators of blast furnace smelting is required).

7.2.8 Efficiency of using washing briquettes made of mill scale

One of the ways of using mill scale briquettes in blast furnace production is their use as a washing material. Evaluation of the effectiveness of the use of scale briquettes for washing the hearth of blast furnaces was carried out in [1].

A pilot batch of washing briquettes was made from various types of scale, supplied for briquetting without averaging from different accumulation and temporary storage sites. In addition to the scale used, also a significant amount of metal concentrate fraction 0–5 mm, obtained during the processing of converter slag. The briquettes produced had a cylindrical shape of 120×90 mm and were accumulated and stored in an open warehouse before being melted in blast furnaces. Briquettes were used as standard washing material in two blast furnaces with a volume of 2000 m^3.

Evaluation of the effectiveness of the use of scale briquettes was carried out by comparing the results of furnace operation under comparable conditions when using briquettes and when using a washing sinter. The results were reduced to the same conditions using the coefficients of the factorial analysis. The chemical composition of the used briquettes from scale and flushing sinter is shown in Table 7.37.

Table 7.37. Chemical composition of flushing sinter and scale briquettes.

Material	Fe_{tot} %	FeO %	Fe_2O_3 %	CaO %	SiO_2 %	MgO %	Al_2O_3 %	CaO/SiO_2
Scale briquettes	50.9	36.2	32.5	13.2	8.52	1.25	3.06	1.55
Washing sinter	60.2	44.3	36.8	7.05	8.57	1.13	0.78	0.82

Table 7.38. Indices of operation of blast furnace No. 3 when using a flushing sinter (Option A) and scale briquettes (Option B).

Blast furnace performance	Option A		Option B
	Period 1 (9 days)	Period 2 (9 days)	Period 3 (9 days)
Productivity, t/day	4422	4335	4248
Coke rate, kg/t	414	423	423
Sinter, kg/t	1494	1417	1452
Washing sinter, kg/t	5	32	0
Washing briquettes, kg/t	0	0	10
Pellets, kg/t	97	192	196
Converter slag, kg/t	2	15	21
Fe content in ore part of charge, %	59,16	59,02	58,92
Natural gas, m/t	104	99	102
Blow flow rate, m^3/min	3337	3311	3316
O_2 content in blow, %	28,4	28,7	28,7
T_{blow} °C	1103	1104	1109
Pressure under the top, kPa	138	138	137
Hot metal composition, %: [Si]	0,56	0,63	0,63
[Mn]	0,08	0,10	0,10
[S]	0,017	0,013	0,015
[P]	0,06	0,06	0,06
Slag yield, kg/t	318	330	330
Slag basicity (CaO/SiO_2)	1,01	1,00	0,98
ΔP_u kPa upper drop	16	17	17
ΔP_l, kPa bottom drop	119	117	117
Gas permeability index (G_n *10^3)	21092,5	21188,4	21367,5
Present productivity, t/day	4422	4565	4507
Present Coke rate, kg/t	414	388	393

Briquettes made of cement-bonded scale were used for flushing the hearth on blast furnaces No. 3 in 450 tons and No. 4 in 200 tons, 5 days. The results of blast furnace operation during the periods of use of flushing materials are shown in Tables 7.38 and 7.39.

The gas permeability index given in the tables was calculated using the equation:

$$G_n = \frac{(V_{tg})^2}{\Delta P_l}$$

where V_{tg}, – tuyere gas output, calculated from the flow rate and composition of the blast and the flow rate of natural gas, m/min; ΔP_l – bottom pressure drop.

Table 7.39. Indices of blast furnace No. 4 operation when using flushing sinter (Option A) and scale briquettes (Option B).

Blast furnace performance	Option A	Option B
	Period 1 (10 days)	Period 2 (11 days)
Productivity, t/day	3908	4563
Coke rate, kg/t	432	431
Sinter, kg/t	1484	1429
Washing sinter, kg/t	18	0
Washing briquettes, kg/t	0	4
Pellets, kg/t	117	228
Converter slag, kg/t	7	14
Fe content in ore part of charge, %	58.74	59.00
Natural gas, m/t	102	107
Blow flow rate, m^3/min	3354	3655
O_2 content in blow, %	25.4	27.3
T_{blow} °C	1194	1192
Pressure under the top, kPa	133	155
Hot metal composition, % [Si]	0.73	0.66
[Mn]	0.09	0.11
[S]	0.011	0.012
[P]	0.06	0.06
Slag yield, kg/t	316	313
Slag basicity (CaO/SiO$_2$)	1.05	0.99
ΔP_u kPa upper drop	15	14
ΔP_l, kPa bottom drop	116	121
Gas permeability index (G_n * 10^3)	20003.0	24520.5
Present productivity, t/day	**3908**	**4381**
Present Coke rate, kg/t	**432**	**421**

Analysis of the results of blast furnace operation during the periods of use of briquettes from scale and washing sinter for washing the hearth showed that these materials are quite comparable in their washing properties. Both the use of flushing sinter and the use of briquettes contributed to the stable and efficient operation of blast furnaces without disturbing the gas-dynamic mode of smelting. At blast furnace No. 3, the maximum reduced productivity and the minimum reduced coke consumption were obtained during the period of increased consumption of washing sinter. The increased productivity of blast furnace # 4 and the minimum consumption of coke occurred during and after the use of the scale washing briquettes.

However, it is not possible to attribute the indicated changes in the performance of the furnaces completely to one or the other flushing material. As can be seen from Tables 7.38 and 7.39, the best performance of the furnaces, both when using a washing sinter and when using washing briquettes, was achieved during periods when the furnaces were operated with an increased (twice) pellet consumption. There were other differences in the operating conditions of the furnaces, which could not be taken into account by the coefficients of the factorial analysis.

The gas permeability index calculated from the gas-dynamic parameters shows a real improvement in the state of coke oven toterman in all furnaces. Moreover, the effect of using briquettes was higher in all cases of their use.

To conduct a deeper analysis and comparison of the washing properties of the test materials, we used such calculation criteria as a change in the degree of direct reduction (r_d) and a change in the "DMI" (Deadman index) [2].

When evaluating according to this criterion, the hourly values of the degree of direct reduction were calculated. The flushing effect was assessed by the change in the degree of direct recovery 5–6 hours after the flushing material.

The DMI index indirectly evaluates the drainage capacity of the hearth by comparing the actual carbon content in the pig iron at the outlets, depending on the surface and the contact time of the pig iron with coke in the hearth and the carbon content in the saturated state. The higher this indicator, the better the drainage capacity of the hearth.

The DMI is calculated using the formula:

$$DMI = 2 \cdot T_{hm} - 120.62 \cdot [Si] - 128.40 \cdot [F] - 155.64 \cdot [S] + 10,89 \cdot [Mn] - 389.11 \cdot [C] - 190 \cdot B_{sl} - 690.46$$

where T_{hm} – hot metal temperature, °C
[Si], [P], [S], [Mn], [C] – mass fraction of these elements in hot metal, %
B_{sl} – slag basicity (CaO/SiO_2).

The results of calculating the criteria for washing are presented in Tables 7.40 and 7.41.

Table 7.40. Results of calculation of flushing criteria for blast furnace No. 3.

Blast Furnace No. 3	Sinter washing		Briquettes washing	
	Day 1	Day 2	Day 3	Day 4
Change in the degree of direct reduction				
Before washing r_d (1)	39.0	45.4	46.5	50.5
during washing r_d (2)	50.1	48.5	49.6	53.8
$\Delta = r_d$ (2)– r_d (1)	11.1	3.1	3.1	3.3
Change in «DMI»				
before washing DMI (1)	170	201	242	201
during washing DMI (2)	227	227	265	233
Δ = DMI (2) – DMI (1)	57	26	23	32

Table 7.41. Results of calculation of flushing criteria for BF No. 4.

Blast furnace No. 4	Sinter washing		Briquettes washing	
	Day 5	Day 6	Day 7	Day 8
Change in the degree of direct reduction				
Before washing r_d (1)	39.8	40.5	37.7	41.8
during washing r_d (2)	45.3	46.5	40.0	47.2
$\Delta = r_d$ (2) – r_d (1)	5.5	6.0	2.3	5.4
Change in «DMI»				
before washing DMI (1)	178	152	162	181
during washing DMI (2)	224	178	181	193
Δ = DMI (2) – DM I (1)	46	26	19	12

Changes in the degree of direct reduction during the penetration of washing materials indicate the presence of a stable washing effect in all cases of using washing sinter and briquettes, but washing sinter had a stronger effect on the degree of direct reduction in most cases.

Also, in all cases of the use of washing materials, an unambiguous change in the calculated DMI index was observed, indicating an improvement in the state of the drainage capacity of the coke packing. According to this criterion, the effect of the use of briquettes is also comparable to the effect of the washing sinter, although in two cases it is significantly lower.

A comprehensive assessment based on the sum of the design criteria shows that the use of both types of washing materials leads to stable washing of the coke packing. However, the washing effect from the use of scale briquettes is somewhat lower than when using sinter. This is largely due to the fact that the absolute amount of the loaded washing sinter in the periods under consideration was more than two times higher than the number of loaded briquettes, which could not but affect the quality of washing the coke packing.

The choice of one of the possible schemes for the utilization of mill scale in the composition of the flushing material, requires an assessment of their energy and environmental efficiency.

The total fuel consumption for smelting pig iron from a washing sinter made from scale, and for smelting pig iron from cement-bonded briquettes, also obtained from scale and dust of ferroalloy furnaces can be compared. The chemical composition of the washing sinter and scale briquettes is shown in Table 7.42.

The results of computer modeling in a blast furnace using high-oxide sinter and briquettes from scale are presented in Table 7.43.

Table 7.42. Chemical composition of scale, washing sinter and cement-bonded briquettes obtained from mill scale, %.

Material	Fe	FeO	Fe$_2$O$_3$	SiO$_2$	Al$_2$O$_3$	CaO	MgO	MnO	LOI
Scale	72.6	53.44	44.40	0.90	0.16	0.42	0.11	0.53	0.0
Sinter	69.7	30.960	65.2	1.3	0.36	1.44	0.16	0.51	0.0
Briquettes	58.8	43.26	35.96	4.33	0.60	4.93	0.35	0.41	10.0

Table 7.43. Main indicators of blast furnace operation during melting of washing sinter (charge A) and scale briquettes (charge B).

No.	Blast furnace performance	Charge A	Charge B
1	Sinter, kg/t	1073	1070
2	Pellets	345	353
3	Converter slag, kg/t	21,5	19,5
4	High-acid sinter, kg/t	152	0
5	Briquettes, kg/t	0	172
6	Coke, kg/t	427,9	429,5
7	Natural gas consumption, m^3/t	92	92
8	Oxygen consumption, m^3/t	88	87
9	Blow rate, m^3/t	886	857
10	O$_2$ content in blow, %	28,35	29,5
9	Slag yield, kg/t	264	277
10	Slag basicity, CaO/SiO$_2$	1,015	1,015
11	Furnace capacity, t/m^2 per day	71,08	71,06

Table 7.44. Total consumption of energy carriers for the production of hot metal when using in blast furnace smelting of flushing sinter (Option A) and scale briquettes (Option B).

Process parameters	Option A	Option B
Coke, kg/t	427.9/431.5	429.5/433.1
Natural gas, m³/t	92/105.4	92/105.4
Oxygen, m³/t	88/8.7	87/8.6
Coke breeze, kg/t	14.8/14.8	0
Mixed gas, m³/t	3.2/0.8	0
Elevtricity, kWh/t	32/3.9	29/3.6
Total energy consumption, kg of fuel equivalent/t	565.1	550.7
Output of secondary energy resources, kg of fuel equivalent/t	211.7	210
Total consumption of equivalent fuel, kg/t	353.4	340.7

Note: the denominator is the consumption in kg of fuel equivalent.

Fuel costs for the production of 1 ton of washing sinter are:

- coke breeze 80 kg/t
- mixed gas – 17 m³/t
- electricity – 43 kWh

The total costs of energy carriers per 1 ton of pig iron for the considered scale recycling schemes are shown in Table 7.44.

When scale consumption (in briquettes or sinter) in blast furnace smelting in the amount of 150 kg/t of pig iron, the difference in the total consumption of equivalent fuel per 1 ton of scale used for smelting pig iron according to the options A and B is: (353.4–340.7): 0.15 = 84.67 kg/t of scale.

Scale recycling by sintering is realized with a higher fuel consumption compared to briquetting. Fuel consumption for scale sintering significantly exceeds the additional consumption of coke when melting briquettes from scale, associated with an increase in the slag yield and the presence of chemically bound water in the briquettes. This excess for the specified amount of disposed scale by fuel type is:

- conventional fuel – 12,700 tons/year;
- coke breeze – 14805 t/year;
- mixed gas – 3195000 m³/year.

Additional CO_2 emissions into the atmosphere during scale recycling through sintering are for a given amount of scale:

$$14805 * 0.86 * 44/12 + 3195 * (0.216 + 0.115) * 44/22.4 = 48762 \text{ t/year}$$

According to option B, there is no increase in CO_2 emissions into the atmosphere.

References

[1] Kurunov, I.F., Bolshakova, O.G., Scheglov, E.M. et al. 2007. Experience of BF hearth flushing with briquettes made of scale. J. Metallurg 6.

[2] Sergeant, R., Bonte, L., Huysse, K., Sergeant R. et al. 2005. Hearth management at Sidmar for an optimal hot metal and slag evacuation [P]. The 5th European Coke and Ironmaking Congress. Stockholm, 2, We1: 3.

7.2.9 De-oiling of mill scale in a magnetic field for further briquetting

In the rolling production of metallurgical enterprises, hundreds of thousands of tons of oily scale are formed, which is of limited use and is dumped into sludge collectors. Briquetting of such material is usually difficult and requires special preparation of an oily scale. Such a scale is heterogeneous in composition and can contain from 10 to 95% scale, 10 to 50% oils and from 3 to 80% water. The greatest difficulties are caused by the disposal of oily mill scale from secondary settling tanks, containing up to 20 ... 30% oils. Recycling of such scale in the sintering process requires its preliminary dehydration and removal or neutralization of oils. The methods used for de-oiling the scale are energy-consuming and environmentally unsafe due to the presence of organochlorine compounds in it. Recycling of oily scale in the sintering process is accompanied by significant emissions into the atmosphere of the strongest toxins - dioxins and furans, and requires the use of expensive gas cleaning systems at sinter plants or special technologies that prevent the formation of dioxins during sintering. In addition, oils that do not completely burn out during the sintering process shorten the life of the exhausters.

The approximate chemical composition and oiliness of the scale are given in Table 7.45.

The content of shavings and metal inclusions is less than 7%, the oil content is less than 14 mg/kg of scale, the mass fraction of moisture is not more than 5.0%. The rest of the scale (oily and with metal inclusions) is sold to external consumers. Bulk mass of mill scale is on average 2.9 ± 0.2 t/m^3 (Table 7.46).

At some factories, due to the complexity and high cost of cleaning scale from oil, it is mixed into sinter charge or briquetted with subsequent feeding into a blast furnace. Scale with a layer of 10 mm is placed on top of the sinter charge layer in front of the incendiary hearth. The combustion temperature of the sinter charge is 1250 ... 1300°C. When ignited, the oil burns out (oils burn out at a temperature of 900°C), and the combustion products are removed along with the exhaust gases. The main disadvantage of this method: a blocking layer appears on the upper layer of the sinter charge as a result of the scale dripping. Therefore, the ignition process of the sinter charge slows down and part of the oil does not have time to burn and is drawn into the charge layer. This method is valid only for fresh scale, and it does not provide environmental protection and is economically ineffective (dioxins are formed during the sintering process). The bulk of the scale goes to the sludge

Table 7.45. Chemical composition of oily mill scale.

No.	Content, %									Oiliness, g/kg
	Fe$_{tot}$	FeO	Fe$_{met}$	CaO	SiO$_2$	Al$_2$O$_3$	C	S	moisture	
1	73.2	-	5.0	-	-	-	0.19	0.009	1.0	100.0
2	71.9	-	5.5	-	-	-	0.48	0.019	2.26	150.0
3	73.5	-	3.4	-	-	-	0.31	0.017	2.0	15.0
4	72.9	-	4.0	-	-	-	0.30	0.017	1.89	20.0
5	73.3	-	3.9	-	-	-	0.25	0.015	2.1	50.0
6	73.1	64.1	-	-	0.9	-	-	-	-	-
7	69.0	29.6	-	0.30	2.08	0.44	-	-	-	-
Average	72.4	47.0	4.4	0.30	1.5	0.44	0.31	0.015	1.85	67.0

Table 7.46. Technical indicators of scale with a fraction of not more than 16 mm.

Material	Commodity moisture, %	Oil, %	Mass fraction of elements, %					
			Fe	Si	Mn	Cr	Total alloying	Total iron and alloying
Scale No. 1	2.98	0	73.3	0.31	0.52	0.38	1.21	74.51
Scale No. 2	14.22	4	71.6	0.5	0.5	0.17	1.17	72.77

ponds, in which there is a large amount of moisture and dirt, which further reduces the efficiency of its processing, in order to feed it into the blast furnace.

A method has been developed for the disposal of oil-containing scale by blowing it into a blast furnace together with shredding dust. Dust from shredding units in terms of particle size distribution is a fraction from fractions to 3 millimeters, i.e., corresponds to the parameters of experimental-industrial research on the injection of fine materials into shaft furnaces. Industrial trials were carried out in 1996 with the participation of the Institute for Iron and Steel Technology of the Technical University of the Freiberg Mining Academy, Eko Stahl GmbH in Eisenhüttenstadt, Stein Injection Technology in Gevelsberg and Carbofer Verfahrenstechnik in Meerbusch.

MISIS (Moscow Institute of Steel and Alloys, Russia) proposed a low-waste technology for injection of combined liquid fuel from oil waste and oily scale into a blast furnace. It has been established that the disposal of oily scale in a blast furnace as part of a combined liquid fuel is environmentally safe. Studies have shown that the process of obtaining this fuel, which has a sufficiently high stability and fluidity to transport it through pipelines, is efficient and easy to implement. It has been established that an injection of combined liquid fuel into a blast furnace makes it possible to reduce the consumption of iron ore agglomerated charge and coke.

In another method of disposal, a certain amount of used oil is added to the oily scale until a liquid consistency is established and the resulting mass is injected into the lance of the blast furnace. But from an economic point of view, this method is unprofitable, since there is a lack of the required amount of oil.

Each of these methods has certain disadvantages. The supply of scale to an open-hearth furnace requires the addition of coke breeze, which is a scarce raw material, the use of additional special equipment for weighing and mixing components, oil loss, the cost of removing moisture, the formation of toxic compounds is not excluded. Technologies associated with briquetting and the subsequent supply of briquettes to the blast furnace and arc steel furnace are also accompanied by the loss of oil, the cost of additional equipment, do not exclude the formation of poisonous gases, and most importantly, briquettes at temperatures above 300°C decompose into dusty materials that are thrown with flue gases.

The Uralmekhanobr Company (Russia, [1]) tested in laboratory conditions, a method for cleaning scale from oil by burning oil without air access in a rotating drum, based on pyrolysis. The oily scale is heated to a temperature not exceeding the metal oxidation temperature (300 ... 400°C). The waste within the plant is heated in a controlled non-oxidizing environment. Contaminants in the form of oil are volatilized into the flue gas cleaning system, where the oil is collected.

The same company tested a method of oil removal by flotation. In this way, it was proposed to clean oil-containing effluents. Oil-containing effluents can be degreased at local treatment plants after separation in settling cones by acidification with sulfuric acid or spent pickling solutions. From the reactor, the emulsion is fed to the skimmer, from where the foam is removed by scrapers and transferred to an oil trap, and then to regeneration or incineration. After neutralization, the aqueous phase enters the settling tank. Mass fraction of organic substances in purified water after this kind of conditioning is 100 ... 150 mg/l. After dilution, the water is sent to the preparation of a new batch of working solutions, for example, emulsions. Spent oil-containing waste can also be used as fuel for boiler plants. This method does not provide high water purification, and also requires a reagent and, as a result, the formation of additional sludge.

In Japan, Germany, and the USA, oily scale is treated with acid-base compounds that dissolve the oil.

Advantages:

• high degree of purification.

Flaws:

- loss of oil occurs;
- the environment is polluted, acid-base compounds together with oil are discharged into the drain, since the treatment facilities are expensive and complex.

Chemical and thermal de-oiling are expensive processes that create additional environmental complications. The safest and most economically profitable can be the developed method of descaling from oil using a vortex layer apparatus [2]. Table 7.47 shows the main results of the experiment for four samples of sludge from the sludge collector. The starting material was a lump with the constitution of butter, black in color with a pungent odor. Even from one polyethylene canister with a volume of 4 liters, four samples were obtained with different contents of oil, moisture and scale. It is important to note that the moisture content in the sludge is relatively low, almost commensurate with the oil content. The bulk of the sludge is scale.

The phase analysis of the scale showed that it contains 5.7% water, 0.1% cutting fluid, the rest is scale. Granulometric composition of the scale after processing: from 3 to 1 mm – 32%, from 1 mm to 20 microns – 68%. The chemical composition of the scale: magnetite – 98.5%, silicon oxide – 0.87%. The latter is very important from the point of view of cleaning the resulting scale from oxides of other materials (SiO_2, Al_2O_3, etc.). This proposal was tested on the same laboratory setup in the selected mode in a glass flask with an opening bottom. After 30 seconds of treatment, 100% magnetite was obtained in the flask. This makes it possible to count on the possibility of heat release from the field of the inductor in the scale of the briquette.

Thus, the cleaning of sludge scale from oil by a magnetic field, unlike other methods, allows the oil to be separated and used in further production, to obtain iron scale in the form of magnetite, to enrich the resulting magnetite up to 100%, separating non-magnetic fractions in the same installation.

The Vortex Layer Apparatus (VLA) is a special installation, the principle of which is based on the creation of a rotating magnetic field. It has been established that technologies with the participation of these devices have a performance often thousands of times higher, consuming energy several times less than traditional technologies for the same purpose [2–4]. A fundamental solution to the principle of operation and design of the apparatus was found back in the 60s of the XX century by D.D. Logvinenko and who built the first industrial apparatus. He called them "devices of the vortex layer".

In such an installation (Fig. 7.59), the processed material 2 with the necessary additives (for example, water and needles) is placed in the steel reactor 1, the three-phase stator 3 is turned on, and the process of separating the oil from the scale into water is carried out for several tens of seconds. The water-cooled stator is hermetically closed with a steel casing 4.

The following order of the main technological operations is proposed. With the lower cover 5 closed and the upper cover 4 open, a certain amount of oily scale is fed from the dispenser into the pipe 2; process water is poured from another dispenser. After that, the cover 4 is closed and cooling water is supplied through the nozzles 6 and 7 of the chamber 1, and the stator is turned on at a rated power. After a certain processing time has elapsed, the power of the inductor is reduced, cover 5 is opened and the "water-oil" emulsion is released onto the conveyor. Scale particles are released into the other chamber of the conveyor at the moment the stator is disconnected from the supply network. Then the process is repeated.

Table 7.47. Results of experiments on the separation of coolant from scale by magnetic field.

Coolant, %	Coolant, g	Water, g	Water, g	Mill Scale, g
5	10	4	8	182
2	24	2.5	5	71
17	34	6	12	154
24	48	4.5	9	43

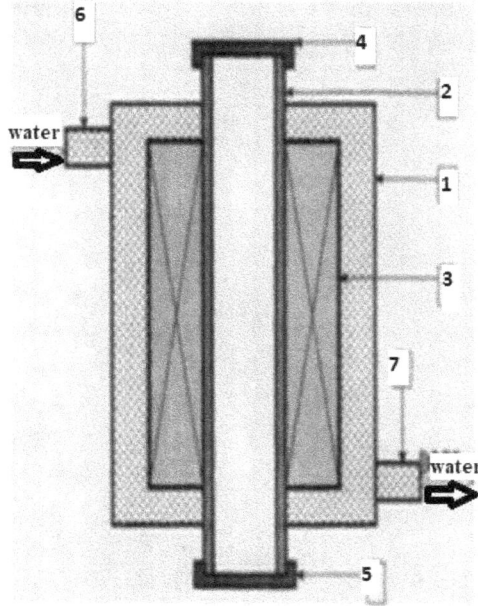

Figure 7.59. Structural diagram of the vortex layer apparatus installation. 1 - steel case, 2 - pipe, 3 - inductor, 4 and 5 - covers, 6 and 7 - branch pipes.

The complex physical, chemical and mechanochemical phenomena occurring in the vortex layer have not been sufficiently studied, although the devices of the vortex layer have already been tested in various industries.

This technology will make it possible to clean the scale from oil (up to 0.01%) and will allow it to be used in the future to obtain finely dispersed iron powder.

VLA was used to remove oil from the sludge generated at the West Siberian Metallurgical Plant (EVRAZ). The consistency of the starting material was in the form of "black butter".

To begin with, the "boat" was weighed into which the scale was subsequently placed – 480 g. Then it was filled with oily scale (400 g), the weight of the filled "boat" was 880 g. Placed in a resistance oven and kept for 2 hours at a temperature of 900°C (white vapor was observed), after weighing, a decrease in weight of about 23 g was found, i.e., 5.75% of 400 g of the original was water. We placed the resistance in the furnace at a temperature of 400°C, withstood it—black smoke was observed, like soot, after about 2 hours its emission stopped. Then, after weighing, a decrease in weight was found – 36 g, i.e., 9% of the 400 g of the original were oils.

The analytical complex ARL 9900 Workstation IP3600 was used for chemical and phase analysis of samples. The results are shown in Tables 7.48–7.52, respectively, for each of the samples.

The given values of the contents correspond to the relative contents of the crystalline phases (normalized to 100%); additionally, as an informational parameter, the estimated content of the amorphous phase is given.

Then three identical experiments were carried out on a vortex layer apparatus to remove this scale from oil. To do this, mixed 100 g of needles, 250 g of scale, 0.5 liters of technical water. The processing time is 2 minutes. The resulting solution was poured into two jars of 0.3 l each. Then, after the solution had settled, the scale was collected from the bottom and dried (Fig. 7.60) The scale showed magnetic properties.

The analysis of the cleaned scale showed that it contains up to 1% oil, i.e., about 2.5 g (Table 7.50). The results of X-ray fluorescence analysis and quantitative X-ray phase analysis of the scale are given in Tables 7.51 and 7.52.

Table 7.48. Results of X-ray fluorescence analysis of scale samples.

Component	Oily mill scale
Fe_{tot}	72.32
SiO_2	1,610
CaO	0.221
MnO	0.711
Al_2O_3	0.468
MgO	0.079
Cr_2O_3	0.061
NiO	0.034
CuO	0.060
S	0.075
P	0.032
TiO_2	0.017

Table 7.49. Results of quantitative X-ray phase analysis of scale samples.

Sample	Oily mill scale
Phase composition (mass fractions, %)	
Hematite (Fe_2O_3)	7.5
Magnetite (Fe_3O_4)	40.1
Wustite ($Fe_{1-x}O$)	40.4*
Amorphous phase	11

* compositionally close to $Fe_{0.91}O$

Table 7.50. Results of the scale cleaning study.

Mass of scale, g	Processing time, min	Oil mass, g	Oil content, %
Raw			
480	-	36	9
Cleaned			
250	2	2,5	1

The results of the experiments showed that the apparatus of the vortex layer can effectively clean the oily scale. In 2012, a patent was obtained for a method for cleaning oily scale [5]. In this method, oily mill scale slime, oily cast iron/steel shavings no more than 15 mm in size and technically pure water are mixed in a ratio of 2:1:6, after which they are processed in a reactor with a magnetic field of 50 Hz and a strength of 200 A/m up to 1100 A/m to obtain scale and shavings free of oil. The invention allows simultaneously with the environmentally friendly return of materials to production, to ensure the safety of the process oil.

The same approach was used for dezincification of sludge. In laboratory conditions, a special design ABC tested the possibility of processing dust from gas purifiers after a blast furnace according to a two-stage scheme: extraction of iron oxide into solution using caustic soda and complete destruction of zinc ferrite, and at the second stage—extraction of zinc oxide into the solution. The duration of each stage did not exceed 1–2 minutes. In this case, the zinc content in the solid residue was obtained below the MPC. Technically pure water and reagent were returned to the beginning of the process. The results formed the basis of another patent [6]. The invention relates to a method for selective extraction of iron oxide and zinc oxide from sludge and dust of metallurgical units. Sludge

Table 7.51. Results of X-ray fluorescence analysis of scale samples.

Component	Cleaned scale
Fe_{tot}	70.36
SiO_2	1,89
CaO	1,31
MnO	0.674
Al_2O_3	0.403
MgO	0.157
Cr_2O_3	0.153
NiO	0.109
Co_3O_4	0.086
CuO	0.072
S	0.066
BaO	0.036
ZrO_2	0.035
P	0.025
TiO_2	0.020

Table 7.52. Results of quantitative X-ray phase analysis of scale samples.

Sample	Cleaned scale
Phase composition (mass fractions, %)	
Hematite (Fe_2O_3)	9.0
Magnetite (Fe_3O_4)	48.9
Wustite ($Fe_{1-x}O$)	39.3*
Calcite ($CaCO_3$)	2.0
Amorphous phase	n/d (< 1 %)

* compositionally close to $FeO_{.91}O$

Figure 7.60. Scale cleaned from oil.

or dust, industrial water, alkali and active bodies in a ratio of 4:7:2:3 are fed in the form of a slurry to the reactor of the vortex bed unit (ABC) and treated with a magnetic field with a given frequency and intensity. Next, the primary solution and sediment obtained in the apparatus are separated. A secondary pulp is obtained by mixing sediment, industrial water, alkali, active bodies, soda and lime in a ratio of 4:7:2.5:3.2:0.3:1.5. Then the secondary pulp is fed into the reactor of the ABC unit, treated with a magnetic field with a given frequency and intensity, a secondary solution is obtained, iron oxides are extracted from the primary solution, and zinc oxide is extracted from the secondary solution. The technical result is a reduction in the duration of the process and the selective extraction of iron and zinc oxides.

References

[1] Electronic resource. www.umbr.ru Official site of the Institute OJSC "Uralmekhanobr".
[2] Logvinenko, D.D. and Shelyakov, O.P. 1976. Intensification of technological processes in the vortex layer apparatus. Kiev: "Technics". S. 157 (in Russian).
[3] Vershinin, N.P. 2004. Process activation settings. - Rostov-on-Don: "Innovator". P. 256 (in Russian).
[4] Vershinin, I.N. and Vershinin, N.P. 2007. Devices with a rotating electromagnetic field. Salsk: LLC "Advanced technologies of the XXI century". P. 134 (in Russian).
[5] Patent No. 2012142543A of the Russian Federation, C23G. Removal of Oil from Oiled Iron/steel Chips and Scale of Rolling Process Slimes/Farnasov, G.A., Kovalev, V.I., Kurunov, I.F., Bizhanov, A.M, Publ. 10/04/2014.
[6] Patent No. 2617086 C1 of the Russian federation, C22B. Method of Selective Iron Oxide and Zinc Extraction from Gas Treatment Sludges and Dusts of Metallurgical units/Farnasov, G.A., Kovalev, V.I., Kurunov, I.F., Bizhanov, A.M. Publ. 19/04/2017.

7.3 The best available technologies in the production of agglomerated materials for blast furnaces

The Best Available Technologies (BAT) concept was for first time used in the 1992 by OSPAR Convention for the protection of the marine environment of the North-East Atlantic for all types of industrial installations.

The best available technology must meet a set of criteria:

- the lowest level of negative impact on the environment per unit of time or volume of products (goods), work performed, services rendered;
- economic efficiency of its implementation and operation;
- application of resource-and energy-saving methods;
- period of its implementation;
- industrial implementation of this technology on two and more objects having a negative impact on the environment.

With regard to the processes of production of agglomerated iron containing materials, the list of the BAT and their technological indicators are presented in the information technology directories of the best available technologies [1, 2].

7.3.1 Production of sinter as a BAT

Sinter production, carried out by burning solid fuel in the air flow filtered through a layer of iron ore charge, remains the primary method of agglomeration of iron ore materials to date. As a result of physical and chemical processes in the sintering layer emissions of pollutants are formed:

- dust emissions are caused by mechanical removal of the fines from the sintering layer or formed in the process of destruction of materials (sinter when coming of the sintering belt; cooling, crushing and screening of the sinter cake after sintering);

- carbon monoxide (CO) emissions which, due to the combustion characteristics of the distributed solid fuel in the sintering layer (due to incomplete combustion), are technologically justified;
- emissions of sulfur dioxide (SO_2), depending on the sulfur content in coke fines, which is used as fuel, as well as the composition of iron ore charge;
- emissions of Nitrogen Oxides (NO_x) from combustion in ignition hoods; partially solid fuel combustion, and also because to the emission of "thermal" nitrogen oxides due to the nitrogen content in coke breeze and iron ore materials.

The sintering process involves a significant impact on the environment. It is suffice to mention [3] that in the structure of pollutant emissions of the integrated metallurgical plant for all the most significant emission components (dust, carbon monoxide, sulfur dioxide, nitrogen oxides) sinter production holds the leading positions (Table 7.53).

A priority source of emissions during sinter production, as can be seen from Table 7.53, is the processes of combustion. To ensure the goals of minimizing the impact on the environment and reducing the consumption of resources, the following technical solutions which can be attributed to the best available in the sinter production are applied [1]:

- reduction of fuel consumption due to rational charge preparation (optimization of granulometric composition of charge components, the component composition of the charge, the size of coke fines, etc.);
- increase in the height of the sintered layer (in most cases, requires some amount of reconstruction of the sintering machine);
- improvement of the combustion conditions of coke fines in the layer by increasing the gas permeability of the charge (optimization of the pelletizing process, the introduction of lime, the use of modulators of combustion (water mist));
- recirculation of waste gas heat (depending on the process requirements, different volumes of recirculation are possible, but under the conditions of moisture content, it cannot exceed 35%; in any case, the processing of agglomeration gases requires an obligatory additional supply of air (oxygen); organization of the flow of agglomeration gases is associated with an additional power consumption).

The above measures, improving the classical sintering technology, essentially provide no more than 20% reduction in emissions; the most radical technical solution (recycling of agglomeration

Table 7.53. Air emissions from major iron and steel industries.*

Production	Share of production in total emissions of components, %			
	Dust	CO	SO$_2$	NO$_x$
Sinter	**31.1**	**77.8**	**61.0**	**26.0**
Steel making	19.7	5.4	0.02	6.5
Refractory (lime-burning)	18.4	0.4	0.4	5.4
Blast furnace	17.3	3.5	0.3	3.0
Thermal power station-steam and air blower station	7.4	n/d	36.7	36.6
Coke making	2.0	7.8	1.0	9.1
Rolling production	1.2	n/d	0.2	10.5
Repair	1.0	4.9	0.02	1.5
Other	1.9	0.2	0.36	1.4
Total:	100	100	100	100

* For specific enterprises can be characterized by different values while maintaining the structure of emissions.

gases) at the maximum possible amount of recycling reduces emissions to 50% (and thus obtain a specific emission at the level not higher than 10 kg/t of sinter).

7.3.2 Production of pellets as a BAT

Pellets production refers to pyrometallurgical technologies, since the strengthening of agglomerated granules (raw pellets) requires high temperature (1250–1350°C) firing.

The emissions generated by the pellets production are due only to the combustion of natural gas to ensure the heat requirements of firing and the chemical transformation of the components of the charge (oxidation of sulfur compounds):

- emissions of NO_x from the combustion process in heated sections of an induration machine hearth;
- emissions of SO_2, determined by the sulfur content in iron ore material (concentrate);
- dust emissions due to mechanical removal of dispersed particles of raw pellets, indurated pellets during unloading from the induration machine, their sieving, loading.

Due to stoichiometric combustion of natural gas, there is no chemical underburning, so carbon monoxide is practically not formed.

Measures to reduce emissions of nitrogen oxides are associated with the use of burners of special design with low formation of nitrogen oxides, or technology of selective catalytic reduction of nitrogen oxides (efficiency up to 60%).

7.3.3 Stiff extrusion briquetting as a BAT

Pyrometallurgical technologies of agglomeration are associated with the formation of essential emissions, which logically leads to the conclusion that for the radical reduction of environmental impact in the technologies of preparation of metallurgical raw materials for smelting, combustion-free processes are promising.

To obtain the proper strength, cold-bonded pellets contain a binding substance (the most applicable is Portland cement) and therefore require heat and moisture treatment followed by strengthening for 7–28 days or drying with heating (the optimum temperature is 200–250°C).

The technology of briquetting by the method of stiff vacuum extrusion is without such disadvantages. The technology does not require heat treatment of raw extrusion briquettes (brex), it allows to obtain a durable material (hot strength of the briquette from magnetite concentrate and coke fines in terms of $RDI_{+6.3}$ exceeds $RDI_{+6.3}$ of sinter with a basicity of 1.2–1.6) of isometric shape and metallurgical dimensions, suitable for loading into a blast furnace (as well as for use in other metallurgical units).

As in the case of cold-bonded pellets, a binder is required to be introduced into the batch for the production of brex, the consumption of which is reduced in comparison with the production of non-ignition pellets, and in some cases (for individual charge) the addition of a plasticizer is required in the batch for the production of brex.

The main sources of environmental impact in the production of brex are the following:

- warehouse of raw materials (unloading operations of charge materials, storage and feed into the technological process);
- Department of dosing, mixing, extrusion;
- brex unloading, stacking and handling of stacks (the attrition of the fine fraction, shipment to consumers);
- aspiration system of brex making unit.

The main component of emissions into the atmospheric air in the production of brex is inorganic dust. According to the assessment, the specific emission of inorganic dust in the production of brex does not exceed 0.05 kg/t.

The technology of briquetting by the method of stiff vacuum extrusion is very effective not only as a method of utilization of anthropogenic materials, but primarily as an environmentally safe alternative to the agglomeration process [1].

7.3.4 *Comparative analysis of technologies for agglomeration of the raw materials for blast furnace*

Comparison of various agglomeration technologies according to the criteria of the best available technologies [1, 2] is presented in Table 7.54.

From Table 7.54 it follows that in comparison with the production of sinter and pellets brex have minimal performance impact on the environment and energy consumption, and require significantly smaller investment, which qualifies the technology of briquetting by stiff vacuum extrusion as the best available technology.

Agglomerating iron-bearing materials by the stiff vacuum extrusion is a highly efficient technology which is capable to compete with sinter production. Brex fully complies with the requirements imposed by the conditions of the blast furnace.

In terms of environmental impact and resource consumption, this method can be qualified as the best available technology.

Smelting of iron and steel, in addition to the well-known negative impact on the environment, is accompanied by the formation of a significant amount of iron-containing waste, the direct processing of which is impossible or difficult without agglomeration.

Table 7.54. Characterization of various agglomeration technologies according to the criteria of the best available technologies (BAT).

#	BAT criterion	Agglomerated iron-containing material		
		Sinter	Pellet	Brex
1	The minimum level of impact on the environment, kg/t:			
	- dust	≤ 1.2	≤ 0.6	0.05*
	- nitrogen oxide	≤ 0.55	≤ 0.535	0
	- sulfur dioxide	≤ 4.0	≤ 0.5	0
	- carbon oxide	≤ 14.0	~ 0	0
	Total emissions, kg/t:	≤ 20	≤ 2	≤ 0.05
2	Resources consumption:			
	- solid fuel, kg/t**	23.6–48.9	0	0
	- gaseous fuel, m³/t	2.45–6.3	9.5–15.0	0
	- Electricity, kWh/t	23.0–48.7	29.0–48.5	15.0–33.0
3	Investments, USD/t***	~ 5000	~ 5500	~ 2000
4	Implementation period, years****	3	3	2

Notes:
*   Estimation (based on following sources: ore warehouse, loadings, dosing and mixing of charge components, the unloading of the finished brex, stacking and stack handling, aspiration system of shop floor space)
**   Refers to the provision of heat treatment
***   Adopted the averaged specific indicator for the implementation on a "turnkey" basis
****   The conditional duration of the investment cycle for productivity of 1 million tons of the product is taken; for the production technology of brex the duration is reduced due to significantly lower metal consumption level (the volume of construction works).

Agglomeration and pellet production are the most common industrial agglomeration methods based on high-temperature processing of iron-containing raw materials and associated with significant emissions of pollutants into the atmosphere (agglomeration – up to 20 kg/t of sinter, pellet production – 2 kg/t of pellets).

References

[1] Information and technical reference on the best available technologies ITS 26-2017. Production of iron, steel and ferroalloys//www.burondt.ru/NDT/NDTDocs.Detail (electronic resource).
[2] Information and technical reference on the best available technologies ITS 25-2017. Iron Ore Mining and Processing//www.burondt.ru/NDT/NDTDocs.Detail (electronic resource).
[3] Frolov, Yu.A. and Sintering, M. 2016. Metallurgizdat, 672 p. (in Russian).

7.4 Synergy of Briquetting and Sintering in Ironmaking

Briquetting for a long time turned out to be uncompetitive in comparison with sinter and pellet production due to insufficient productivity of briquette equipment and inconsistency of briquette properties with the requirements of metallurgical processing.

The situation changed with the advent of the stiff extrusion technology adapted for the production of briquettes (briquettes), which is widely used in more than 60 countries around the world for the production of ceramic bricks, on the agglomeration market. The success of the first implemented projects made it possible for this technology to compete with the agglomeration.

In order to substantiate this possibility, the considered agglomeration technologies from the point of view of metallurgical properties of products (sinter and extrusion briquettes), productivity of sintering and extrusion equipment, and, no less important, from the point of view of environmental impact, need to be compared.

It is advisable to compare the mechanical strength of sinter and extruded briquettes during tumble tests in accordance with ISO 3271 (GOST 15137-77 "Iron and manganese ores, sinter and pellets. Method for determining the strength in a rotating drum"). The method is based on the mechanical processing of briquettes in a rotating steel drum (diameter 1000 mm, length 500 mm, 200 revolutions at a drum rotation frequency of 25 ± 1 rpm) and subsequent determination by sieve analysis of changes in the particle size distribution of the sample (with dimensions less than 0.5 mm, from 0.5 mm to 6.3 mm and over 6.3 mm), which characterizes the ability of briquettes to resist impact and abrasion during transportation and overloading. The relative proportion of the fraction with particle sizes less than 0.5 mm characterizes the resistance of the briquette to abrasion (AI, Abrasion Index), and the proportion of the coarse fraction (+6.3 mm) characterizes its impact strength (TI, Tumble Index). The quality of the sinter is considered satisfactory if its impact strength according to this method is not lower than 70%. Table 7.55 shows the results of measuring the

Table 7.55. Strength of brex on the basis of mill scale in a drum sample.

Brex sample, No.			1	2	3	4
Initial mass	m_0	g	15012	15001	14999	15000
+ 6,30 mm	m_1	g	12917	12948	12943	12909
+ 0,5 mm	m_2	g	239	215	229	221
– 0,5 mm	m_3	g	1690	1658	1685	1735
m_0-m_1-m_2-m_3	d	g	166	180	142	135
d/m_0	D	%	1.1	1.2	0.9	0.9
(m_1/m_0)	TI	%	86.0	86.3	86.3	86.1
$(m_0$-m_1-$m_2)/m_0$	AI	%	12.4	12.3	12.2	12.5
		TI (average)	86.2			
		AI (average)	12.3			

strength of brex from mill scale on a cement bond. It can be seen that, in terms of the impact strength parameter, the brex exceed the rejection limit for sinter. Slightly higher values of abrasion values were explained in [1] based on the results of numerical simulation of a drum sample of brex. The results obtained associate the increase in abrasion with the features of falling briquettes on the steel surface of the drum and create the basis for revising the rejection limits of abrasion strength of briquettes in a drum sample, which will make it possible to adequately compare this parameter with that for the sinter.

A comparison of the hot strength of sinter and brex of various compositions was carried out in [2]. We investigated the metallurgical properties of the brex made from blast furnace and converter sludge, as well as iron-carbon-containing brex from the iron ore concentrate of Stoilensky GOK and coke breeze (Table 7.56), which were also considered as a promising component of the blast furnace charge.

Hot strength testing of slurry and concentrate brex according to ISO 4696 showed a significant advantage of concentrate brex (RDI + 6.3 = 96.5%) compared to slurry brex (RDI + 6.3 = 61.8%), which is explained by the absence of any hematite in the first brex and, on the contrary, by the presence of secondary hematite in the sludge brex (in the particles of both blast furnace sludge and converter sludge), the crystal lattice of which is rearranged on low-temperature and slow reduction with an increase in geometric parameters, which causes the occurrence of mechanical stress and disintegration of pieces of material containing hematite.

For comparison, the hot strength of sinter with basicity $(CaO + MgO)/(Al_2O_3 + SiO_2)$ 1.2, 1.4, and 1.6, obtained from the usual charge for NLMK PJSC, was simultaneously evaluated according to the same standard (Table 7.57). The hot strength of the sludge brex is comparable to the hot strength of sinter with a basicity of 1.2 and 1.4 (64 and 60%), which is explained by the similar contents of secondary hematite in these materials. The hot strength of sinter with a basicity of 1.6 (77%) is higher than the hot strength of a sludge brex due to the fact that, with a basicity of 1.6, a new phase already appears in the sinter—calcium ferrites, which strengthens the structure of the sinter and reduces its disintegration during low-temperature reduction. It has been experimentally established that, due to the strength of the cement bond and the absence of disintegration of ore particles during low-temperature reduction, the hot strength of brex from magnetite iron ore concentrate and coke breeze (RDI + 6.3) significantly exceeds the hot strength of sinter with basicity (B4) 1.2, 1.4 and 1.6.

Table 7.56. Component composition of sludge and ore-coke brex.

Brex components	Mass share of components, %	
	Brex No. 2	Brex No. 4
Portland cement PC 500	9.1	9.0
Coke breeze	-	13.5
Bentonite	-	0.9
BF sludge	54.5	-
BOF sludge	36.4	-
Iron ore concentrate	-	76.6

Table 7.57. The results of determining the hot strength of brex and sinter of various basicity.

Agglomerated material	RDI (+6.3), %
Brex No. 4 (basicity 0.75)	96.5
Brex No. 2 (basicity 1.93)	61.9
Sinter (basicity 1.2)	64
Sinter (basicity 1.4)	60
Sinter (basicity 1.6)	77

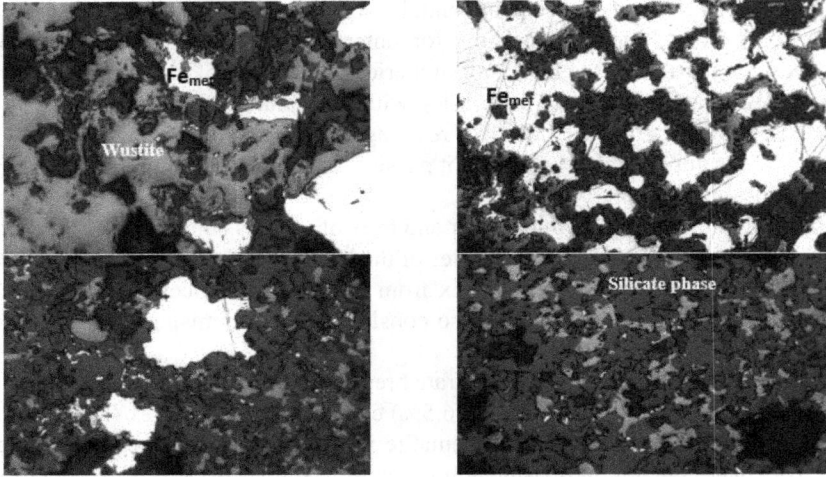

Figure 7.61. Light microscopy of brex after heating in a reducing atmosphere at 1400°C (magnification 200).

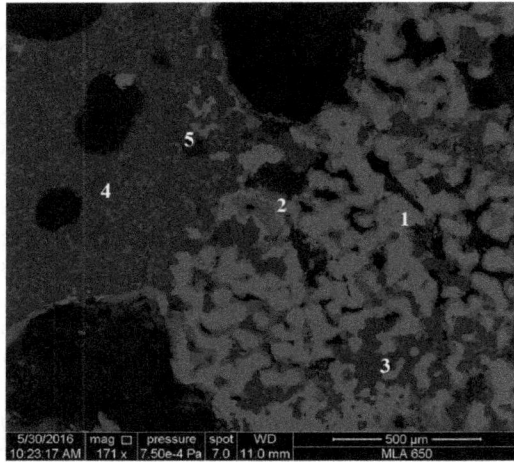

Figure 7.62. Scanning microscopy of brex sample after reduction heating at 1400°C; 1 - α-iron, 2 - wustite, 3, 4 - glass phase, 5 - wustite.

The mechanism of hot strength is qualitatively illustrated in Figs. 7.61 and 7.62, which show images of the brex structure after reduction at 1400°C, showing that metallic iron forms a kind of mesh, within which wustite microzones are found. The metal precipitated from the glass phase crystallized to silicates.

Thus, the provision of the hot strength of the brex after the loss of the binding capacity of the cement bond is achieved due to the formation (even during gas reduction) of a surface metal shell-frame and because of the formation of an iron silicate matrix, from which iron is reduced at temperatures above 1100°C due to carbon of coke breeze.

At the beginning of 2021, tests were carried out for the metallurgical properties of extrusion briquettes based on iron-containing dust from cement-bonded electrostatic precipitators (8%) and with bentonite as a plasticizer (0.8%). An independent laboratory (CCI, part of the COTEGNA GROUP) carried out a set of studies, including:

1. Determination of the relative reducibility according to ISO 7215. During the test, the sample is exposed to a mixture of reducing gas, consisting of CO (30%) and N2 (70%), for 180 minutes at a temperature of 9000C. The final recovery index is determined by the loss of mass during the recovery process. Reducing gas consumption 15 l/min. The size of the test sample: diameter 19 mm, length from 19 to 120 mm.

2. Determination of the free swelling index according to ISO 4698 (for pellets). The sample is exposed to a mixture of a reducing gas consisting of CO (30%) and N2 (70%) for 60 minutes at a temperature of 900°C. The free swelling index is defined as the difference in sample volumes before and after reconstitution. Reducing gas consumption 15 l/min. The size of the test sample: diameter 19 mm, length from 19 to 120 mm.

3. Determination of indicators of low-temperature reduction - grinding by a dynamic method according to ISO 13930. The sample in a rotating tube is exposed to a mixture of a reducing gas consisting of CO (20%), CO_2 (20%), H_2 (2%) and N_2 (58%), within 60 minutes, at a temperature of 500°C. The low-temperature fracture index is determined depending on the size class of more than 6.3 mm, less than 3.15 mm and less than 0.5 mm. Reducing gas consumption 20 l/min. The size of the test sample: diameter 19 mm, length from 19 to 120 mm.

The results of the measurements of metallurgical properties are shown in Table 7.56 and in Figs. 7.63–7.66.

Table 7.56. Results of independent tests of metallurgical properties of briquettes.

The name of indicators	Regulatory documents	Results
Total iron Fe_t, %	ISO 2597-1	47.61
Ferrous oxide Fe (II), %	ISO 9035	10.23
Ultimate reduction degree R_{180}, %	ISO 7215	88.8
Free Swelling Index FSI, %	ISO 4698	9.4
Low temperature disintegration indices (+ 6.3 mm) $LTD_{+6.3}$, %	ISO 13930	86.1
Low temperature disintegration indices (−3.15 mm) $LTD_{-3.15}$, %		13.8
Low temperature disintegration indices (−0.5 mm) $LTD_{-0.5}$, %		13.6

Figure 7.63. Appearance of the sample before testing.

Figure 7.64. Appearance of the sample after measuring the relative reducibility (ISO 7215).

Figure 7.65. Appearance of the sample after the swelling test (ISO 4698).

Figure 7.66. Sample appearance after low temperature recovery test - grinding (ISO 13930).

According to CCI, the test results for iron briquettes are in the range of values obtained with good quality iron ore pellets and sinter samples.

Thus, both the mechanical strength of the brex and their hot strength ensure that the level of similar properties of the sinter is achieved (and exceeded). Practical and long-term experience of operating a number of briquette factories as part of blast-furnace shops (Russia, Japan, India, Chile, etc.) confirm the high metallurgical properties of brex, which are achieved by a notably lower binder content than in alternative briquette technologies.

From the point of view of comparing the productivity of sinter plants and briquette plants, the undoubted superiority of sinter plants in this parameter should be recognized. Nevertheless, the already accumulated logistic experience of large-scale factories for the production of brex indicates the possibility of achieving a productivity level of 2000–7000 tons of brex per day. Sintering belt capacity: 1500–10000 tons per day. Productivity of the roasting machine: 2500–9000 tons per day. The world's largest ceramic brick factory, Saudi's Stiff Extrusion Red Brick Factory, produces 1 million bricks per day.

The sintering process is carried out due to the combustion of solid fuel in an air stream filtered through a layer of iron ore charge. As a result of physical and chemical processes in the agglomeration layer, emissions of pollutants are formed:

- dust emissions caused by mechanical removal of a finely dispersed charge from the layer or formed in the process of destruction of materials (agglomerate when coming off the sinter belt, during cooling, crushing and sieving of the sinter cake after sintering);
- emissions of carbon monoxide CO, which, due to the features of combustion of distributed solid fuel in the agglomeration layer (due to incomplete combustion), are technologically justified;
- emissions of sulfur dioxide SO_2, depending on the sulfur content in coke breeze and coal, which is used as a fuel, as well as on the composition of the iron ore charge;
- emissions of nitrogen oxides NOx from the combustion process in the burners of the incendiary furnace, partly from the combustion of solid fuel, as well as due to the emission of "thermal" nitrogen oxides caused by the nitrogen content in coke breeze and iron ore materials.

Sintering has significant environmental impacts. In the structure of pollutant emissions of an integrated metallurgical enterprise, the agglomeration occupies a leading position in all the most significant emission components (dust, carbon monoxide, sulfur dioxide, nitrogen oxides) (Table 7.53, Section 7.5.3, [3]).

Unlike the sintering process, when agglomeration by the method of stiff vacuum extrusion (pressure range of pressure 1.5–3.5 MPa and higher, the moisture content of the mixture is 12–16%), no heat treatment of raw briquettes is required, which makes it possible to obtain a durable material with an isometric shape and dimensions suitable for loading into a blast furnace, withstanding stacking, multiple handling and long-term storage in ambient conditions. As in the case of cold bonded pellets, a binder is necessarily introduced into the charge for the production of brex, the consumption of which is reduced in comparison with the production of non-fired pellets [4], and in some cases (for individual charges) the addition of a plasticizer is required in the charge for the production of brex.

The main sources of environmental impact during the production of brex are the following (in aggregate):

- warehouse of raw materials (operations of unloading charge materials, storage and supply to the technological process);
- department of dosing, mixing, extrusion;
- unloading of brex, stacking and processing of a stack (screening of fines, shipment to consumers);
- workshop space aspiration system. Inorganic dust is the priority component of emissions into the atmospheric air during the production of brex.

According to estimates, the specific emission of inorganic dust in the production of brex does not exceed 0.05 kg/t.

Comparison of the main industrial agglomeration technologies in terms of the degree of environmental impact are given in Table 7.54 (Section 7.3.4, [5–6]).

Extrusion agglomeration, due to the lack of energy consumption for high-temperature processing of raw brex, is distinguished by minimal indicators of environmental impact, which is its undoubted competitive advantage.

Thus, the possibility of an effective partial replacement of sinter with brex is substantiated according to the criteria listed above (metallurgical properties, equipment performance and environmental impact).

When fine-dispersed waste (dust and sludge) is used in the sinter production, additional energy and material resources are re-expended. Not all such materials can be effectively used as part of sinter batch, which leads to a constant increase in dumps (in fact, technogenic deposits) on the territory of enterprises. In some countries, the disposal of such waste is prohibited, while in others it is allowed only after its neutralization, which requires high costs.

Flue dust is a very valuable material, since its composition is close to that of a blast furnace charge. However, due to the high carbon content and dispersion, the flue dust is poorly agglomerated. Poor agglomeration of the material constantly hindered its use in the production of sinter. At present, metallurgists have, to one degree or another, solved the problems of using blast furnace dust in the sinter batch, although some experts believe that for the above reasons, a significant part of the dust does not enter the sinter, is carried away with sinter gases and again passes into the sludge. The flue dust carbon does not work during the sintering process. At the same time, in the composition of the briquette, carbon of the blast furnace dust is effectively used for the reduction of oxides, which leads to a decrease in coke consumption.

Utilization of converter dust and sludge as an additive in sinter production is also possible, but not rational. In addition to a significant decrease in the productivity of sintering machines associated with the deterioration of pelletization and gas permeability, large amounts of alkalis and zinc enters the blast furnace, which has a negative effect on the operation of blast furnaces. The use of man-made raw materials of various genesis in sinter batch due to the instability of its chemical and granulometric compositions without the possibility of dosing at sinter plants that do not have a blending warehouse, negatively affects the stability of the composition and quality of the sinter.

For the first time, the concept of partial replacement of sinter with brex was expressed in [7] in the form of a hypothesis about the possibility of achieving high values of the proportion of briquettes in the blast furnace charge (up to 100%). In this work, the concept is revised and is developed in the form of a statement about the possibility and feasibility of achieving synergy between agglomeration and extrusion briquetting, achieved as a result of a 50% reduction in sinter production, leading to a twofold decrease in emissions and a proportional decrease in the share of agglomerate in the blast furnace charge, compensated for by adding brex made from materials removed from the sintering process.

As a result, the strength of the renewed charge (sinter + brex) increases, firstly, because of the use of stronger brex, and secondly, due to the production of a stronger agglomerate (without or with a reduced proportion of secondary materials, higher basicity). This will lead to a decrease in the formation of fine fractions in the agglomeration and blast-furnace processes.

From calculations for the conditions of one of the plants (Table 7.57), a decrease in the proportion of agglomerate in the blast furnace charge requires an increase in the basicity of the agglomerate. The calculation was carried out under the condition of a stable proportion of fluxed pellets and the use of brex and while maintaining the basicity of the blast furnace slag, as well as the exclusion of raw fluxes from the blast furnace process.

Figure 7.67 shows the graphical dependence of the proportion of sinter in the charge on its basicity.

As can be seen from Table 7.59 and Fig. 7.49, with the basicity of blast furnace slag at the level of 1.00–1.10, which is characteristic of almost all blast furnace workshops in Russia, a decrease in the share of sinter below 0.5 (50%) in the charge will require its basicity of at least 1.7, and with a fraction below 0.3, the basicity is already closer to 3.

For values of the fraction of sinter in the blast furnace charge less than 50%, the basicity of the sinter is illustrated in Table 7.58.

Table 7.57. Fraction of sinter in the charge depending on the basicity of the slag.*

Sinter basicity		Slag basicity			
	1,00	1,05	1,10	1,15	1,20
1.04	1.000				
1.20	0.796	0.873	0.951		
1.30	0.718*	0.790	0.860	0.930	1.000
1.40	0.648	0.710	0.773	0.836	0.899
1.50	0.591	0.648	0.705	0.762	0.819
1.60	0.544	0.596	0.648	0.700	0.753
1.70	0.503	0.551	0.599	0.647	0.696
1.80	0.469	0.514	0.558	0.604	0.649
1.90	0.438	0.480	0.522	0.564	0.606
2.00	0.411	0.450	0.490	0.529	0.569
2.10	0.388	0.425	0.462	0.499	0.537
2.20	0.367	0.402	0.437	0.472	0.507
2.30	0.348	0.381	0.414	0.447	0.481
2.40	0.331	0.362	0.394	0.425	0.457
2.50	0.315	0.345	0.375	0.405	0.436
2.60	0.301	0.330	0.358	0.387	0.416
2.70	0.289	0.316	0.343	0.371	0.398
2.80	0.277	0.303	0.329	0.356	0.382
2.90	0.266	0.291	0.316	0.341	0.367
3.00	0.256	0.280	0.304	0.328	0.353

* - basic version without brex, the rest (0.282) - pellets. In variants with a higher basicity, the proportion of pellets is lower, in the rest it is preserved, the rest is brex.

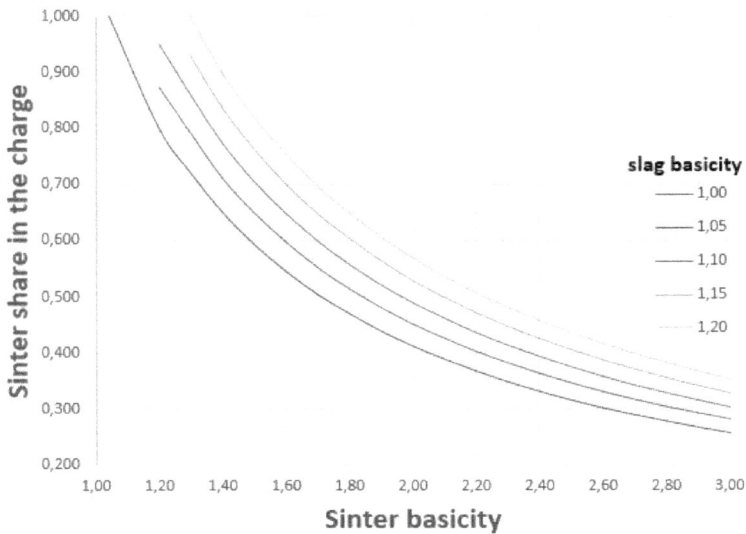

Figure 7.67. Dependence of the fraction of sinter in the blast furnace charge on its basicity and basicity of the slag (Table 7.59).

Table 7.58. Basicity of the sinter depending on its share in the blast furnace charge.

Operating parameters while maintaining the basicity of the sinter B = 1.315				
Consumption, kg/t of hot metal	**Consumption of briquettes, % of iron ore mixture**			
	0 (base)	**10**	**20**	**30**
Sinter	1193	1037	879	719
Pellets	443	449	450	454
Briquettes	0	165	332	503
Coke (dry skip)	426	426	425	423
Limestone	0	40	60	80
Dolomite	0	0	10	20
Correction of the basicity of the sinter				
Recommended agglomerate basicity to exclude wet flux from blast-furnace smelting	1,32	1,53	1,79	2,21
Consumption, kg/t of hot metal		167.01	169.94	173.08
Sinter	1193	1051	899	743
Pellets	443	453	461	469
Briquettes	0	167	340	519
Coke (dry skip)	426	417	409	401
Limestone	0	0	0	0
Dolomite	0	0	0	0

Notes: basicity of the slag 0.99, the proportion of pellets in the charge 27.1%, the degree of gas utilization and the composition of the slag are stable: Composition of briquettes: 80% mixture of concentrates (Mikhailovsky and Lebedinsky 3: 1), 13% coke breeze, 6% cement, 1% bentonite.

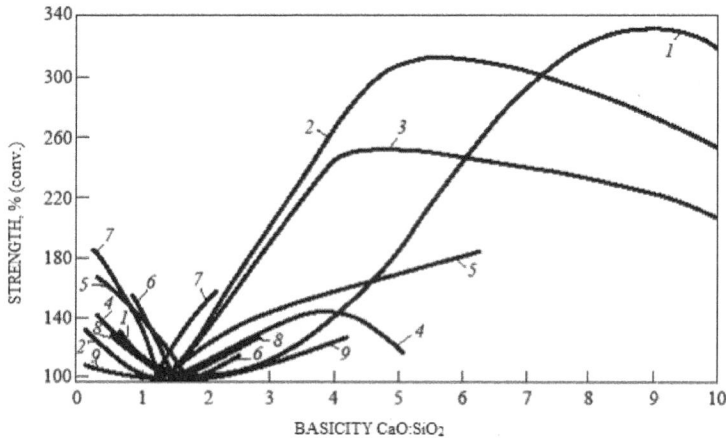

Figure 7.68. The influence of the basicity of the agglomerate on its strength, depending on the quality of raw materials [3]. 1–3 - finely divided concentrate, respectively, rich, medium and poor in iron; 4 - mixture of ores; 5 - a mixture of fine concentrate and dusty ores; 6 - magnetite ore; 7 - hematite ore; 8, 9 - mixture of ores.

In turn, an increase in the basicity of the sinter (above the minimum values of 1.3–1.5 [8]) leads, as is known, to an increase in its strength for almost all types of iron ores (Fig. 7.68).

According to Fig. 7.68 for the calculation conditions, the results of which are shown in Tables 7.59 and 7.60 and Fig. 7.49, replacing 50% of the sinter with brex with a corresponding increase in the basicity of the sinter will lead to an increase in its strength (only by increasing the basicity) by 60–80%. An additional increase in strength due to the removal of dust and sludge from the sinter charge makes it possible to count on a total of at least a twofold increase in strength.

When using "acidic" pellets, the sinter will initially be more basic; nevertheless, an increase in its basicity to a level of at least 2.0–2.5 will also allow the use of brex in the blast furnace charge, which will ensure the removal of secondary materials from the sintering.

Thus, in fact, reducing the proportion of sinter in the blast furnace charge by half will lead, in addition to a twofold decrease in harmful emissions, to an increase in the strength of the produced sinter, which, in turn, will lead to a significant, to a minimum, decrease in the dropout rate in the blast furnace workshop. For old workshops, where there is no sinter screening operation, this will minimize the harmful effect of this factor on the process indicators (a decrease in the content of a fine fraction of 5–0 mm in an iron ore charge by every 1% will lead to a decrease in coke consumption by 0.5%, an increase in productivity by 1.0% [3]). In addition, the following changes in the sinter batch will occur:

1) secondary materials such as blast furnace and aspiration dust, partially or completely sludge is removed from the agglomeration;
2) the consumption of limestone and dolomite increases;
3) the consumption of solid fuel increases;
4) the specific consumption of iron ore concentrate increases with a decrease in the absolute consumption.

In the blast-furnace process, as a result of the implementation of the concept of partial replacement of the agglomerate with brex, the following will be achieved:

• reducing the consumption of skip agglomerate with a significant increase in strength characteristics;
• the possibility of increasing the proportion of pellets in the charge;
• a decrease in the specific consumption of coke and an increase in productivity due to a decrease in the proportion of fines.
• the mass fraction of iron will vary depending on the composition of the brex and the fraction of the components of the charge.

Conclusion

1. In general, the partial replacement of sinter in the blast furnace charge by brex will improve the quality of the sinter, both due to the necessary increase in basicity and to the possible removal of secondary materials from the sinter batch, in total, leading not only to a reduction in coke consumption and an increase in the productivity of the blast furnace, but also to a decrease in the total emissions and environmental impact.

2. The possibility of achieving high values of the proportions of briquettes in the blast furnace charge was shown in [9–12], to analyze the results of pilot melting at PJSC NLMK in 2003.

3. When melting a pilot batch of briquettes (2475 tons) from a mixture of iron ore concentrate, coke breeze and Portland cement (15% of the briquette mass), the level of consumption of briquettes exceeded 303 kg/t of pig iron. The results of the smelting confirmed that such briquettes are a full-fledged self-reducing component of the blast-furnace charge, the use of which ensures a decrease in coke consumption in blast-furnace smelting, in proportion to their consumption. The share of such a component in the blast furnace charge is only slightly limited by a decrease in the furnace productivity due to a decrease in the iron content in the charge and can reach, according to the authors of the work, 50% or more. The maximum achieved value of the share of briquettes in the ore part of the charge was about 19%. Note that the achievement of this level required a serious "dilution" of the charge due to the use of 15% cement. For extrusion briquettes, the required hot strength is achieved by 6–9% cement. Further reduction in binder

consumption achieved with the use of polymer binders will increase the degree of technically feasible replacement of the agglomerate with brex.

References

[1] Bizhanov, A.M. and Zagainov, S.A. 2021. Testing of briquettes for mechanical strength. Metallurg. 3: 11–18 (in Russian).

[2] Kurunov, I.F., Bizhanov, A.M., Tikhonov, D.N. and Mansurova, N.R. 2012. Metallurgical properties of brex. Metallurg 6: 44–48 (in Russian).

[3] Yu, A. and Frolov, M. 2016. Agglomeration: technology, heat engineering, management, ecology. Metallurgizdat. 672 p. (in Russian).

[4] Lotosh, V.E. 2009. Non-firing agglomeration of finely dispersed materials and minerals. Ekaterinburg. Publishing House "Philanthropist" 525 p. (in Russian).

[5] ITS NDT 26 "Production of pig iron, steel and ferroalloys"//www.burondt.ru/NDT/NDTDocsDetail/ (electronic resource).

[6] ITS NDT 25 "Extraction and beneficiation of iron ores" /www.burondt.ru/NDT/NDTDocsDetail/ (electronic resource).

[7] Bizhanov, A.M. and Kurunov, I.F. 2017. Extrusion briquettes (brex)—a new stage in the agglomeration of raw materials for ferrous metallurgy. Moscow: OOO Metallurgizdat 236 p. (in Russian).

[8] Vegman, E.F., Zherebin, B.N., Pokhvisnev, A.N. and Yusfin, Yu.S. 2004. M. Iron metallurgy. Metallurgy, pp. 184–204 (in Russian).

[9] Kurunov, I.F., Shcheglov, E.M., Kononov, A.I., Bolshakova, O.G. and others. 2007. Research of metallurgical properties of briquettes from technogenic and natural raw materials and assessment of the effectiveness of their use in blast-furnace smelting. Part 1. Bulletin of Scientific, Technical and Economic Information "Ferrous Metallurgy" 12: 39–48 (in Russian).

[10] Kurunov, I.F., Shcheglov, E.M., Kononov, A.I., Bolshakova, O.G. and others. 2008. Investigation of the metallurgical properties of briquettes from technogenic and natural raw materials and assessment of the effectiveness of their use in blast-furnace smelting. Part 1. Bulletin of Scientific, Technical and Economic Information "Ferrous Metallurgy" 1: 8–16.

[11] Kurunov, I.F., Malysheva, T.Ya. and Bolshakova, O.G. 2007. Study of the phase composition of iron ore briquettes in order to assess their behavior in a blast furnace. Metallurg 10: 41–46 (in Russian).

[12] Kurunov, I.F., Bolshakova, O.G., Shcheglov, E.M. et al. 2007. Metallurgist 6: 36–39 (in Russian).

CHAPTER 8

Briquetting of Natural and Anthropogenic Raw Materials for Ferroalloys Production

Significant volumes of finely-dispersed, effectively-recyclable materials are created in the process of extraction and beneficiation of raw materials for ferroalloy production during their transportation to ferroalloy plants and ferroalloy smelting.

In particular, the mining and processing plants have accumulated large reserves of manganese ore fines with manganese content below the marketable value.

One of the most common anthropogenic raw materials is dust and sludge from gas cleaning of ferroalloy furnaces. The specific dust emissions of ferroalloy furnaces are 8–30 kg/ton of ferroalloys. The content of manganese oxides in dust of the production of silico-manganese is 21–34.3%, in ferromanganese – 30–35%; chromium oxides in dust in the production of ferrochrome – 22–30% [1]. The use of finely dispersed raw materials without their agglomeration in Submerged Electric Arc Furnaces (SEAF) disrupts the normal course of the process, leading to an increase in dust generation, a decrease in productivity and metal extraction rate, increased risks of accidents and number of hot down time, and as a result, deteriorates technical and economic indicators of the furnace generally.

The traditional method of agglomeration of such fine materials is sintering. Until 1973, ferroalloy plants did not have sinter plants [2]. The first sinter plant was commissioned at the Nikopol Ferroalloy Plant in 1973, and in 1979, the sinter plant at the Zestafoni Ferroalloy Plant was commissioned. The average productivity of sintering machines is 100 t/h (specific – 1.2 t/m^2 per hour).

Another industrial method of agglomeration is pellet calcination at a temperature of 1150–1220°C. With this calcination the hardening of the pellet is provided by the resulting liquid phase. At the "Kashima" plant (Japan), for high-carbon ferromanganese smelting in SEAF with a capacity of 40 MVA, manganese concentrates are agglomerated in a 110 t/h capacity pelletizer and then calcined in a 75 m long rotary kiln with a diameter of 3.5 m [3].

The experience of using briquettes in the charge of ferroalloy furnaces has confirmed the effectiveness of this kind of agglomeration technology. The first industrial experience of using briquettes based on manganese ore concentrates in the charge of the ore-smelting furnace was obtained in 1961, when 160 tons of roll briquettes made of manganese ore concentrate (30.9% Mn; 4% Fe; 0.0577% P; 26.95% SiO_2; P/Mn = 0.0018; 0.33% S; 1.9% CaO; 4.05% BaO) were used in a 2500 KVA furnace of the Zestafoni Ferroalloy Plant (Georgia) [4]. The results of the heats showed that this type of charge component is quite effective.

In 1966, full-scale industrial testing was carried out at the same plant for the smelting of ferrosilicon manganese in industrial ore-smelting furnaces using ore briquettes of manganese concentrate from the Chiatura deposit in the charge [32]. For industrial experiments, briquettes were made on a roller press with a capacity of 5 tons per hour from ore with a particle size of 5–0 mm on a binder of Sulfite-Alcohol Bard (SAB) with a density of 1.2 g/cm^3. With an ore moisture content of 4% and an addition of 8% binder, the mixture was stirred for 15 minutes. Next, the mixture was subjected to heat treatment. The fact is that the binding properties of the SAB are due to the occurrence of polymerization processes in it, leading to the formation of long chains of molecules in the body of the briquette. The polymerization reaction in the presence of manganese is most completely realized at temperatures of 160–180°C. For iron ore briquettes, the polymerization temperature reaches 200°C. Thus, the achievement of the required values of the strength of the briquettes required their drying at temperatures of 50–300°C. Silicomanganese was smelted on a charge with ore briquettes in a three-phase open ferroalloy furnace with a capacity of 16.5 MVA. The furnace worked normally and stably, the gas permeability of the charge was good, the flame was distributed evenly throughout the furnace. After 112 hours of experimental melting, it was concluded that sufficiently strong briquettes can be obtained from manganese ore of this size suitable for use in the charge of ferroalloy furnaces. When working on ore briquettes, the furnace productivity increases, the power consumption and the consumption of reducing agent are reduced.

In 1970, in the Dnepropetrovsk Metallurgical Institute, indicators of the smelting of marketable silicomanganese on a briquetted mixture of a mixture of manganese oxide concentrates of grades I and II of the Nikopol deposit in the 1:1 ratio and on sinter made from the same mixture were studied [6]. Before briquetting, the mixture of concentrates of a fraction of 10–0 mm was ground to a size of 3–0 mm. Briquettes were made on a semi-industrial roller press at a pressure of 500 kg/cm², the binder was a mixture of bitumen, fuel oil and SAB in an amount of 10% by weight of the charge. The charge was mixed in tanks with steam heated (to activate the polymerization of the SAB). Briquettes of two compositions were prepared and melted: with an excess of reducing agent (coal) in the amount of 50%, introduced to create a skeleton in the briquette and to increase its strength (mixture of concentrates – 54.5%, river sand – 9.1%, coal – 27.3%, a mixture of bitumen and fuel oil – 3.6%, SAB – 5.5%), with a stoichiometric amount of reducing agent required for the reduction of silicon and manganese (a mixture of concentrates – 60.6%, river sand – 10.1%, coal – 20.2%, a mixture of bitumen and fuel oil – 3.6%, SAB – 5.5%). Laboratory tests showed that the physical properties of raw briquettes were superior to those for roasted briquettes, thus they refused to roast the briquettes. The values of mechanical strength and heat resistance were higher for raw briquettes. Commodity silicomanganese was decided to be smelted on raw briquettes. Semi-industrial smelting on briquettes of the above composition and on sinter was carried out in a three-phase open ore-smelting furnace with a capacity of 1.2 MVA at a voltage of 81.60–85.89 V and a current of 6000–6500 A with a continuous process with a closed top chamber.

Briquettes with an excess of reducing agent had a high electrical conductivity that, when melted, led to a significant increase in the load current, and as a result to the rise of the electrodes and the opening of the top. To eliminate this phenomenon, 10% manganese concentrate was added to the briquette charge. Thus, 14 tons of briquettes with an excess of reducing agent, 1.2 tons of concentrate and 200 kg of dolomite were melted. Melting silicomanganese on briquettes containing a stoichiometric amount of reducing agent, passed without any complications. The load was steady; except for briquettes, nothing else was added to the furnace. Comparative data of heats at different loads (briquettes, sinter) showed that the melting of silico-manganese on briquettes should be more preferable than melting on the sinter.

The results of pilot heats formed the basis for the construction and commissioning in 1976 of a briquette factory at the site of the Zestafoni Ferroalloy Plant. Manganese ore briquettes (with the addition of gas cleaning dust and without it) were produced by roller presses of Sahut Conreur Company (France).

Briquetted charge was also used for smelting ferrosilicon. At the Zaporizhia plant of ferroalloys (Ukraine) briquettes were produced of sand, coking coal and iron ore concentrate. The mixture was heated and stirred in a steam mixer and then pressed with a roller press. Raw briquettes were subjected to roasting at a temperature of 600–800°C for 10–13 hours to achieve a reduction degree of 70–80%. Smelting of ferrosilicon on the briquetted charge was carried out in a 3.5 MVA furnace. When working on briquettes, the furnace productivity increased, the power consumption and reducing agent rates decreased [7]. The smelting of ferroalloys from briquetted charge became widespread in the United States, France, Japan and Germany [8].

Melting technology of low-phosphorus carbon ferromanganese with 100% briquetted charge was described in [9]. In 1987, experimental tests of this technology were carried out under the conditions of the Nikopol Ferroalloy Plant in comparison with the traditional technology. A three-phase ore-smelting furnace with a capacity of 3600 KVA with coal lining was used as a melting unit. The essence of the technology was the smelting of carbon ferromanganese with the required low phosphorus content using the fluxless method on the charge, the ore part of which, in addition to graphite concentrate, contains a concentrate obtained by chemical enrichment of sludge. Low-phosphorus manganese slag obtained in the smelting of carbon ferromanganese by the fluxless method is used as a raw material in the smelting of low phosphorus silicomanganese, refined ferromanganese and metallic manganese. The results of the experimental heats with 100% briquettes showed that the briquetted charge provided an economically acceptable solution to the problem of smelting high-quality ferromanganese from highly phosphorous manganese ores.

In 1988, the first vibropress briquetting factory with a capacity of 110 thousand tons of briquettes for ferrochrome smelting was built at Vargön Alloys in Sweden [10]. Since 1991, this method of briquetting has been used in France by Eurometa SA for the agglomeration of the fines of the crushing of ferroalloys [11]. Ferroalloy fines briquettes were used in the foundry business.

The technology of smelting manganese ferroalloys using ore-dust briquettes was developed at the Zestafoni ferroalloy plant [12]. The need for disposal of dust and sludge from furnace gas cleaning was due to the large volume of its formation (50 thousand tons annually). Taking into account the metallurgical evaluation of waste, their dispersed state and physicochemical properties, it was proposed to mix dry dust with previously dehydrated sludge, and make briquettes from such a mixture with the addition of manganese ore concentrate. It has been established that to obtain mechanically strong briquettes (specific crushing force of 8–12 MPa), the optimum parameters of briquetting are: moisture content of the charge 4–6%, binder content (SAB) 6–8%, amount of fine component (dust, sludge) 30% and minimum pressure of roller-press 19.6 MPa. The comparative kinetics of the reduction of dust-sludge and ore briquettes and of manganese sinter was investigated. It was found that briquettes gave the greatest degrees of reduction at different temperatures. The utilization of manganese and silicon from wastes increased due to the presence of closely bound carbon and the inhibition of the formation of silicates, in contrast to sinter. The results of high-carbon ferromanganese heats showed that the furnace productivity in the case of using briquettes increased from 73.33 to 75.67 tons/day, and the specific power consumption decreased by 90 kWh/ton. Coke consumption decreased by 34 kg/ton.

The first successfully implemented project with the use of SVE technology in the ferroalloy industry was the production of brex and their use in the charge for ferronickel smelting at BHP Billiton in Cerro Matoso (Montelibano, Colombia [13]) in the framework of the RKEF (*rotary kiln - electric furnace*) process [14]. The nickel content in this ore is usually small and ranges from 0.8–3.0%. During the RKEF process, laterite ores are sieved, ground and added into a charge with some iron, nickel, SiO_2 and MgO content. After that, the charge material is roasted and partially metallized with coal or coke fines. This semi-product and residual coke are then delivered to the electric furnace where ferronickel is smelted. In Cerro-Matoso, the RKEF process entails the use of briquetted fines of nickel laterite ore.

SVE technology has not previously been used for the agglomeration of manganese and chromium ores. The first attempt to use it for such purposes took place in 2010. A full-scale pilot testing of this technology was conducted using brex as a charge component of industrial Submerged EAF. The possibility of obtaining metallurgically valuable brex based on various combinations of manganese ore fines and dry gas cleaning dust formed during the production of silicomanganese was studied in [15].

As in the case of blast furnace briquettes, the metallurgical properties of ferroalloy briquettes must ensure their integrity before entering the workspace of the reduction unit (cold strength) and in complex physical and chemical processes that occur at high temperatures under the reducing conditions of ore-smelting furnaces (hot strength).

Analysis of the combination of such processes allows one to distinguish four zones in the vertical section of the furnace, differing in the nature of the processes [16] occurring in them.

In the first zone (in the upper horizons of the top of the furnace) drying and calcination occurs (25–438°C):

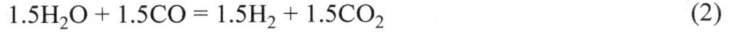

$$11H_2O = 11H_2O_{(g)} \tag{1}$$

$$1.5H_2O + 1.5CO = 1.5H_2 + 1.5CO_2 \tag{2}$$

$$9.5MnO_2 + 4.7CO = 4.7Mn_2O_3 + 4.7CO_2 \tag{3}$$

In the second zone, Mn_2O_3 is reduced by gas to Mn_3O_4 (438–773°C).

$$4.7Mn_2O_3 + 1.6CO = 3.2Mn_3O_4 + 1.6CO_2 \tag{4}$$

$$4.8Mn_3O_4 + 4.8CO = 14.4Mn_3O_4 + 4.8CO_2 \tag{5}$$

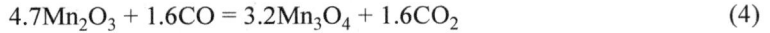

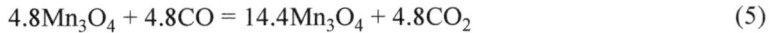

In the third zone (773–1250°C), direct reduction of Mn_3O_4 and carbon gasification reactions (Boudoir reaction) occur simultaneously:

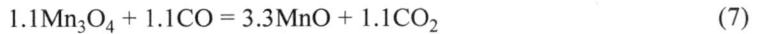

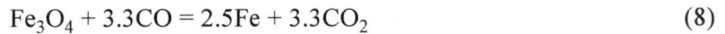

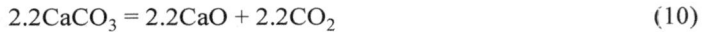

$$8.0C + 8.0CO_2 = 16.0CO \tag{6}$$

$$1.1Mn_3O_4 + 1.1CO = 3.3MnO + 1.1CO_2 \tag{7}$$

$$Fe_3O_4 + 3.3CO = 2.5Fe + 3.3CO_2 \tag{8}$$

$$1.4MgCO_3 = 1.4MgO + 1.4CO_2 \tag{9}$$

$$2.2CaCO_3 = 2.2CaO + 2.2CO_2 \tag{10}$$

In the fourth zone (1250–1500°C), MnO is melted in the slag and partially reduced to a liquid metal. In the same zone, SiO_2 is reduced and carbon is dissolved in the metal.

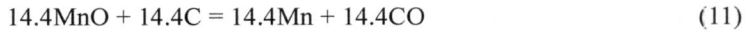

$$14.4MnO + 14.4C = 14.4Mn + 14.4CO \tag{11}$$

$$0.014SiO_2 + 0.028C = 0.014Si + 0.028CO \tag{12}$$

$$C = \underline{C} \tag{13}$$

The metallurgical properties of ferroalloy briquette must ensure that the "melting rate" and "reduction rate" are equal. If the melting rate of ore materials is greater than the rate of formation of the metal and the final slag, then partially reduced slags are formed. Such slags penetrate into the lower horizons of the furnace, the reduction process turns into an unfavorable slag regime, the slag boils and, in general, the normal operation of the furnace is disturbed. Otherwise, there is a need to intensify melting by increasing the input of electric power into the unit. The equality of the rates of melting and metallization ensures the entry of metal and non-aggressive slag from the layer of charge materials into the furnace hearth.

The character of the processes of reducing oxides of iron and manganese in the ore-reducing electric furnace is associated with the passage of a portion of the electric current through charge materials. When using briquettes with a substantial content of carbonaceous reducing agent in its composition, the current begins to flow through the body of the briquettes, heating them, which significantly speeds up the reduction processes inside the briquette, and the direct passage of current through the briquette can lead to its destruction. The choice of the optimal content of reducing agent in the briquette should also take into account the possibility of reducing the cold strength of the briquette by adding a carbonaceous reducing agent. At the same time, the presence of a carbonaceous reducing agent largely raises the temperature at which softening of briquettes begins (1250–1400°C versus 750–850°C for ore briquettes) and leads to an increase in the permissible critical current density [17]. In case of close contact of particles of manganese ore with carbon in briquettes, the reduction of higher oxides can occur with the participation of carbon:

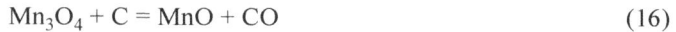

$$2MnO_2 + C = Mn_2O_3 + CO \tag{14}$$

$$3Mn_2O_3 + C = 2Mn_3O_4 + CO \tag{15}$$

$$Mn_3O_4 + C = MnO + CO \tag{16}$$

8.1 Metallurgical Properties of Brex on the Basis of Manganese Ore Concentrate

Brex from primary oxidized manganese ore concentrate were tested for the purpose of this study. They were manufactured on the industrial SVE line in India. Portland cement (5% mass) was used as a binder; bentonite (1% mass) was used as a plasticizer. Table 8.1 shows the chemical composition of the primary oxidized manganese concentrate.

The share of particles larger than 0.071 mm is 22%; the manganese content of these particles is 35.4%. The share of particles smaller than 0.071 mm is 78%; the manganese content of these particles is 38.77%. Iron particles distributed in the samples of the concentrate in both size ranges evenly – 3.81–3.86%. The surface area of cement particles exceeds 4000 cm^2/g. The chemical composition of cement is (%): CaO – 62–64; MgO – 2.5 (max); SiO$_2$ 22–24; Al$_2$O$_3$ – 4.5; Fe$_2$O$_3$ – 4.5; SO$_3$ – 1.8; alkali – 0.6. The chemical composition of bentonite (%) is: SiO$_2$ – 58.4; Al$_2$O$_3$ – 12.6; CaO = 0.8; Fe$_2$O$_3$ – 10.6; alkali – 4.4.

Physical and mechanical properties of the brex were measured after three-day strengthening in an open warehouse. The compressive strength was equal to 30 kgF/cm^2 and was measured on the Tonipact 3000 equipment (Germany) in accordance with the DIN 51067 standard. The open porosity was determined in accordance with DIN 51056 and amounted to 14.39%. Brex density averaged 2.59 g/cm^3.

A specially designed unit, described in 5.2 (Figs. 5.19 and 5.20) was used to study the behavior of industrial brex samples from primary oxidized manganese concentrate during heating in a reducing atmosphere.

The physical and mechanical properties of brex after heating at different temperatures are given in Table 8.2.

The results of the phase analysis of the samples of initial brex and brex after heating for 2 hours in a reducing atmosphere are presented in Table 8.3.

It is seen that in the presence of iron oxides an iron-manganese melt was formed, from which manganese was subsequently recovered.

The technology for the production of manganese ferroalloys is currently based on the use of manganese concentrates only of the highest grades, however, in this case, the phosphorus content

Table 8.1. Chemical composition of primary oxidized manganese concentrate.

Elements	Mn	Fe	SiO$_2$	Al$_2$O$_3$	CaO	MgO	P	S	Zn
Concentrate, %	38	3.84	10	1.36	17	0.76	0.07	0.10	0.084

Table 8.2. Physical and mechanical properties of brex after heating at different temperatures.

Properties/T, °C	1000	1100	1200	1300
Porosity, %	43.48	39.10	32.22	15.02
Density, (kg/cm^3)	2.31	2.49	2.82	3.28
Compressive strength (kgF/cm^2)	34	43	47	61

Table 8.3. Main phases of brex from primary oxidized concentrate.

Temperature, °C	Phases
20	Hausmannite (Mn_3O_4); Bixbyite (($Mn,Fe)_2O_3$); Rhodonite ($MnSiO_3$); Bixbyite C ($FeMnO_3$)
1000	Jacobsite ($MnFe_2O_4$); $Fe_{0.798}Mn_{0.202}$ O.
1100	$Fe_{0.798}Mn_{0.202}O$
1200	$Fe_{0.798}Mn_{0.202}$ O; $Fe_{0.664}Mn_{0.336}O$.
1300	$Fe_{2.08}Mn_{0.92}O_4$; $Fe_{0.798}Mn_{0.202}O$.
1400	Manganese

in manganese ferroalloys is at the permissible limit. The technology of double metallurgical processing, based on the electrometallurgical method of dephosphorization of manganese concentrates, adopted for the production of low-phosphorous carbon ferromanganese, has a number of significant disadvantages. Low-phosphorous slag obtained by the electrometallurgical method of dephosphorization contains 28–30% silica, therefore, carbon ferromanganese is smelted from low-phosphorous slag by the flux method. The smelting process by the flux method is characterized by low technical and economic indicators—significant losses of manganese with dump slag and high power consumption, which necessitated the creation of a more efficient technological scheme for smelting low-phosphorous carbon ferromanganese.

At the Institute of Metallurgy of the Russian Academy of Sciences (IMET) a technology was developed [18], which allows to obtain low-phosphorous carbon ferromanganese with high technical and economic indicators. According to this technology, ferromanganese carbonaceous with the required low phosphorus content is smelted by a non-flux method on a charge, the ore part of which, in addition to the graphite concentrate, contains a concentrate obtained by chemical enrichment of waste sludge. Low-phosphorous manganese slag, obtained in the smelting of carbon ferromanganese by the flux-free method, is used as a feedstock for the smelting of low-phosphorous silicomanganese, refined ferromanganese and metallic manganese.

The above-described technology for smelting low-phosphorous carbon ferromanganese has passed a pilot-industrial test under the conditions of the Nikopol Ferroalloy Plant in comparison with the existing technology. A three-phase ore-thermal furnace with a capacity of 1600 kVA with a coal lining was used as a melting unit. The composition of the used manganese-containing charge materials is given in Table 8.4.

The dithionate concentrate was produced at the experimental industrial complex for the chemical enrichment of sludge of the Marganets mining and processing plant and delivered in the form of briquettes. Low-phosphorous slag was used current, smelted at the Nikopol Ferroalloy Plant.

Low-phosphorous carbon ferromanganese grade FMn78A was smelted according to two options: by the flux method from low-phosphorous slag—the current technology and by the flux-free method from the dithionate concentrate—the IMET technology. When melting by the non-flux method, compared with the flux method, the furnace stroke was significantly smoother, and the electrodes in the charge were more stable. This is due to the fact that there are practically no fluxing additives in the charge of heats using the IMET technology, the consumption of the reducing agent is lower and, therefore, the electrical resistance of the charge is higher.

The composition of the metal and slag of both types of heats is given in Table 8.5. The phosphorus content in the metal smelted by the current technology exceeds the permissible level due to the

Table 8.4. Chemical composition of manganese-containing charge materials, %.

Material	Mn	P	Fe	SiO_2	CaO	Al_2O_3	M_gO	S	P/Mn
Low phosphorous slag	40.3	0.021	0.84	28.04	5.68	3.83	2.90	0.24	$5*10^{-4}$
Dithionate concentrate	51.7	0.012	1.71	2.76	7.04	3.03	0.83	2.53	$2*10^{-4}$

Table 8.5. Chemical composition of metal and slag (mass fraction, %).

Shot number	Metal					Slag							CaO/SiO$_2$
	Mn	C	Si	P	S	Mn	SiO$_2$	CaO	Al$_2$O$_3$	MgO	S	P	
Existing technology													
4	75.69	5.63	2.79	0.110	0.014	7.40	32.55	48.70	5.07	3.03	0.62	0.009	1.50
5	80.23	6.38	1.56	0.098	0.013	7.40	31.97	48.91	5.15	3.60	0.66	0.011	1 53
6	78.85	5.93	1.76	0.087	0.011	7.59	32.59	46.21	7.78	2.68	0.59	0.012	1 42
7	80.53	5.83	2.42	0.096	0.016	8.75	33.07	45.97	5.84	2.90	0.61	0.016	1 39
8	77.30	5.37	3.25	0.127	0.015	11.70	33.87	41.48	5.24	2.63	0.63	0.011	1 27
9	76.34	5.18	3.09	0.160	0.013	11.51	32.58	41.66	5.54	4.31	0.64	0.013	1 28
10	77.24	5.02	3.80	0.179	0.014	11.70	32.67	37.68	5.86	3.80	0.63	0.011	1 15
11	78.52	5.19	3.72	0.163	0.011	12.47	32.57	38.46	4.91	4.37	0.61	0.014	1.18
12	79.03	5.33	3.70	0.129	0.014	11.51	32.97	37.85	5.56	4.45	0.62	0.012	1 15
13	77.74	4.90	3.38	0.124	0.011	14.0	33.33	34.80	3.62	4.35	0.62	0.005	1.04
IMET technology													
15	78.15	4.48	3.92	0.058	0.022	38.9	26.90	16.36	1.46	1.11	4.01	0.005	0.61
16	80.57	5.31	1.24	0.060	0.022	39.3	25.66	16.74	1.24	0.70	3.61	0.005	0.65
17	80.47	5.16	1.61	0.058	0.015	43.2	25.61	14.33	1.16	1.38	3.87	0.005	0.56
18	80.62	5.37	0.67	0.052	0.019	42.4	24.90	14.58	1.31	0.48	3.82	0.005	0.59
19	81.07	5.71	0.33	0.047	0.014	42.6	23.16	14.33	1.29	0.98	4.19	0.005	0.62
20	81.53	5.38	0.43	0.046	0.022	40.3	23.36	15.55	0.72	2.05	4.43	0.005	0.67
21	81.42	5.25	1.41	0.052	0.017	35.4	27.25	19.39	2.00	0.94	3.48	0.005	0.71

significant amount of phosphorus supplied with the beads of the high-phosphorous associated metal contained in the low-phosphorous slag [19]. The phosphorus content in the metal smelted using the IMET technology meets the requirements of GOST. The increased sulfur content in the dithionate concentrate does not lead to a notable increase in the sulfur content in the metal due to the low solubility of sulfur in carbon ferromanganese [20]. The high silicon content in the metal smelted according to the current technology is explained by the high silica content in the low-phosphorous slag.

In the slag of smelts according to the current technology (smelting No. 8–13), the manganese content was 10–12%, the slag is dump, therefore, the manganese contained in the slag is lost. The manganese content in the slag can be reduced by increasing its basicity (smelting No. 4–7), but this does not lead to an improvement in technical and economic indicators, since the slag multiplicity and its viscosity increase, which leads to a deterioration in the furnace operation. In the slag of smelts using the IMET technology, the manganese content was 38–42%, the conversion slag is used as a low-phosphorous manganese-containing charge material in the smelting of silicomanganese, refined ferromanganese and metallic manganese.

Technical indicators of heats are given in Table 8.6. In the case of the IMET technology, the slag ratio is two times lower, and the useful use of manganese is 20% higher than in the case of the existing technology due to the subsequent use of the conversion slag. The energy costs required for the process, in the case of smelting carbon ferromanganese by the flux-free method, compared to the flux method, are significantly lower, therefore, when smelting using the IMET technology, the specific power consumption is 1.5 times lower, and the furnace productivity is 1.5 times higher. The specific consumption of manganese-containing materials in terms of the basic content of manganese 48% is slightly lower in the case of the current technology, the specific consumption of the remaining charge materials, primarily coke, is significantly lower in the case of the IMET technology.

Table 8.6. The main indicators of the process of low-phosphorous carbon ferromanganese smelting.

Parameter	Existing technology	IMET technology
Mass fraction in metal, %:		
Mn	78.10	80.55
C	5.26	5.24
Si	3.34	1.37
P	0.140	0.053
S	0.013	0.019
Mass fraction in slag, %:		
Mn	11.66	40.16
SiO_2	33.01	25.26
CaO	39.70	15.90
Slag basicity CaO/SiO_2	1.20	0.63
Slag ratio	2.21	1.15
Mn distribution, %:		
into metal	65.75	55.38
into slag	21.77	32.88
LOI	12.48	11.74
Useful Mn utilization, %	65.75	88.26
Capacity, bas.t/day	4.10	6.44
Specific material consumption, kg/bas. t:		
low phosphorous slag (48% Mn)	2470.9	—
dithionate concentrate (48% Mn)	—	2934.3
Coke nut	662.4	476.8
limestone	1704.7	—
quartzite	—	272.4
Iron chips	132.5	109.0
Specific power consumption, kW * h/base t.	5556.5	3485.5

The analysis of the obtained results shows that the IMET technology based on the briquetting of the charge provides a solution to the problem of production of high-grade manganese ferroalloys from Nikopol manganese ores, characterized by an increased phosphorus content. The introduction of this technology allows: to provide the steel industry with manganese ferroalloys with the required low phosphorus content; to significantly increase the end-to-end extraction of manganese; largely reduce energy costs for alloy smelting and, accordingly, increase the productivity of ferroalloy furnaces; mainly reduce the cost of low-phosphorous carbon ferromanganese; to provide low-phosphorus manganese-containing raw materials (conversion slag of flux-free smelting of carbon ferromanganese) smelting processes of silicomanganese, refined ferromanganese and metallic manganese; significantly improve the use of mineral resources.

References

[1] Tolochko, A.I. 1990. Utilization of Dust and Residues in Ferrous Metallurgy. Tolochko, A.I., Slavin, V.I., Suprun, Y.M., Khairudinov, R.M. Moscow: Metallurgy, 143 p. (in Russian).
[2] Gasik, M.I., Lyakishev, N.P. and Emlin, B.I. 1988. Theory and Technology of Ferroalloys Production. Moscow: Metallurgy, 784 p. (in Russian).

[3] Zhuchkov, V.I., Smirnov, L.A. and Zayko, V.P. 2008. Technology of Manganese Ferroalloys. Yekaterinburg: UrB RAS, 442 p. (in Russian).

[4] Khazanova, T.P. 1961. Production of manganese alloys from poor oxide and carbonate ores. Khazanova, T.P., Shearer, G.B. and Lyakishev, N.P. Development of the USSR Ferroalloy Industry. - Kiev. P. 122 (in Russian).

[5] Khvichiya, A.P. 1970. Smelting of silico-manganese from ore briquettes in a 16.5 MVA furnace. Khvichiya, A.P., Mazmishvili, S.M. Steel. 2: 138 (in Russian).

[6] Sukhorukov, A.I., Sosedko, P.M. and Khitrik, S.I. 1970. Smelting of marketable silicomanganese on briquettes and sinter. Steel. 2: 135 (in Russian).

[7] Kozhevnikov, I.Yu. and Ravich, B.M. 1991. Agglomeration and the foundations of metallurgy. M.: Metallurgy, 296 p. (in Russian).

[8] Ravich, B.M. 1982. The Briquetting of Ores [in Russian], Nedra, Moscow (in Russian).

[9] Dashevskyi, V.Ya., Kashin, V.I., Lyakishev, N.P., Velitchko, B.F. and Ishutin, V.I. 1992. Improving the technological processes of production of manganese ferroalloys. Izvestiya VUZov. Ferrous Metallurgy 12: 45 (in Russian).

[10] Electronic resource. http://www.vargonalloys.se/index_eng.html.

[11] Electronic resource. http://www.eurometa.fr/en/.

[12] Mazmishvili, S.M. 1992. Development and industrial development of technologies for producing dust briquettes and smelting manganese ferroalloys from them. Mazmishvili, S.M., Simongulashvili, Z.A. Izvestiya VUZov. Ferrous Metallurgy 12: 43 (in Russian).

[13] Duarte, A. and Lindquist, W.E. 1999. Recovery of Nickel Laterite Fines by Extrusion. Proc. 27th Biennial Conference. IBA, USA, pp. 205–217.

[14] Christian Redl, Matthias Pfennig, Roland Kristl, Siegfried Fritsch, Heinz Muller Refining of Ferronickel. 2013. The Thirteenth International Ferroalloys Congress Efficient Technologies in Ferroalloy Industry 229 June 9–13, 2013 Almaty, Kazakhstan.

[15] Bizhanov, A., Kurunov, I., Podgorodetskyi, G., Dashevskyi, V., Pavlov, A. and Chadaeva, O. 2014. Extruded briquettes—new charge component for the ferroalloys production. ISIJ International, vol. 54(10): 2206–2214.

[16] Olsen, S.E., Tangstad, M. and Lindstad, T. 2007. Production of Manganese Ferroalloys, P. 247, SINTEF and Tapir Academic Press, Trondheim.

[17] Gasik, M.I. 1992. Manganese. Gasik, M.I. Moscow, 608 p. (in Russian).

[18] Dashevsky, V.Ya., Dashevsky, Ya.V., Kashin, V.I. et al. 1983. Russian Patent. 1002390 (USSR). Publ. in BI 9: 92.

[19] Ovcharuk, A.N., Chepelenko, Yu.V., Kucher, A.G. et al. 1974. In the Book: Metallurgy and Coke Chemistry, Kiev: Technique 39: 17–20.

[20] Dashevsky, V.Ya. and Kashin, V.I. 1973. Izv. USSR Academy of Sciences Metals. 5: S. 85–88.

8.2 Metallurgical Properties of Brex on the basis of Manganese Ore Concentrate and Baghouse Dusts of Silicomanganese Production

Chiatura manganese ore concentrate (Republic of Georgia) and baghouse dust from silicomanganese production was used as raw material for brex production. The chemical composition of the manganese ore concentrates and silicomanganese production baghouse dust is given in Table 8.7. The main minerals of the concentrate are pyrolusite, psilomelan and manganite. Pyrolusite from Chiatura manganese ore has a finely-spherulitic (oolitic) and clastic brecciate structure. Spherulites fragments do not exceed 0.25 mm in size. Pyrolusite is partially recrystallized up to hundredths of a millimeter during grain coarsening. The Chiatura oxide concentrate softening point is 1049°C according to the data in [1]. Eighty-five percent of baghouse dust particles are smaller than 10 μm. Primary particles have a spherical shape due to phase transitions.

Table 8.7. Chemical composition of manganese ore concentrate and baghouse dust.

Material	Mn	MnO	Fe_2O_3	FeO	Al_2O_3	SiO_2	CaO	MgO	P
Concentrate	35.0–45.0	-	1.2–2.2	-	2.0–3.5	14.0–25.0	2.0–3.5	0.86	0.20
Baghouse dust	-	29.74–31.41	-	0.30	0.01	30.0–34.0	1.67–2.42	1.50	-

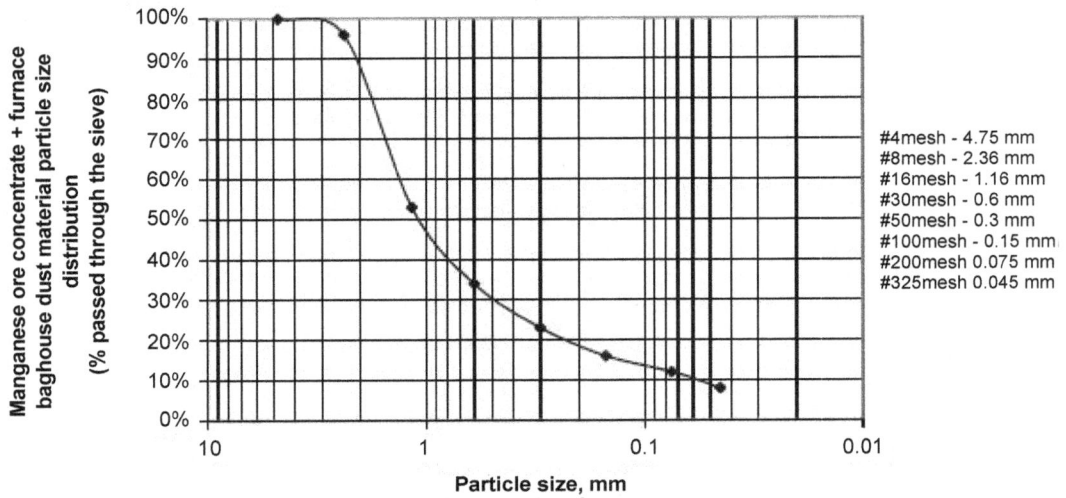

Figure 8.1. Granulometric composition of mixture for brex production.

Table 8.8. Composition of brex.

Component	No. 1	No. 2	No. 3
Manganese ore concentrate	47.6	66.7	56.0
Coke breeze	-	-	15.0
Baghouse dust	47.6	28.6	24.0
Portland cement	4.8	4.7	5.0

Two kinds of mixtures from the above materials were prepared for brex making: (1) a mixture of an equal mass (50:50) of concentrate and dust, and (2) a mixture of 70% concentrate and 30% dust. The granulometric composition of the second mixture is shown in Fig. 8.1.

Coke breeze was used as a reducing agent; it was provided by the owner of the ferroalloy smelter. The coke breeze has the following composition ratio (%): carbon – 68.72; volatiles – 7.53; ash – 23.75. 88.46% of coke breeze particles are smaller than 0.635 mm in size. Its moisture content is 9.86%. Three brex mixtures with low binder content were tested in the study (Table 8.8).

Test samples of brex were produced using a computerized laboratory extruder that simulates the processing of a briquetted mixture in commercial even feeders, pug mills and the extruder itself. The brex had a circular cross section of 2.5 mm in diameter and 1.5–2.0 mm in length. The average moisture content of freshly formed brex was 11%. The vacuum level in the working chamber of the extruder was maintained at 38–48 mm Hg.

Due to the cylindrical shape of brex, its cold strength was determined by measuring both compressive strength and tensile splitting strength, since these types of stress would be exerted on brex during actual transport and stacking. These properties were measured using an Instron 3345 system (USA). The compressive and tensile splitting strength tests were performed using six specially-prepared brex samples with a diameter of 25 mm and a length of 20 mm which were subjected to compression and tensile splitting.

Figures 8.2 and 8.3 illustrate the behavior of brex and the mechanism of brex destruction during compressive and splitting strength tests. At the initial stage (1), the elastic loading of brex shows a linear dependence between the applied force and the displacement of the active pressing surface. The beginning of the second stage is characterized by a sharp decrease in the applied force due to the formation of surface fractures. At this stage, the "near-surface" layers of the brex are destroyed and peeled (15–20% of the radius). However, the "core" (central part) of the brex continues to endure

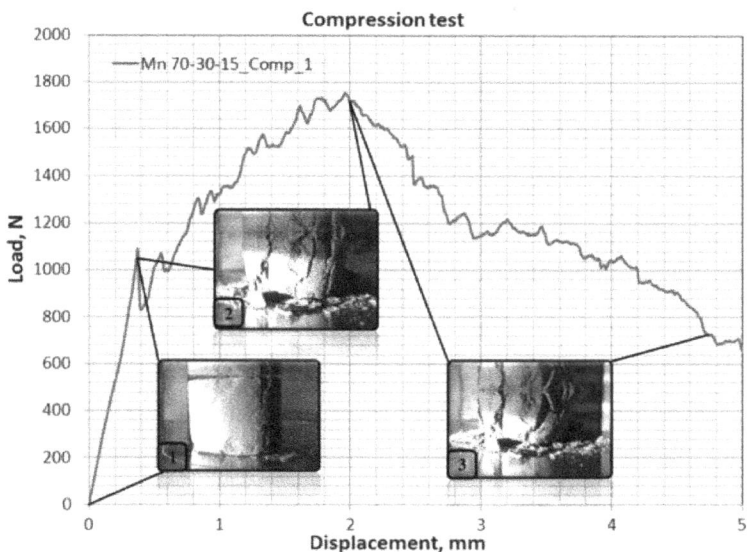

Figure 8.2. Compression test of brex No. 3.

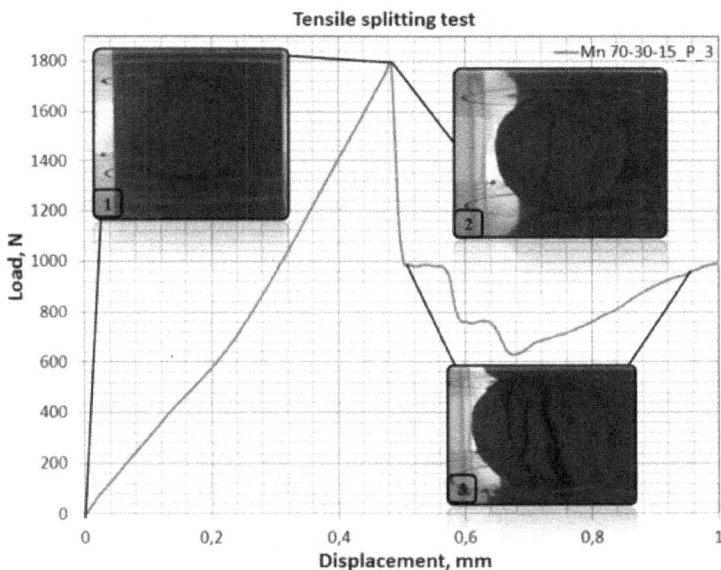

Figure 8.3. Tensile splitting test of brex No. 3.

the pressure, as evidenced by the core's growth. A clear zonal structure of the brex can be observed in the radial direction. The near-surface layers of the brex have a relatively higher stiffness than its "core". This statement is supported by the difference in slopes of the curves at Stage 1 and Stage 2.

In the third stage, once the applied force reaches its maximum, the "core" of brex is destroyed with a complete strength loss.

Table 8.9 illustrates the average compressive strength and splitting strength.

Subsequently, compressive strength of samples of industrially-manufactured brex No. 2 was tested by an independent laboratory L. Robert Kimball & Associates, Inc (USA). The compressive strength of a sample with a diameter of 32.3 mm and a length of 57.7 mm was 206 kgF/cm^2. It is clear that the results obtained for compressive strength significantly exceed the compressive strength of briquettes required for Submerged EAF.

Table 8.9. Mechanical strength of brex, kgF/cm².

Brex No.	Tensile splitting strength	Compressive strength
1	18.3	160.8
2	28.6	291.2
3	13.1	124.6

Brex porosity was examined by electron microscopy using an LEO 1450 VP microscope (Carl Zeiss, Germany) with a resolution of 3.5 nm in combination with X-ray computed tomography using a Phoenix V | tome | X S 240 tomograph (General Electric, USA). The computed tomography method was used to determine the sample porosity in the pore size range above 100 μm. The electron microscopy method was used to study the porosity fraction represented by pores with sizes less than 100 μm. Interpretation of these microscopic studies was carried out using the computer program STIMAN [2]. It was found that the porosity of the brex was 18.5% (5.6% is the porosity measured by computed tomography and 12.9% by electron microscopy). Subsequently, the porosity of industrially manufactured brex with the same content of major components (with 3% Portland cement) was measured at the mentioned above independent laboratory and amounted to 18.7%.

Thermal studies using powder samples with a mass of 50–70 mg, carried out on an STA 449 C installation (Germany) in an argon atmosphere, in the temperature range of 20–1400°C at a heating rate of 20°/min, allowed to determine the nature of phase transformations in brex during their heating and showed that the thermal effects on the DSC curve for brex No. 1 and No. 2 are almost identical (Fig. 8.4).

The manganite ($MnO(OH)$) dehydrates to form pyrolusite or β-kurnakit ($2MnOOH = Mn_2O_3 + H_2O$) in the temperature range 300–450°C. At temperatures above 400°C, psilomelane of manganese ore transforms to hollandite or Hausmannite. Exothermic peaks with a maximum temperature of 795.8°C in brex No. 1 and 801.0°C in brex No. 2 are most likely associated with the dissociation of pyrolusite and the formation of β-kurnakit which is accompanied by oxygen release and mass loss. The same peak in brex No. 3 at a temperature of 742.8°C is not followed by a loss of mass; it can be explained by the effects of recrystallization of amorphous phases. β-kurnakit decomposes in the temperature range 900–1050°C, and the formation of β-Hausmannite takes place [3]. In the range 1080–1250°C, β-Hausmannite is polymorphically converted to γ-Hausmannite.

In brex No. 3, the endothermic peak in the 150–200°C temperature range is accompanied by mass loss due to sorption moisture removal. In the 300–500°C temperature range there are no pronounced endothermic peaks, but there is a loss of mass associated with the reduction of MnO_2 to Mn_2O_3. A phase transformation of manganite ($MnOOH$) into kurnakit occurs in the same temperature range. The endothermic effect with a maximum at 1031.7°C is associated with the reduction of Mn_3O_4 to MnO due to mass loss; the effect at 1056.1°C is determined by the ferric phase change.

The endothermic peak with a maximum at 1122.7°C is determined by forming manganese carbide Mn_3C, where carbon recovers manganese from MnO. The coke breeze in brex provides for intensive iron and manganese reduction, marked by endothermic peaks at 1122.7°C and 1213.2°C.

Figure 8.5 shows the microstructure of brex No. 2.

Portland cement is known to lose its binding properties at 750–900°C. Therefore, it is important to understand what provides for the hot strength of brex at temperatures higher than the above-mentioned temperature range. An earlier study [4] explored the behavior of a pyrolusite sample of manganese concentrate from the Chiatura mine when heated in a reducing atmosphere (helium 40% and hydrogen 60%; heating at a rate of 20°C per minute). A cubic sample measuring 5 × 5 × 5 mm was observed to gradually expand by a considerable amount up to 0.8 mm; the thermal expansion was accompanied by sample cracking. Cracks were first observed at 500°C, with the number of cracks increasing with heating. The fractures formed during the tests are characterized by a certain orientation pattern. Radial and concentric cracks are found within the spherulites. Cracks in random directions are found in the inter-spherulites cementing mass. Despite this fracture pattern,

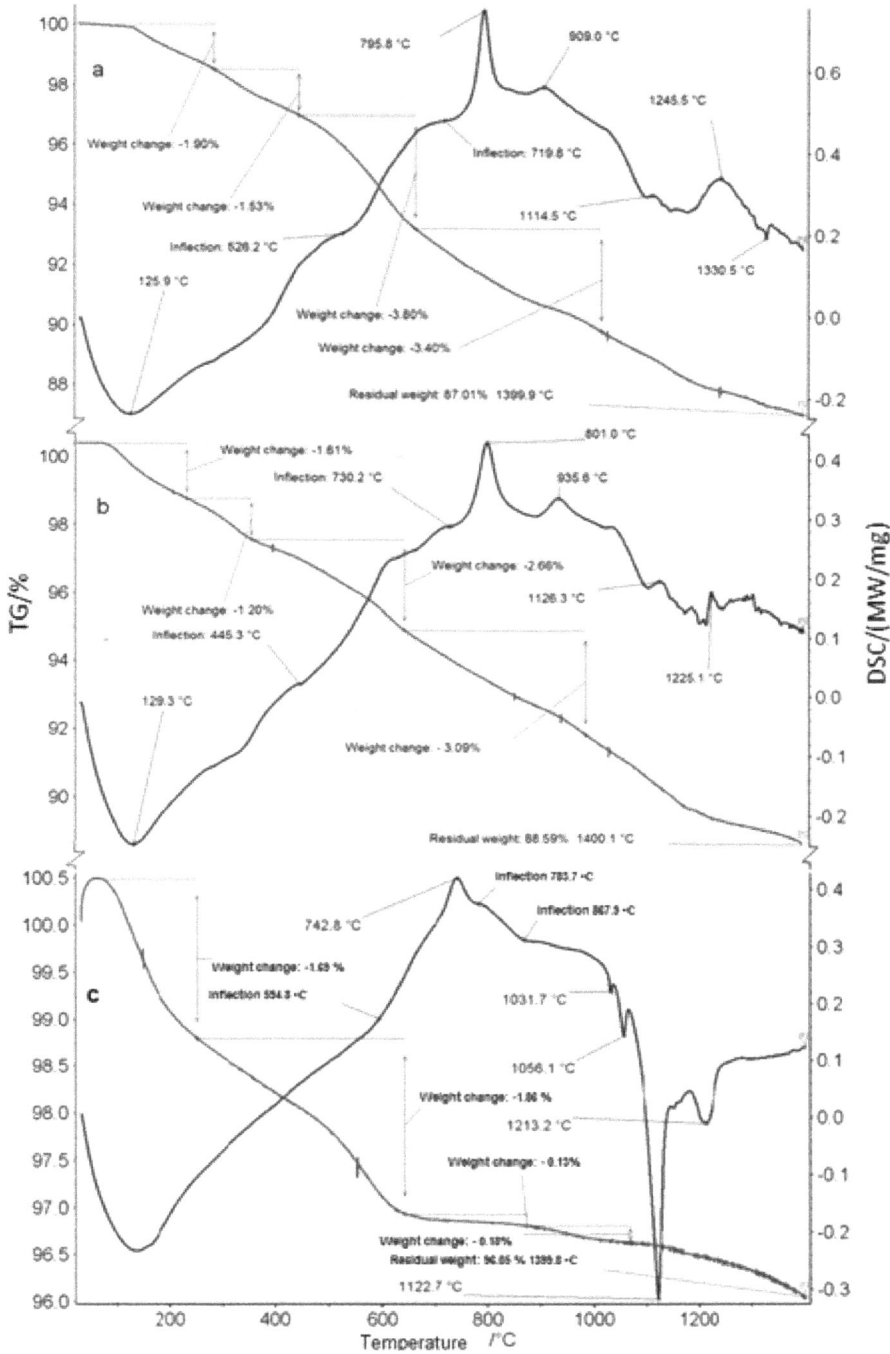

Figure 8.4. Thermograms of the brex samples No. 1 (a), No. 2 (b), No. 3 (c) (left hand vertical TG/%; right hand vertical - DSC/(MWt/mg).

the sample maintained an integrity at 800°C. After that, a slow and irreversible compression process began. In our opinion, a dense silicate phase (olivine or Wollastonite) protected the sample from destruction. The formation of the phase in the indicated temperature range corresponds to the well-known phase diagrams of systems Ca_2SiO_4-Mn_2SiO_4 and $CaSiO_3$-Mn_2SiO_3 (Figs. 8.6 and 8.7 [5]).

Figure 8.5. Microstructure of brex No. 2 structure. (1 - pyrolusite, 2 - gaussmanite, 3 - diopside, 4 - fayalite, 5 - mullite, 6 - pyrolusite, 7 - solid solution of Ca_2SiO_4-Mn_2SiO_4).

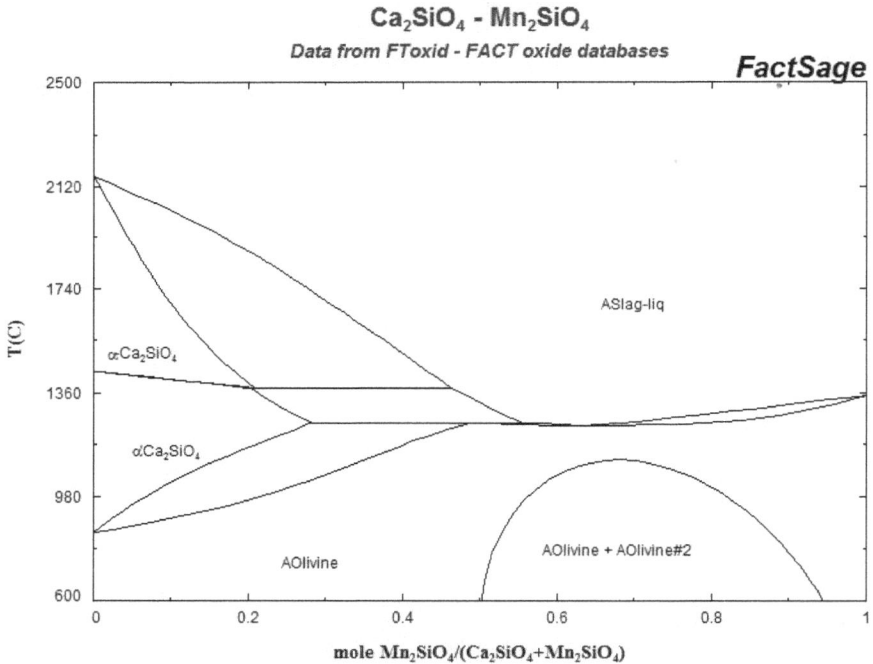

Figure 8.6. Diagram of system Ca_2SiO_4-Mn_2SiO_4.

Manganese silicates (tephroite-Mn_2SiO_4 and rhodonite-$MnSiO_3$) were discovered in the intervals between the oxide phase's grains of the Chiatura concentrate sample.

Up to 1250°C, the strength of brex is provided by the dense structure of the mentioned silicate phases, and above this temperature two eutectics are formed in the MnO-SiO_2 system: tephroite + rhodonite + liquid (1251°C) and tephroite + manganosite + liquid (1315°C) which contribute to preserving the strength of the brex (Fig. 8.8, [6]), due to the appearance of a liquid binder.

An important property of brex for smelting ferroalloys in Submerged EAF is the specific electric resistance. The measured specific electrical resistance of brex samples at room temperature is given in Table 8.10.

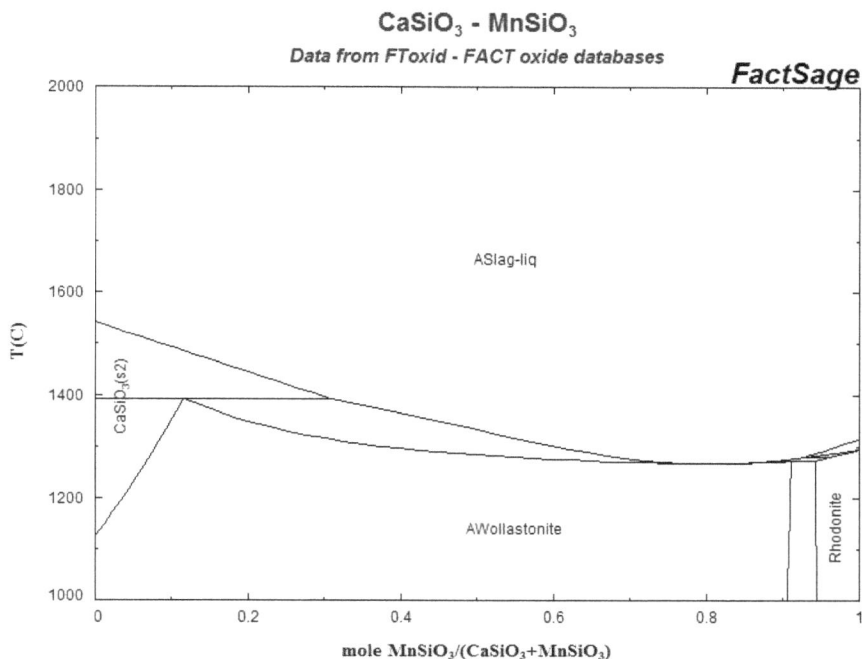

Figure 8.7. Diagram of system $CaSiO_3$-$MnSiO_3$.

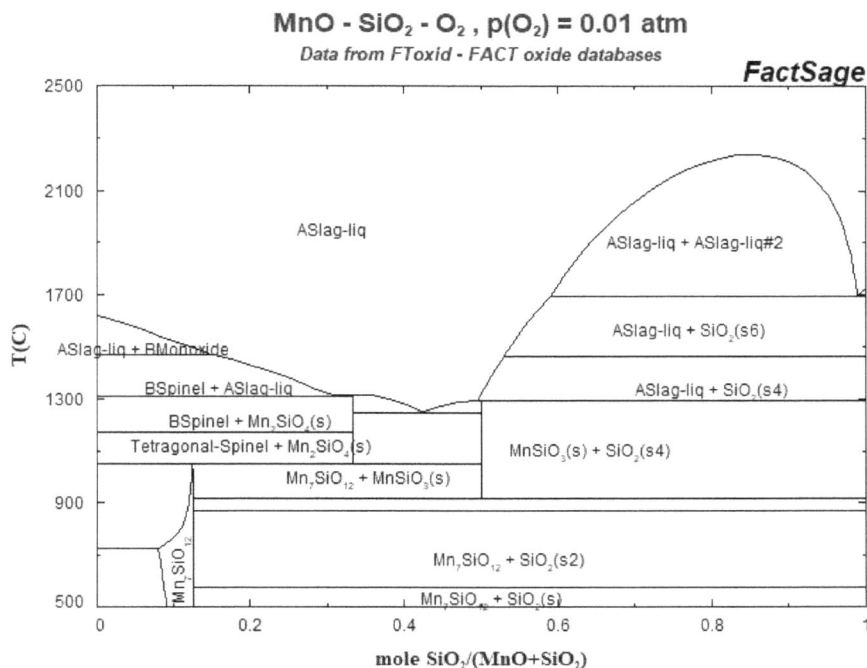

Figure 8.8. Phase diagram of the MnO-SiO_2 system [6].

Table 8.10. Specific electric resistance of brex No. 1–No. 3 at room temperature.

Brex, No.	ρ, MOhm·mm	
1	26.3	25.5
2	32.3	28.9
3	9.42	12.4

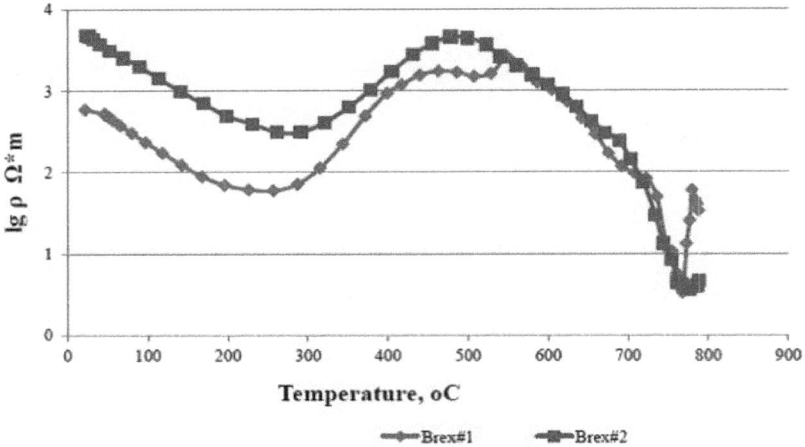

Figure 8.9. Specific electric resistances of brex No. 1 and brex No. 2.

The measuring system used in the study made it possible to investigate the dynamic pattern of brex resistance at a temperature up to 800°C. As a result, the behavior of brex resistance was studied up to the first exothermic peak described in Fig. 8.4. It corresponds to a relatively deep brex immersion in the furnace to levels where lower manganese oxide is formed. Due to the low electric resistance of brex No. 3 (with 15% coke breeze in the mass of brex), which is lower than the resistance of lump manganese ore of the same deposit [7], this type of brex was excluded from further study.

Figure 8.9 shows the results of resistance measurements in brex No. 1 and brex No. 2 when heated to 800°C.

The increase in resistivity of brex in the 300°C–500°C temperature range coincides with the above-mentioned endothermic effects of mass loss, which is associated with manganite dehydration and with kurnakit formation.

It is clear that the resistivity of brex No. 1 is lower than that of brex No. 2 because of the higher carbon content in brex No. 1 resulting from a higher baghouse dust content that adds carbon to brex No. 1.

Based on the above-mentioned conclusions regarding the physical and mechanical properties of brex, the behavior of brex during heating, and measurements of the specific electrical resistivity, brex No. 2 was selected as a component of the Submerged EAF charge.

References

[1] Zhdanov, A.V. 2007. Study of reducibility of manganese ore raw materials. Zhdanov, A.V., Zayakin, O.V., Zhuchkov, V.I. Electrometallurgy. 4: 32–35 (in Russian).
[2] Sokolov, V.N., Yurkovets, D.I. and Razgulina, O.V. 1997. Determination of tortuosity coefficient of pore channels by computer analysis of SEM images. Proceedings of Russian Academy of Sciences, Physical Series. 61(10): 1898–1902 (in Russian).
[3] Ivanova, V.P., Kasatov, B.K. and Krasavina, T.N. 1974. Thermal analysis of minerals and rocks. Leningrad: Resources Publishing House, 399 p. (in Russian).

[4] Tolstunov, V.L. and Petrov, A.V. 1989. Study of the processes of phase and microstructural transformations in manganese ores during their reducing heating. Izvestiya VUZov. Ferrous Metallurgy 4: 9–14 (in Russian).

[5] Glasser, F.P. 1962. The ternary system CaO-MnO-SiO$_2$. Glasser, F.P. Journal Amer. Ceram. Soc. 45(5): 242.

[6] Gasik, M.I. 1992. Manganese. Gasik, M.I. Moscow, 608 p. (in Russian).

[7] Zhdanov, A.V. 2007. Study of electric resistance of materials and batches used for ferromanganese production. Zhdanov, A.V., Zayakin, O.V. and Zhuchkov, V.I. Electrometallurgy 6: 24.

8.3 Full-scale Testing of Silicomanganese Smelting with Brex in the Charge of Submerged EAF

For the full-scale testing an experimental batch of 2000 tons of brex of 1400 tons of Chiatura manganese oxide and 600 tons of aspiration dust from production of silicomanganese were manufactured using the Steele 75 extruder (Fig. 8.10). The components of the mixture were mixed in the ore yard by a front loader and fed to the extrusion line.

In the manufacture of this batch, it was decided to limit the proportion of binder (Portland cement) in the charge mixture of 3%, which was sufficient to ensure the strength of brex that withstood 20 overloads on the way from the briquetting factory to the ferroalloy plant: extruder - conveyors - dump truck - piles - front loader - truck - stack at the port of loading - front loader - bunker - grab - barge - grab at the port of discharge - bunker - conveyors - dump truck - stack in the ore yard of the ferroalloy plant wheel loader - hopper - oven. At the same time, the cumulative formation of fines (less than 6 mm) did not exceed 10%. Such a decrease in the mass content of the binder was made possible due to the well-known binding properties shown by aspiration dust from the production of manganese ferroalloys.

The pilot campaign was conducted in a stably operating industrial Submerged EAF with a capacity of 27 MVA, with a productivity of 85 tons/day. The specific power consumption averaged 4,200 kWh/t. The degree of recovery of manganese 80%. The manganese content in waste slag is 12–14%.

The chemical composition of the manganese components of the mixture is given in Table 8.11.

Figure 8.10. Production of an experimental batch of brex No. 2.

Table 8.11. Chemical composition of charge components.

Material	Components composition						
	Mn	**SiO$_2$**	**CaO**	**MgO**	**Al$_2$O$_3$**	**Fe**	**P**
Ore-1	49.5	13.0	0.7	0.5	1.0	4.0	0.05
Ore-2	29.0	20.2	5.9	5.2	2.1	0.9	0.06
Brex	31.37	24.32	6.10	2.05	2.79	1.36	0.12
Scraps	23.3–38.6	N/A	N/A	N/A	N/A	N/A	N/A
Silicomanganese fines briquettes (vibropressed)	53	18	4	N/A	N/A	10.5	0.15

Table 8.12. Composition of the charge (net tons).

Component of the charge	Reference and full-scale trial periods						
	Reference period	**1**	**2**	**3**	**4**	**5**	**6**
Mn Ore-1	0.526 (30%)	0.525 (30%)	0.525 (30%)	0.525 (30%)	0.525 (30%)	0.525 (30%)	0.525 (30%)
Mn Ore-2	1.205 (70%)	1.120 (65%)	1.030 (60%)	0.855 (50%)	0.705 (41%)	0.600 (35%)	0.525 (30%)
Brex	-	0.087 (5%)	0.175 (10%)	0.350 (20%)	0.500 (29%)	0.605 (35%)	0.690 (40%)
Estimated weight of charge	1.730	1.732	1.730	1.730	1.730	1.730	1.740
Incoming manganese with:							
Ore-1	0.262 (43%)	0.262 (42.6%)	0.262 (42.6%)	0.262 (42.4%)	0.262 (42.2%)	0.262 (42.1%)	0.262 (41.8%)
Ore-2	0.350 (57%)	0.325 (53%)	0.299 (48.6%)	0.224 (40.1%)	0.205 (32.8%)	0.174 (27.9%)	0.152 (24.2%)
Brex	-	0.027 (4.4%)	0.054 (8.8%)	0.109 (17.5%)	0.160 (25%)	0.188 (30%)	0.214 (34%)
Manganese charge weight	0.612	0.614	0.614	0.619	0.622	0.624	0.628
Average manganese content, %	35.4	35.5	35.6	35.8	36.0	36.1	36.1

Scraps are the internal wastes generated while deslagging the surface of metal in a ladle, cleaning buckets and furnace notch, sorting at slag dumps, and casting silicomanganese. Only the total manganese content in slags was determined; these values are given in Table 3.31. Briquettes made from silicomanganese fines are produced by the process of vibropressing (unconditioned fines fraction 0–6 mm). To impart the necessary strength to the briquettes made from these fines, Portland cement was used (10% mass).

A preliminary calculation of the composition of the charge was carried out to achieve stable manganese content in the ore part of the charge (Table 8.12).

For the accuracy of the comparison of the results of the furnace operation with and without brex, a month-long period of furnace operation was observed. A weeklong period of the furnace operating without brex immediately preceding the beginning of the pilot period was taken as a reference.

It was decided to begin with 5% of the brex share in the ore part of the charge to get a first experience with the agglomerated burden. Over the course of three days of observation, no visible changes had been registered in the process of silicomanganese smelting; the furnace worked smoothly, with a constant and uniform current load during the period of brex use. The slight 0.3% decrease in manganese extraction during this period was due to furnace downtime associated with the tap-hole unit repair (1.5 hours downtime). Then, for four consecutive days, the brex share in the charge was maintained at 10%. Improvement in the furnace top functioning was visually apparent. Gas flames throughout the furnace's top area confirmed an improvement in gas permeability of the column of the charge and the uniformity of the temperature distribution over the surface of the furnace top. Deeply submerged electrodes functioned without producing surface blowholes in the electrodes circle and in the surrounding area. No sintering in the electrodes circle was observed. During the next three days, Start brex share in the charge was increased up to 20%. The furnace operated smoothly, the electrodes were submerged deeply, the current load had no visible abnormalities or jolts, and the gas permeability of the charge was good for the entire area of the furnace top. Throughout the next week, the brex share of the charge was increased to 29%. The furnace worked well. An increase in the rate of the charge descent in the electrodes circle, especially during the tapping, indicates an increase in the melting rate of the charge in the active zone of the furnace. The furnace top functioned smoothly without surface blowholes with a constant load current, without bumps and drops. Melt out was good, metal and slag were sufficiently warmed up. However, at the end of this period, there was a furnace downtime for 4 hours due to reasons not related with the brex presence in the charge (electrical problem in the transformer).

The beginning of the next phase of the pilot period, when the brex share rose to 35% substituting the ore-2, was associated with a problem caused by a disruption in the tap-hole electrode bypass, resulting in its shortening and the deterioration of the melt output. The bulk of the slag remained in the furnace. This led in an increased coke rate and scraps withdrawal from the charge along with briquettes of silicomanganese fines. During this period, a deterioration in the technical and economic performance of the furnace had been registered, namely an increase in specific energy consumption and specific consumption of manganese ore raw materials, as well as a reduced extraction of manganese. To restore its normal functioning, the furnace heating rate was increased so as to warm up the slag and ensure its normal tapping.

In the final phase (week) of the pilot operation period, the furnace worked with the brex share in the charge up to 40%. The furnace operation was characterized by a good current load, top gas release was smooth, with no surface blowholes and furnace charge downslides, and with a normal tapping.

As the share of brex in the ore part of the charge increased, the share of ore 2 was proportionally reduced, which is an equivalent substitute for the amount of manganese introduced, since the average manganese content in brex was 31% and in ore - 29% (Tables 8.13 and 8.14).

Substitution of the essential part of manganese ore in the charge with brex based on ore fines and aspiration dust led to an improvement in the technical and economic indicators of the process as a whole. The pilot industrial campaign itself went without any visible changes in the technological process: the furnace worked smoothly, with a constant current load, the melt was produced according to the schedule, there were no significant changes in the chemical composition of the metal and slag, and the gas permeability of the furnace throat improved.

The specific energy consumption during the experimental period significantly decreased. In the base period, the power consumption per 1 ton of alloy was 4091 kWh. When 40% of manganese-containing brex were fed into the ore part of the charge, the specific energy consumption decreased to 3727 kWh per 1 ton of alloy (Fig. 8.11).

The next positive factor of the campaign was an increase in extraction of manganese from the ore component of the charge. If in the comparative period of the furnace operation without brex, manganese extraction from the ore part of the charge averaged 80%, while using 30% of the brex in the ore part of the charge, manganese extraction reached 83.5% (Fig. 8.12).

The deterioration of the manganese extraction parameters in the period preceding the final one was not related to the presence of brex in the charge and was due to the stopping of the furnace and problems with the electrode. The reduction in coke consumption is illustrated in Fig. 8.13.

Table 8.13. Furnace parameters during the reference and the full-scale trial period.

Parameter		Reference and Full-scale trial periods						
		Reference period	1	2	3	4	5	6
Actual metal production over a period of time, t	ton	816.323	298.4	277.1	196.6	570.6	397.2	757.2
	b.t. (basic ton)	839.670	300.7	285.8	199.5	584.75	393.2	767.8
Actual furnace performance, %		98.9	97.6	97.4	99.7	96.2	98.6	93.6
Power consumption, MW		3,339.27	1,078.2	1,085.4	734.7	2,120.8	1,536.3	2,821.48
Specific power consumption, kW*h/b.t.		3,977	3,586	3,798	3,682	3,627	3,908	3,675
Ore-2	t(29% Mn)/b.t.	1.106	0.933	0.892	0.714	0.635	0.626	0.492
	t(48% Mn)/b.t.	0.668	0.563	0.539	0.431	0.383	0.378	0.297
Ore-1	t(49.5% Mn)/b.t.	0.565	0.505	0.482	0.509	0.480	0.524	0.484
	t(48% Mn)/b.t.	0.582	0.520	0.497	0.525	0.495	0.540	0.499
Brex	t(31.37% Mn)/b.t.	0	0.077	0.164	0.273	0.387	0.571	0.605
	t(48% Mn)/b.t.	0	0.050	0.107	0.178	0.252	0.373	0.395
The total consumption of raw manganese ore	t/b.t.	1.671	1.515	1.538	1.496	1.502	1.721	1.581
	t(48%Mn)/b.t.	1.250	1.133	1.143	1.134	1.130	1.291	1.191
Coke, t/b.t.		0.446	0.339	0.420	0.405	0.395	0.415	0.413
Quartzite, t/b.t.		0.419	0.499	0.524	0.465	0.529	0.456	0.475
Silicomanganese fines briquettes t/b.t.		0.158	0.082	0.103	0.092	0.120	0.110	0.094
Scrap (Mn content in scrap, %), t/b.t.		0.358 (23.3)	0.601 (29.9)	0.462 (33.0)	0.477 (35.3)	0.461 (32.0)	0.455 (25.8)	0.373 (38.6)
Electrode mass, t/b.t.		0.034	0.032	0.030	0.035	0.028	0.033	0.028
Manganese extraction from the ore component, %		80.1	79.8	80.7	80.7	83.6	79.1	79.9

Table 8.14. Main components composition during the reference and pilot operation period.

Phase	Main components composition								
	in metal				in furnace charge				
	Mn	Si	Fe	P	MnO	MgO	SiO$_2$	CaO	Al$_2$O$_3$
Base	66.64	17.71	14.06	0.127	10.50	7.38	44.28	21.43	15.29
1	66.33	16.30	14.25	0.128	12.70	7.14	45.33	20.31	13.46
2	66.90	17.69	14.18	0.098	12.89	6.52	44.78	18.67	13.80
3	66.77	16.45	14.21	0.12	11.30	6.77	44.57	19.86	15.07
4	68.02	16.02	14.01	0.14	11.79	6.39	45.11	19.42	15.13
5	65.98	15.22	14.30	0.16	15.08	5.73	46.20	26.11	15.71
6	66.08	17.07	14.25	0.18	10.8	5.17	45.30	19.83	15.35

The obtained results formed the basis for the design and construction of several briquetting factories based on stiff vacuum extrusion in different countries around the world. In particular, a briquette factory with a capacity of 260 thousand tons of briquettes per year was put into operation at the Chelyabinsk Electrometallurgical Plant in Russia. Briquettes are made from a mixture of manganese ore and baghouse dusts (Fig. 8.14). The ratio of these components in the briquette charge depends on the type of imported ore fines. At 30% of the dust content in the mass of the briquette,

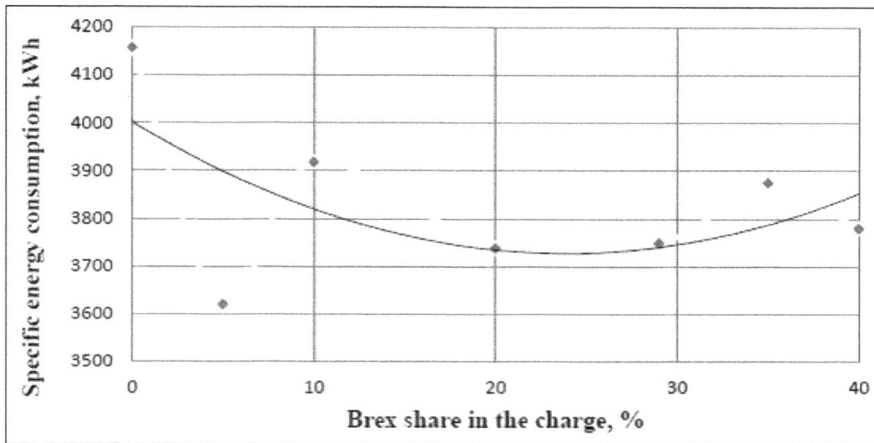

Figure 8.11. Dependence of specific energy consumption on the share of brex in the furnace charge.

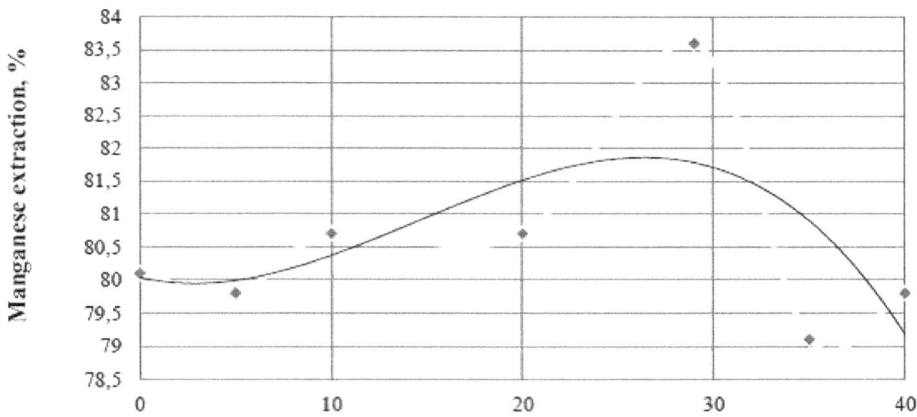

Figure 8.12. Manganese extraction as a function of the brex share in the furnace charge.

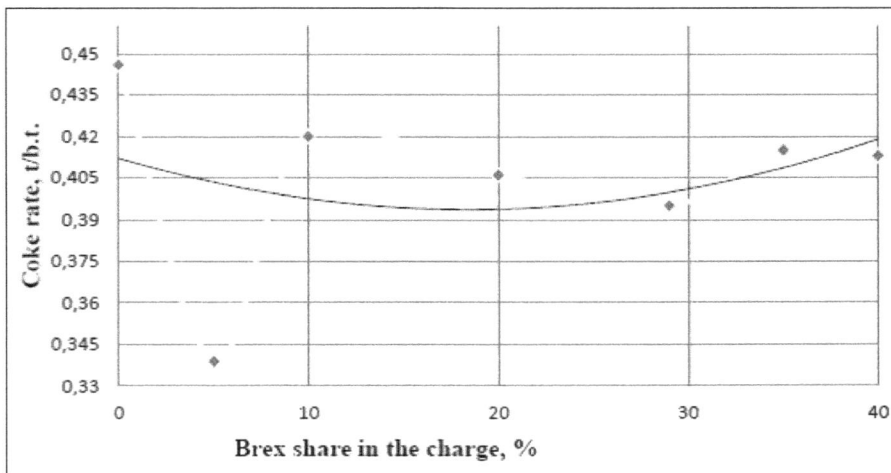

Figure 8.13. Dependence of coke consumption on the share of brex in the furnace charge.

Figure 8.14. Production of brex in Chelyabinsk (Russia).

only 1.5% of the cement by mass of the briquette is required. The share of cement increases to 3 at 15% of the content in the dust in the briquette.

8.4 Metallurgical Properties of Briquettes on the Basis of Chromium-containing Materials

The possibility of obtaining inexpensive and efficient agglomerated charge materials for smelting ferrochrome without earlier calcination has always attracted ferroalloys technologists. However, cold briquetting of chromium ore fines and concentrates is not widespread and is still significantly inferior to sinter and pellet production. Practical experience of using roller presses and vibropressing confirmed the possibility of achieving satisfactory metallurgical properties of briquettes obtained by such methods. It is suffice to mention the experience of the Serov Ferroalloy Plant [1], which confirmed the feasibility of replacing lumpy chromium ore with briquettes due to productivity growth (by 4%) without deterioration of the slag mode with a calm and even furnace running. The specific energy consumption (kWh/t) decreased by 1%. The decrease in chromium extraction when working on briquettes (by 2.3%) is due to the additional consumption of ore to increase the ore layer.

The successful results of the full-scale industrial testing of Globe Metallurgical Inc. are well known [2] using briquettes made of Turkish chromite ore. A mixture of lime and molasses was used as a binder (5% by weight). The share of briquettes in the ore part of the charge reached 100%. Since 1988 in Sweden, as noted earlier, the plant for the production of briquettes by vibropressing has been successfully operating at Vargön Alloys.

However, roller briquetting is accompanied by the formation of a significant amount of waste. Achieving hot strength requires the use of significant quantities of expensive binder. Vibropressing technology also involves the heat treatment of raw briquettes and inevitably leads to the "dilution" of the charge component thus obtained due to the need to use a substantial amount of binder. A significant amount of cement binder also inevitably leads to an increase in the sulfur content in the alloy. This, in particular, caused the closure of the vibropress briquetting factory at the Serov Ferroalloy Plant.

The SVE method opens up new opportunities for the widespread commercialization of cold briquetting chromium ore materials due to its higher performance, sufficiently high strength of raw briquettes and mainly lower binder content required to achieve the required level of metallurgical properties and the lack of thermal processing of raw briquettes.

In [3], the metallurgical properties of two types of briquettes made of chrome-containing materials—ore and ore-coal (including ore-coke) briquettes—were studied.

The choice of cement as a binder was also based on the results of studies of hot strength of laboratory briquettes made of chromium ore concentrate, described in [4]. The hot strength of briquettes was determined by the method of linear heating of the sample under load (2 kg) at a rate of 17°C/min under reducing conditions. The choice of the load value corresponds to the pressure of a briquette column with a height of 2 meters. Reducing conditions are provided by conducting an experiment in a stagnant furnace atmosphere of gaseous reaction products (CO) and placing the test sample of the briquette between two graphite plates, which simulates the contact of the briquette with pieces of coke in an electric furnace.

An experimental setup for determining the hot strength of briquettes based on fine-grained chromium ore when heated in a reducing atmosphere in the temperature range 20–1800°C is shown in Fig. 8.15. Before the start of the experiment, the briquette was installed in the isothermal zone of the furnace between graphite plates. The top plate was prevented by the load and the alundum displacement indicator with a total mass of 2 kg. The furnace was closed, pumped out and filled with argon. In the course of linear heating at a rate of 17°C/min, set by the furnace programmer, the gas from the furnace was vented, maintaining the pressure in the furnace equal to the atmospheric pressure. Changes in the sizes of the briquette were fixed according to the movement of the alundum pointer. To improve the accuracy of the readings, an Intel digital microscope was used, which made it possible to conduct linear movements on the computer screen with an accuracy of 0.01 mm. If the absolute value of the displacement is 1 mm (typical for the first heating period of the briquette, when the usual thermal expansion of the briquette prevails), then the relative measurement accuracy is 1% (rel.), which is sufficient for technical measurements. After the experiment, a metallographic section was prepared from the "crushed" briquette and a metallographic analysis of the phase components formed in the briquette was performed.

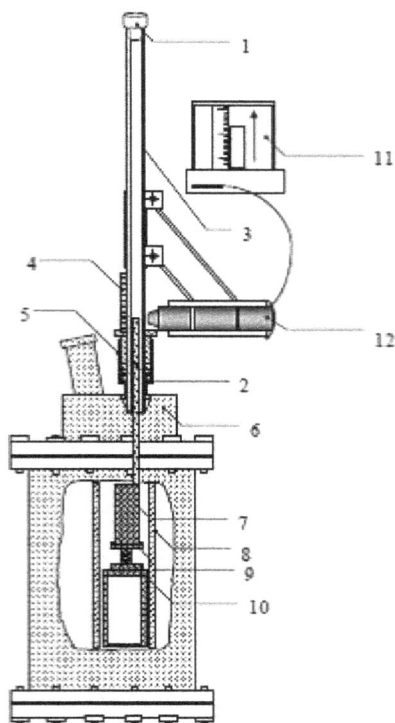

Figure 8.15. Installation diagram for determining the hot strength of briquettes in reducing conditions with a temperature change from 293 to 2073 K. 1 - rubber stopper; 2 - alundum index pointer; 3 - transparent quartz tube; 4 - ruler; 5 - vacuum tight sleeve; 6 - cover; 7 - load weighing 2 kg; 8 - graphite heater; 9 - pressure graphite plates; 10 - tested briquette; 11 - the computer; 12 - digital microscope.

Comparison of softening curves (dilatometric curve) and the results of metallographic analysis of briquettes after melting makes it possible to qualitatively compare the melting and reduction rates for a given briquette composition and thus simulate their behavior in the ore-smelting furnace. The use of the method of linear heating under load can be illustrated by the example of briquette No. 1 from Kazakhstanian concentrate of chromium ore with cement as a binder (6% by mass) in comparison with briquette No. 2 with the addition of coke breeze (8% by weight) and with the synthetically modified liquid glass substance (Penolite) as a binder (6% of the mass) and with the briquette No. 3, which differs from the briquette No. 2 by the absence of coke in its composition.

Photos of briquettes and polished sections from such briquettes are presented in Figs. 8.16–8.18. Figure 8.19 shows the softening curve of briquettes when heated under load.

Briquette No. 1 has a high final softening temperature - 1620°C, which allows intensive processes of formation of ferrochrome to develop at the places of contact of the ore briquette with carbon pressure plates. The bulk of the briquette at the end of the experiment turns into a pasty solid-liquid state. In general, it can be assumed that for this case the melting rate corresponds to the rate of reduction of the components of the briquette by carbon.

Interestingly, the smallest drops of metal are distinguishable in the body of briquette No. 1, although in this case this process does not get developed. The most likely reason for this is the transfer of the reducing agent from carbon pieces to ore grains in the form of gaseous carbon-containing molecules of the type CHx along cracks and pores of the briquette. Such a mechanism for the carbothermic reduction of chromites was proposed in [5] to explain the results of a study on the reduction of Ural chromium ores by carbon. Creating conditions in the ore-smelting furnace for the formation of gaseous carbon-containing compounds with increased reducing ability can be a significant reserve for the intensification of reduction processes.

Adding 8% of coke to the composition of the briquette No. 2 leads to a decrease in the softening end temperature to the level of 1550°C, i.e., less than for cement bonded briquettes (1620°C). However, the reason for the decrease in this temperature here is not the melting of the ore material, but the intensive formation of liquid metal inside the briquette. The appearance of metallic liquid in the body of the briquette leads to its weakening and crushing, although the remainder of the ore material of the briquette is predominantly in the solid state. In this case, the melting rate lags behind the rate of reduction of the components of the briquette by carbon, which allows increasing the heating rate of such a charge without disrupting the process conditions.

Briquette No. 3 shows a record high softening end temperature of 1765°C. At such a high temperature, there is an intensive reduction of chromic ore with carbon at the interface of briquette-graphite; there is a vigorous outflow of droplets of ferrochrome from the briquette. The remaining substance of the briquette ("slag residue") is squeezed out under the load from the pressure plates,

Figure 8.16. Appearance and thin sections of briquette No. 1 with a cement binder after determining the hot strength (on the left is a briquette; on the right is a thin section, magnification 10). The composition of the briquette - Kazakhstanian chrome ore and cement grade M500 in the amount of 6% by weight of the ore, drying at room temperature for 24 hours. The experimental conditions are linear heating of the sample at a rate of 17°C/min to a temperature of 1680°C under a load of 2 kg in a reducing atmosphere.

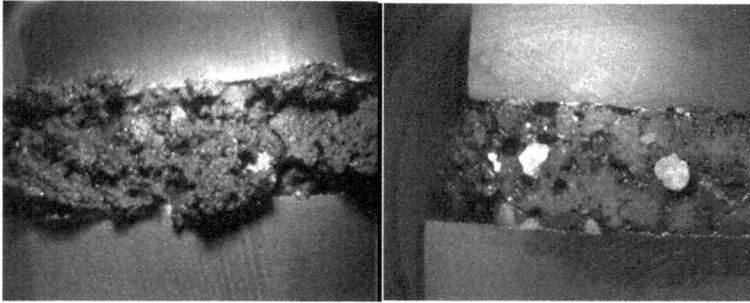

Figure 8.17. Appearance and thin sections of ore-coke briquette No. 2 with a Penolite binder after determining the hot strength (on the left is a briquette; on the right is a thin section, magnification 10). The composition of the briquette - Kazakhstanian chrome ore, coke breeze - 8% by weight of ore and Penolite in the amount of 6% by weight of the ore, drying at room temperature for 24 hours. The experimental conditions are linear heating of the sample at a rate of 17°C/min to a temperature of 1780°C under a load of 2 kg in a reducing atmosphere.

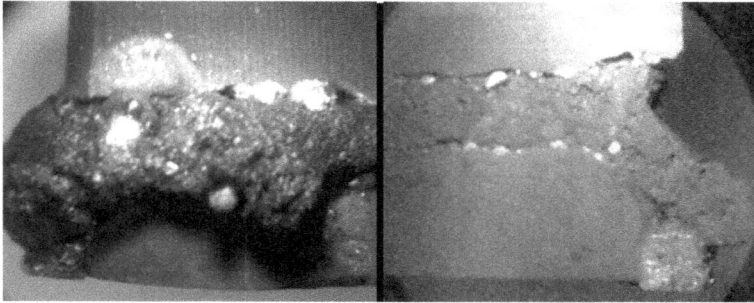

Figure 8.18. Appearance and thin sections of briquette No. 3 with a Penolite as a binder after determining the hot strength (on the left - briquette, on the right - thin section, magnification 10). The composition of the briquette is Kazakhstanian chrome ore and Penolite in the amount of 6% by weight of the ore, drying at room temperature for 24 hours. Experimental conditions - linear heating of the sample at a rate of 17°C/min to a temperature of 1780°C under a load of 2 kg in a reducing atmosphere.

but retains the solid-liquid state Here, the melting rate also corresponds to the reduction rate of the components of the briquette with carbon, but the reduction process until the softening of the briquette goes much further in terms of the amount of ferrochrome produced. This is natural, because this binder (Penolite) allows "overheating" of the briquette by more than 100°C than the cement binder, with the corresponding larger development of reduction processes.

The addition of a carbonaceous reducing agent leads to an intensification of the reduction processes, which leads to the possibilities for increasing the productivity of the electric furnace. But at the same time, the choice of the optimal content of carbon-containing reducing agent should be based on the analysis of the totality of the consequences of its introduction into the composition of the briquette, including the reduction of mechanical strength and the growth of electrical conductivity.

In [6], based on the results of the study of hot strength of laboratory briquettes No. 1–3, the metallurgical properties of extrusion briquette (brex) were studied on the basis of screenings and concentrates of Kazakh chrome ore with the addition of coal.

At the first stage, the metallurgical properties of the ore-coal briquettes of the following composition were investigated: briquette No. 4 - concentrate of chromium ore (71.4%), coal (23.8%) and cement (4.8%); briquette No. 5 - concentrate of chromium ore (76.2%), coal (19%) and cement (4.8%).

The following average values of mechanical strength in tensile and compression tests for these briquettes were obtained: briquette No. 1 - 28.8 kgF/cm² and 80.56 kgF/cm², briquette No. 2 -

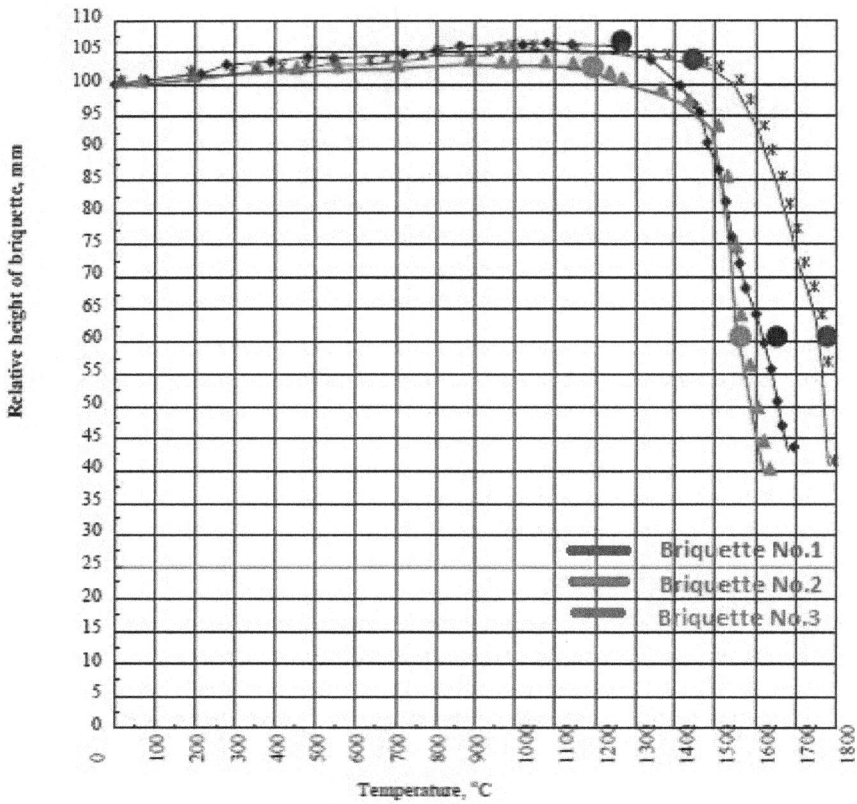

Figure 8.19. Softening curves for briquettes No. 1–3.

Figure 8.20. Briquette No. 5 after splitting (left) and compressing (right).

47.8 kgF/cm^2 and 84.6 kgF/cm^2, respectively. The appearance of the briquette No. 2 after splitting and compression tests is shown in Fig. 8.20.

It can be seen that when splitting a briquette sample brex up into larger fragments than during compression and in both cases the number of fines generated is insignificant, which confirms the expediency of using cement as a binder.

To determine the heat resistance of briquettes, a cylindrical sample (diameter 26.4 mm) weighing 33.6 g was placed in a melt heated to a temperature of 1500°C, which was further heated to 1650°C after lowering the briquette into it. A sample of briquette was removed from the melt after 10 minutes. The mass loss of the sample of briquette No. 4 was 31.25%, its diameter decreased to 21.5 mm (Fig. 8.21).

The dependence of the conductivity of briquette No. 4 on the temperature is shown in the Fig. 8.22. It is seen that almost exponential growth of electrical conductivity closer to 800°C begins, which can lead to the above mentioned negative consequences in the operation of the furnace. The conductivity of such a briquette at 788°C is almost six times higher than that of a briquette based on manganese ore fines and aspiration dust (briquette No. 2, Section 8.2) and 6 orders of magnitude higher than the conductivity of lumpy chromium ore [7].

The properties of chromium and chromium-ore brex on a bentonite binder were also studied. The main component of brex is the gas cleaning dust of carbon ferrochrome production (97% of the mass of brex). Bentonite has been used as a plasticizer and binder, providing high impact strength. The fraction of the fine fraction (particles less than 5 mm in size) formed after four drops of brex samples onto a concrete floor from a height of 2 meters did not exceed 1% (Fig. 8.23).

Figure 8.21. Sample briquette No. 4 before (left) and after holding for 10 minutes at 1650°C.

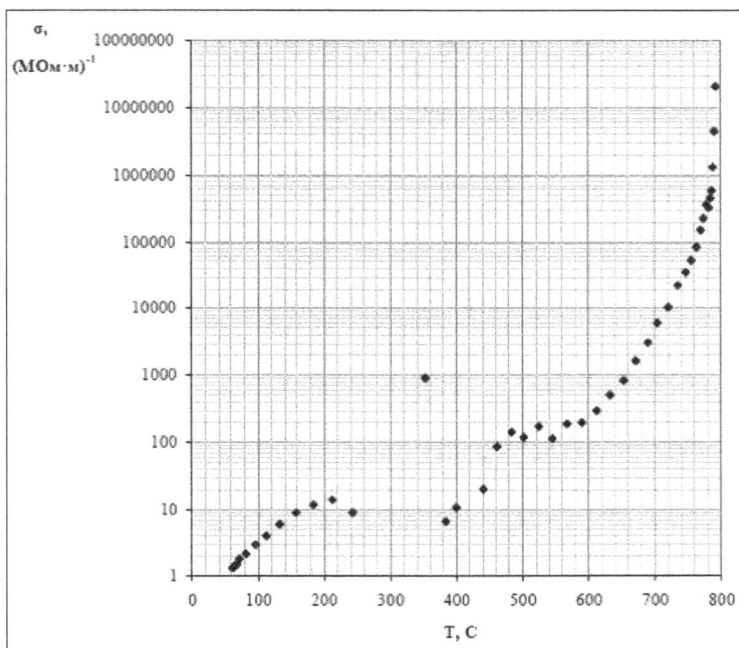

Figure 8.22. Electrical resistivity of briquette No. 4 when heated.

Figure 8.23. Test results for dropping brex with bentonite.

Figure 8.24. Making brex on a laboratory extruder.

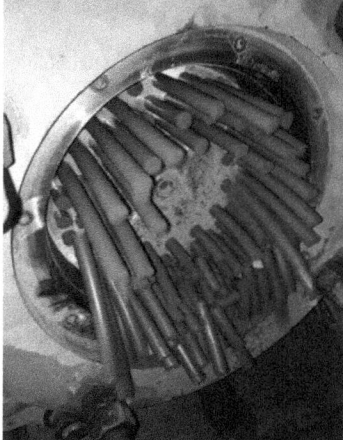

Figure 8.25. Industrial production of chromium-based brex at Aktobe Ferroalloy Plant. Aktobe Ferro Alloys Plant Report on BREX Smelting.

The strength of the raw brex is also shown in Fig. 8.24, which shows the time when the raw brex on the bentonite binder is squeezed out of the die of the laboratory extruder. Raw brex was so strong that it easily moved the cuvette with previously made samples.

The results obtained formed the basis for designing an industrial line for the production of brex in Kazakhstan (TNK Kazchrome, Aktobe), which was commissioned in 2017 (Fig. 8.25). Currently, brex are successfully used as a component of the charge of ore-smelting furnaces for smelting high carbon ferrochrome [8].

Table 8.15. Dust of the aspiration systems of the Aktobe Ferroalloy Plant.

Dust source	Dust formation, t/year	Content of chemical elements, % (wt.)			
		Cr_2O_3	C	S	P
Raw material unloading station	787.3	23.5	16.27	0.25	0.018
Raw material drying department	2737.3	15.5	66.85	0.32	0.019
Dry raw materials warehouse	4499.9	16.0	47.80	0.27	0.018
Storage compartment for prepared batch and its dosage of furnaces No. 41, 42	5753.6	36.5	15.20	0.18	0.017
Department of storage of prepared charge and its dosage of furnaces No. 43, 44	3828.8	30.5	11.00	0.11	0.016
Bag filter dust	23332.3	20.0	5.10	0.68	0.018

At the Aktobe Ferroalloy Plant, the accumulation of the dust from bag filters of ore-smelting furnaces (material 1) takes place. This type of dust differs greatly in its chemical composition, since it is formed as a result of metallurgical processing, and dust from aspiration technological units (material 2) is captured in aspiration systems located above conveyor belt conveyors and drying drums. It has the same chemical composition as the original chromium ore. Adding them to the charge for the extrusion process satisfactorily affects the quality characteristics of the extrusion briquettes. In connection with the ban on the placement and storage of metallurgical redistribution products in the territory of the slag dump, the management of the enterprise made a decision to introduce materials 1 and 2 into the extrusion charge.

Table 8.15 shows the values of annual dust formation, as well as average values for the content of chromium oxide, carbon, sulfur and phosphorus in them.

The study of the composition of polydisperse dust, as well as the selection of binders were carried out in laboratory conditions [9–10].

The cost of processing 1 ton of material is US\$ 11–14, electricity consumption is 33 kWh [11].

Moistening of dry mass was carried out according to the consistency of the mixture. As a result, the optimal ratio of the charge components was determined experimentally:

- 75% material 2;
- 25% material 1;
- + 6% (by weight of the dust mixture) bentonite;
- +3% (by weight of the dust mixture) polymer binder.

The finished extrusion briquettes were unloaded into the body of a dump truck by means of a belt conveyor, after which they were taken out to an open area, where they were dried in natural conditions. Each day, for 6 days, samples of extrusion briquettes were taken for moisture to determine the dynamics of drying at an average daily atmospheric temperature of 25°C, as well as to determine the quality characteristics (in the winter period of time, extrusion briquettes are not dried, since the moisture present in the composition of extrusion briquettes, accelerates freezing of the charge, which is a favorable factor for transportation).

During the tests, a pilot batch of extrusion briquettes weighing 600 tons was produced for remelting in an ore-thermal furnace. Comparative data on the mechanical characteristics of the obtained brex with the required strength values of extrusion briquettes at Aktobe Ferroalloys Plant are presented in Table 8.16.

According to the results of measurements of the strength of experimental extrusion briquettes, it can be seen that, in terms of mechanical properties, extrusion briquettes with the addition of polymer binders showed generally higher values in comparison not only with the current production, but also with the required strength values that were not achieved.

Table 8.16. Strength indicators under basic and experimental conditions.

Option	Binder, %		Mechanical strength of dry briquettes, %					
	Bentonite	Polymeric	CCS, MPa	2 m drop test		Impact	Abrasion	
				+5 mm	-5 mm	+5 mm	-0,5 mm	
Experimental	6	3	10	97.3	2.7	67.7	19.1	
Base	6	-	4.7	88.1	11.9	30.0	40.0	
Requirements	-	-	9	85.0	15.0	60.0	30.0	

The metal smelting test using experimental extrusion briquettes was carried out in an AC ore-smelting furnace RKO-22.5 MVA. A batch of extrusion briquettes dried on a special site was sent to the charge warehouse of chromium ore materials. When stored in bins, spontaneous destruction was not observed within three weeks. Before starting work, the thermodynamic and material-heat balance were calculated [12].

The main raw material for smelting high-carbon ferrochrome was rich chromium ore with a content of at least 48% Cr_2O_3. The number of extrusion briquettes fed into the furnace was increased in several stages. In the base period, only bentonite was used as a binder; then, in combination with a polymer binder, three stages were distinguished depending on the proportion in the sample. At the first stage of experimental smelting, the share of extrusion briquettes in the ore charge was 10%, at the second stage it was increased to 12.8%, at the third stage - up to 18.3%.

The dosing was carried out through the receiving hoppers of the dosing department of the workshop together with the main part of the ore material and the reducing agent. The top temperature was 1000°C. As the charge descended along the height of the furnace, new portions of material were poured onto the top.

Table 8.17 shows the generalized indicators of the furnace operation during the test period. Average composition of metal during testing, %: 69.05 Cr_{met}; 0.75 Si; 8.54 C; 0.026 S; 0.027 P.

Average slag composition, %: 2.2–5.5 Cr_2O_3; 45–47 MgO; 27–31 SiO_2; 0.7–1.1 CaO; 16–17 Al_2O_3; 0.5–0.8 FeO.

The performance of the furnace during the experimental period is close to the baseline. At the most stable experimental stage, the specific power consumption for the smelting of 1 ton of chromium is 3.1% lower than the baseline values. The obtained average extraction of chromium for the period of experimental heats is 1.7% higher than the base period.

Table 8.17. Indicators of operation of the furnace No. 12 in the baseline and experimental periods.

Work period	The share of brex in the ore sample, %	Cr_2O_3 content in ore sample, %	Ore materials supply to the furnace, t/day	Chromium production in ingots, t/day (including downtime)	Daily consumption of electricity, MWh	Average Cr_2O_3 content in slag, %
Base	11,1	49,71	292,51	64,66	414,86	3,9
In general, during the tests		46,5	242,67	49,32	327,81	3,67
Stage 1	10	47,12	305,56	60,78	406,2	3,89
Stage 2	12,8	46,63	163,6	28,63	225	4,08
Stage 3*	18,3	45,96	233,01	49,93	317,38	3,37
Stage 3 in the absence of stoppages of the furnace	18,3	47,17	284,51	62,78	390,75	3,72
* During stage 3 of the test, the furnace stopped for several hours due to technical reasons.						

The improvement in the performance of the furnace was due to a decrease in the content of fine fractions and moisture content in the charge. Earlier, when extrusion briquettes prepared according to the basic method were fed into the furnace, a decrease in lumpy chrome ore and the presence of a large amount of fines in the charge were repeatedly observed, which was the main reason for limiting their share in the sample when smelting high-carbon ferrochrome.

The results of industrial tests showed that the obtained high-carbon ferrochrome meets the requirements of GOST 4757 (ISO 5448-81) [13] and has an average chemical composition, which allows shipping products to consumers in accordance with contracts. Analysis of slag with Cr_2O_3 content in the range of 2.2–5.5% shows the maximum possible recovery of chromium-containing raw materials in industrial conditions.

In the process of manufacturing extrusion briquettes, the addition of a polymer binder led to an increase in their strength, while technological difficulties were not noted.

During the operation of the ore-thermal furnace, deviations in the operation of the furnace were not recorded. The resulting high-carbon ferrochrome met the chemical composition requirements. A small positive effect was observed due to increased strength of extrusion briquettes and improved gas permeability of the charge layer in the furnace. The percentage of the use of chrome raw materials significantly increased - up to 5% for the whole plant.

The ecological problem was solved by using the generated dust in a closed technological cycle.

References

[1] Ravich, B.M. 1982. The Briquetting of Ores, Nedra, Moscow (in Russian).
[2] Ravich, B.M. 1975. Briquetting in Ferrous and Non-Ferrous Metallurgy, Metallurgy, Moscow, 232 (in Russian).
[3] Bizhanov, F.V., Steele, R.B., Podgorodetskyi, G.S., Kurunov, I.F., Dashevskyi, V.Ya. and Korovushkin, V.V. 2013. Extruded briquettes (bricks) for ferroalloy production. Metallurgist. March 2013, 56(11-12): 925–932.
[4] Pavlov, A.V. et al. 2000. Research of briquetting process of the fine chromium ores. 12th International Ferroalloys Conference, Helsinki, Finland, June 6–9, 2000.
[5] Mikhailov, G.G., Pashkeev, I.Yu., Senin, A.V. et al. 2001. Intensification of Chromite Carbothermic Reduction. Metallurgy and Metallurgists of the 21st Century. International Conference-Debate, Moscow: MISIS, pp. 83–98 (in Russian).
[6] Bizhanov, A.M., Podgorodetskyi, G.S., Kurunov, I.F., Dashevskyi, V.Ya. and Farnasov, G.A. 2013. Experience of use of extrusion briquettes (BREX) to make ferrosilicomanganese. Metallurgist. May 2013, 57(1-2): 105–112.
[7] Akimov, E.N., Mal'kov, N.V. and Roschin, V.E. 2013. The electrical conductivity of high-alumina chromic ore. Bulletin of the South Ural State University. Series "Metallurgy" 13(1): 186–188 (in Russian).
[8] Kazchrome News (Vestnik Kazchroma, in Russian) No. 5(406), February 2, 2018, p. 5.
[9] Alimbaev, S.A., Shotanov, A.E. and Myrzagaliev, A.A. 2018. Thermal strength of ore-coal briquettes. Proceedings of the Scientific and Practical Conference. Prospects for the Development of Metallurgy and Mechanical Engineering using Completed Fundamental Research and R&D: Ferroalloys—"R&D - 2018". Ekaterinburg, pp. 172–175.
[10] Kornev, A.V., Kuskov, V.B. and Bazhin, V.Yu. 2019. Iron ore agglomeration by extrusion. ChernyeMetally 11: 4–10.
[11] Bizhanov, A.M. and Chizhikova, V. 2020. Agglomeration in Metallurgy. Springer Nature Switzerland AG. Cham. 454 p.
[12] Myrzagaliev, A.A., Almagambetov, M.S., Alimbaev, S.A., Ulmaganbetov, N.A. and Khalitov, T.V. 2020. The use of low-grade chromium-containing briquettes for smelting high-carbon ferrochrome. Vestnik nauki. 1(1): 342–349.
[13] GOST 4757 (ISO 5448-81). Ferrochromium. Specification and conditions of delivery. Introduced: 01.01.1993.

8.5 Stiff Extrusion Briquetting of the Ferroalloys Fines

Before shipping to consumers, ferroalloys are crushed and screened to obtain a required piece size. During such processing ferroalloys fines and aspiration dust are generated. One of the ways to utilize the ferroalloy fines and dusts is briquetting. The basic requirements for quality of commercial briquettes made from fines of ferroalloys are reduced to the necessity to provide the required mechanical strength at lowest possible binder content and high thermal stability, i.e., the ability to maintain integrity when placed in the environment with high temperature. Briquetted ferroalloys used as components of the Submerged EAF charge must have sufficiently high values of hot strength and of specific electrical resistance. In order to increase the specific electrical resistance, dust of aspiration systems can be added to metallized ferroalloys fines. In addition, aspiration dusts of ferroalloys production can exhibit binding properties which can largely reduce binder consumption.

The field experience in using roller presses for agglomeration of manganese-containing, and hence abrasive ferroalloys fines, confirmed that this method has a number of disadvantages. In August 2003, Nikopol Ferroalloys Plant (Ukraine) launched the experimental production of ferroalloys-fines-based briquettes using an organic binder. The capacity of the briquetting line allowed producing 40000 tons of briquettes per year [1]. It was projected that 70% of the briquettes would be produced from silicomanganese fines and 30% from ferromanganese fines. In the process of the production of silicomanganese-fines-based briquettes, the following parameters were achieved: productivity, t/h – 4.5 ... 5.0; briquette volume, cm^3 – 22 ... 25; briquette density, g/cm^3 – 4.5 ... 4.8; briquette moisture, % –0.8 ... 1.7; moisture content in briquetted charge, % – 2.5 ... 4; temperature of charge before pressing, °C – 60–80; drop strength of briquettes (+5 mm size fraction when tested in accordance with the Russian Standard GOST 25471-82), % – 94 ... 96.

When implementing the briquetting technology, an increased wear of the sleeves surface was noted. As a result, the first set of sleeves was worn out after briquetting 1100 tons of high abrasive silicomanganese fines.

The Assmang plant in Cato Ridge (South Africa) used vibropress briquetting technology to agglomerate manganese-containing anthropogenic materials, including ferroalloys fines [2]. The initial composition of the tested briquettes consisted of 69.2% ferromanganese fines, 26.8% of aspiration dust and 4% of cement binder. Mechanical strength was tested using a drop test procedure, which showed that less than 10% of fines with smaller than 0.5 mm size particles are formed. However, the tests that were carried out in a muffle furnace with heating in the range of 200–800°C showed that the briquettes of the above-mentioned composition do not have hot strength. As the temperature increased with the interval of 200°C, the briquettes were removed from the furnace and subjected to a strength test. It was noted that the briquettes completely lost their strength at 400°C. At the next stage, briquettes with alumina cement as a binder were tested, which allowed achieving specified hot strength. However, the mechanical strength of this type of briquettes did not meet the requirements. In addition, the use of alumina cement would lead to a great increase in the briquette cost. The mixture of cement and clay rich in alumina (3% by weight) was decided to be used as a binder, which eventually resulted in the production of briquettes with the sufficient strength for a 12-hour long stacking and with acceptable mechanical strength after 48 hours of storing. In addition, the briquettes had sufficient hot strength. Since the cost of clay is 50% lower than the cost of alumina cement, the final composition of the briquettes is the following: 67.0% ferromanganese fines, 26.0% aspiration dust, 4% cement binder and 3% aluminum-rich clay. The chemical composition of the briquette is shown in Table 8.18.

After that, the briquettes were used as charge components for Submerged EAF (11.5 MVA). The briquettes replaced two grades of manganese ore, namely Nchwaning and Gloria. The share of briquettes in the furnace charge was 10%. During this full-scale testing, no rise of electrodes was observed; no charge sintering occurred. The use of ferromanganese fines-based briquettes as charge components resulted in an improvement in the technical and economic parameters of the furnace operation.

In order to study possible ways of using the SVE technology for agglomeration of similar materials, the following types of brex were tested: brex from ferroalloy crushing fines and brex with the addition of aspiration dusts of the ferroalloys crushing units.

The briquettes were manufactured on a laboratory extruder with the hole size of 15 mm and the cylindrical shape of brex. The vacuum was 37.5–62.5 mmHg. The brex composition and the

Table 8.18. Chemical composition of Assmang briquette.

Elements	Content, %
Mn	56.2
Fe	9.84
SiO_2	4.46
CaO	3.63
MgO	0.58
C	6.28
Al_2O_3	1.83
Mn/Fe	6/1

Table 8.19. Composition of brex based on silicomanganese fines.

Composition of Brex	No. 1	No. 2	No. 3	No. 4*
Silicomanganese fines, %	94.0	93.0	92.0	94.0
Portland cement, %	5.0	5.5	6.0	5.0
Bentonite, %	1.0	1.5	2.0	1.0
Breaking load for brex (after 24 hours), kgF	18.4	29.4	32.8	20.9
Fines formed during the drop test after 24 hours (less than 4.75 mm), %	6.9	3.8	3.3	8.9
Failure load on brex (after 7 days), kgF	62.4	66.3	79.9	64.7
Fines formed during the drop test after 7 days, (less than 4.75 mm), %	3.5%	2.5%	3.1%	3.0%
Density, g/cm³	4.13	4.10	4.08	4.02

* with homogenization during 24 hours

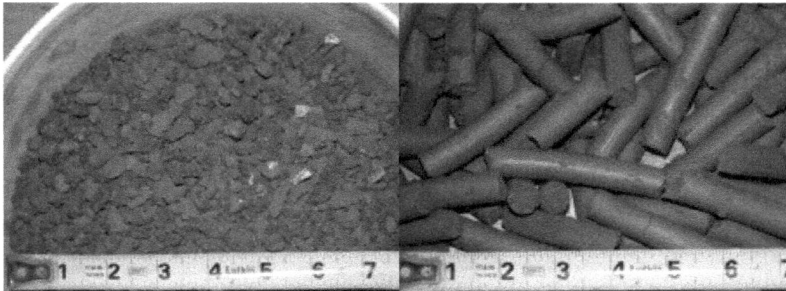

Figure 8.26. Silicomanganese crushing fines (left) and brex based on Silicomanganese crushing fines (right).

physical and mechanical properties of brex are given in Table 8.19. The mixture for Brex No. 4 was subjected to a 24-hour homogenization. Figure 8.26 illustrates silicomanganese crushing fines and brex based on them.

It is evident that the brex achieved sufficiently high compressive strength and impact strength, and homogenization did not have an obvious effect. The increase in cement content from 5–6% and in bentonite content from 1–2% improved compressive strength by 28% and impact strength by 12%. Applicable strength level was achieved with a minimum content of cement (5%) and bentonite (1%).

Briquette sample No. 3 was melted at a temperature of 1450°C for 15 minutes in a flowing argon atmosphere in a heated resistance furnace. In Fig. 8.27 shows an image of the appearance of the fusion products after cooling (metal nuggets and slag).

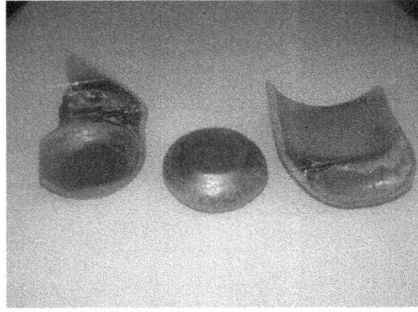

Figure 8.27. Appearance of the products of melting of briquettes from fines of Silicomanganese. Slag and metallized product (middle).

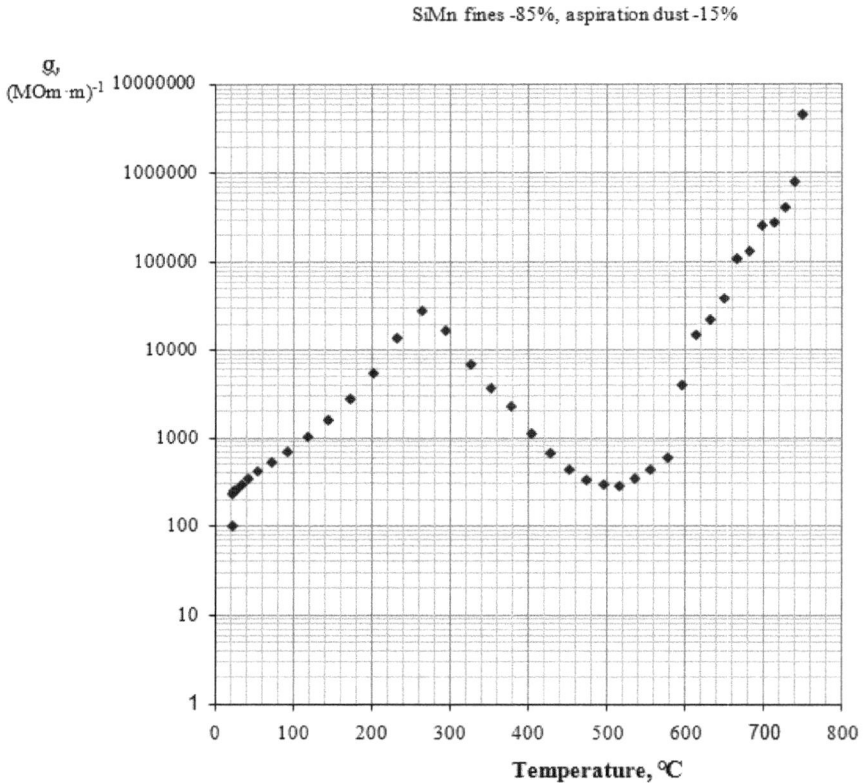

Figure 8.28. Dynamics of changes in the electrical conductivity of brex based made of silicomanganese fines when heated.

The composition of the metal phase, %: Mn – 80.8; Fe – 11.53; P – 0.10; Ni 0.19; Slag composition, %: MnO – 13.57; Al_2O_3 30.16; SiO_2 – 30.87; CaO 20.63; FeO – 0.2; NiO – 0.024.

The composition of brex based on ferroalloy fines with addition of the aspiration dust collected at the ferroalloys crushing unit was tested as well: (1) silicomanganese fines – 80.2%, aspiration dust of silicomanganese production – 14.2%, Portland cement – 4.7%, bentonite – 0.9%; (2) silicomanganese fines – 66.1%, aspiration dust of silicomanganese production – 28.3%, Portland cement – 4.7%, bentonite – 0.9%.

The crushing strength of these brex was, on average, according to the results of measurements on 10 samples 244.6 kgF/cm^2 and of 116.8 kg/cm^2, respectively.

The dynamics of the electrical conductivity of the brex from the first batch was examined during the heating of the sample till 800°C (Fig. 8.28).

For this type of brex the room temperature electrical conductivity is 101.2 (MOhm·m)$^{-1}$. For the brex used in the above-described industrial pilot tests of silicomanganese smelting, the electrical conductivity was 207.7 (MOhm·m)$^{-1}$ at 20°C. When heated to a temperature of 750°C, the conductivity of both types of brex were 4463670.5 (MOhm·m)$^{-1}$ and 120328.2 (MOhm·m)$^{-1}$, respectively. The electrical conductivity of the silicomanganese fines-based brex was 37 times higher than that of the brex from manganese oxides. To reduce the electrical conductivity, it is necessary to increase the content of aspiration dust in the brex.

The thermal stability tests of the brex were carried out in a Tamman furnace by holding the brex above the melt for 15 minutes at a temperature of 1200°C. All samples passed the test and maintained their integrity. In addition, the brex were also smoothly melted after being dropped into a slag melt at a temperature of 1580°C. It took 10 minutes to melt a single brex unit (with a diameter and length of 25 mm) (Fig. 8.29).

In recent years, a number of ferroalloys producers have fostered efforts to develop an efficient technology for briquetting ferrosilicon crushing fines. The paper [3] examines the results using liquid glass (sodium silicate) as a binder to produce strong briquettes from this material. The required strength of the briquettes was achieved by adding liquid glass in the amount of about 6% of the weight of a given briquette batch; however, use of the same binder (2%) in combination with starch (about 3.0%) resulted in strength gain. It has been stated that the results of the study provided the basis for the design and construction of three industrial lines for briquetted ferrosilicon production. The briquetting line of Silingen Polska [4] is known to produce about 18000 tons of these briquettes per year. The maximum compression strength of such briquettes is 14 MPa. The briquette dimensions are 66 mm × 55 mm × 35 mm and composition is 68% silicon and up to 2% carbon, due to the use of an organic binder, which exceeds the carbon content limit in ferrosilicon. The content of liquid glass-supplied Na_2O is not indicated.

In tests of the efficiency of using stiff extrusion for agglomeration of metallized ferrosilicon fines, fines of 75% FeSi grade were used. FeSi dust has also been used in the briquetted charge (up to 10% mass). Microsilica, starch and different lignins were used as binders. Bentonite was used as the plasticizer. FeSi (75%) crusher fines (Fig. 8.30) had a density of 2.1 g/cm^3. The proportion of particles with sizes less than 2.36 mm was 96%; 47% less than 0.3 mm. The maximum particle size was 4 mm.

Microslica – density 0.26 g/cm^3, wet basis moisture 1.5%; 99% of particles are smaller than 0.045 mm.

Three material mixes were evaluated and were run with and without 4 hours souring. Each mix was comprised of a base mix containing 85% FeSi (75%) Crusher Fines + 10% FeSi Crusher Dust + 5% Microsilica. To all these base mixes 1% of Bentonite was added. In samples designated 01 and 04 a liquid lignin was used at 3.5% on a dry solid basis. In samples 02 and 05 a powdered lignin was added at 3.5% on a dry solid basis. In samples 03 and 06 a starch binder was added at 2.5% on a dry solid basis (compositions and testing results in Table 8.20, Figs. 8.31–8.32).

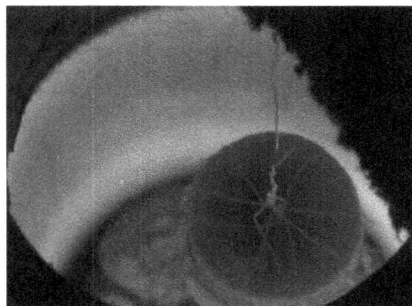

Figure 8.29. Brex tested in Tamman furnace.

Figure 8.30. 75% FeSi crusher fines.

Table 8.20. Brex compositions and strength testing results.

Sample	01	04	02	05	03	06
Base Mix Proportions FeSi/FeSi Dust/MS	85/10/5	85/10/5	85/10/5	85/10/5	85/10/5	85/10/5
75% FeSi Screened 8 Mesh	81.1%	81.1%	81.1%	81.1%	81.9%	81.9%
FeSi Dust	9.6%	9.6%	9.6%	9.6%	9.8%	9.8%
Microsilica	4.8%	4.8%	4.8%	4.8%	4.8%	4.8%
Bentonite	1.0%	1.0%	1.0%	1.0%	1.0%	1.0%
Liquid Lignin-Dry Solids	3.5%	3.5%				
Dry Lignin Solids			3.5%	3.5%		
Starch					2.5%	2.5%
Souring hours	0.0	4.0	0.0	4.0	0.0	4.0
Extrusion wet basis moisture	5.6%	5.9%	5.1%	5.7%	7.1%	6.8%
Ambient temperature (°C)	16	16	16	16	16	16
Extrusion temperature (°C)	37	40	39	42	32	31
Brex properties						
density (g/cm³)	2.45	2.39	2.44	2.40	2.38	2.31
Green brex drop test (CPFT 4.75 mm)	23.0%	19.0%	31.0%	26.0%	2.0%	1.0%
Dried at 150 °C brex drop test (CPFT 4.75 mm)	3.0%	4.0%	3.0%	4.0%	3.0%	3.0%
Dried at 150°C compressive strength (MPa)	12.6	9.0	9.6	8.4	14.0	12.8
Brex properties after 24 hours curing at 20°C						
Wet bas moisture, %	3.4%	3.4%	3.5%	3.2%	6.3%	4.6%
Drop test (CPFT 4.75 mm)	8.0%	21.0%	17.0%	13.0%	3.0%	1.0%
Compressive strength: (MPa)	0.26	0.23	0.4	0.31	0.83	0.63
Brex properties after 3 days curing at 20°C						
Wet bas moisture, %	1.2%	2.0%	0.9%	1.7%	3.7%	1.6%
Drop test (CPFT 4.75 mm)	2.0%	11.0%	1.0%	3.0%	1.0%	2.0%
Compressive strength: (MPa)	0.7	0.6	3.3	2.0	1.9	2.1
Brex properties after 8 days curing at 20°C						
Wet bas moisture, %	0.7%	0.7%	0.5%	0.7%	0.6%	0.7%
Drop test (CPFT 4.75 mm)	0.5%	1.2%	0.7%	1.3%	0.6%	0.7%
Compressive strength: (MPa)	13.8	11.1	15.2	10.5	20.9	19.2

CPFT stands for Cumulative Percent Finer Than

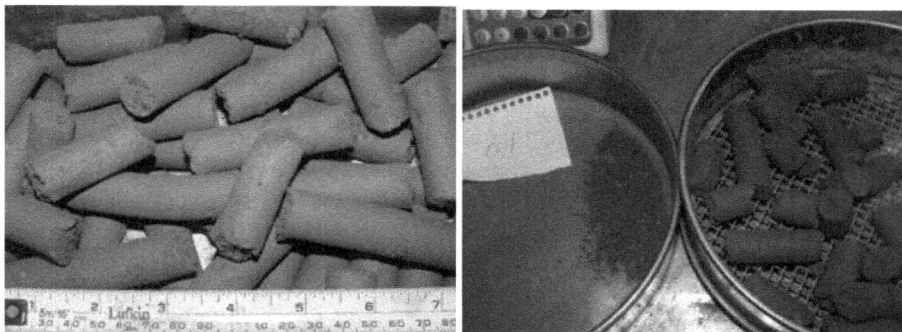

Figure 8.31. Brex 01: Green extruded and after dryer drop test (3% of fines generation).

Figure 8.32. Brex 02 (soured): dryer drop test (4% of fines generation).

The material preparation procedure involved blending all the components with water in the laboratory Hobart mixer. A double batch of material mixtures were prepared and then run one time through a shearing die on the laboratory extruder. One half of the batch was then pelletized through the laboratory extruder, while the other half of the batch was allowed to sit (sour) for 4 hours. After 4 hours this half of the batch was then run through the extruder. The extruder was equipped with a steel extrusion plate die with four ~ 20 mm diameter holes. The pertinent extrusion data is presented in the data table and includes extrusion moisture, die temperature vs ambient temperature.

The extruded briquettes were allowed to break off at random on exit from the die. Samples were immediately taken for initial testing which included moisture content and density determination. Samples were also subjected to drop testing. This drop test criteria involves taking about 500 g of extruded noodles and putting them into a plastic zip lock bag. The samples in the bags were then subjected to 12 drops from one meter onto a concrete surface. The number of fines generated was measured. Fines are defined as being particles smaller than 4.75 mm.

In order to compare the effect of active drying versus ambient drying over time, a portion of each batch were put into a dryer at 150°C until the brex were less than 1% moisture. The rest of each batch of brex were set out in the laboratory at a constant temperature and humidity for ambient curing. It should be noted that "curing" in this case should actually be considered air drying. The temperature in the laboratory stays about 20°C year around. The dried brex were immediately subjected to the prescribed drop test and axial compression test. The ambient cured brex were subjected to the drop test and axial compression tests after 24 hours, 3 days and 7 days.

The extrusion parameters indicate some major differences between the samples with lignin binders versus the samples with starch binder. The extrusion amps for the lignin samples are in the high end of the normal range at 3.2–3.3 for the lignin samples. For starch samples there was a distinct drop to 2.9%. This drop is attributable to the moisture content at extrusion. The Wet Basis Moisture contents (WBM) at extrusion for the lignin samples were higher for the starch samples, in the range of 7% WBM versus about 5.5% WBM for the lignin samples. This is probably due to the fact that starch has a much higher water absorption capacity than the lignin. The starch also appears to impart more internal lubrication to the material mixtures as evidenced by the extrusion temperatures. The starch samples had much lower temperatures which would correlate with less friction.

The axial compression strengths for all the dried samples were greater than any of the samples through 3 days of curing. Infact the axial strengths of the 24 hours and 3 days samples are poor. After 8 days of curing, however, a drastic increase in strengths is observed. These cured strengths uniformly exceed those of the actively dried samples. These samples exhibit compressive strengths that would be considered good for samples which have more than 3.5% of Portland cement as a binder.

The green drop test results show a vast difference in the green strength of the lignin samples versus the starch sample. Drop test results for the lignin samples varied from about 20–30% whereas the results for the starch samples were 1–2%. The base material mix is inherently non-plastic with low cohesion. It is therefore not surprising that it exhibits very poor green strength. The amount of green strength that can be imparted is more dependent on the binders. It is apparent that the lignin binders did not impart nearly as much green strength as the starch.

The drop test results for the actively dried samples are uniformly good for all samples. All the cured sample results, with the exception of the starch samples, are generally poor after 24 hours. After 3 days of curing the lignin samples exhibited good drop test results with the exception of sample 16-0105-04 which had a drop test result of 11%. This may correlate with the WBM% of the samples at this point. The unsoured samples had lower WBM% and had better drop test resistance than the soured samples. The sample 04 noted above had a WBM of 2.0% whereas the 01 and 02 samples had WBMs close to 1% and had drop tests of 1–2%. After 8 days of curing the WBMs drop below 1% and all the drop test results are in the range of 1% or less.

These FeSi material mixes are non-plastic and non-cohesive and therefore need binders that can impart green strength. The lignin binders did not impart sufficient green strength for any kind of stressful handling immediately after extrusion. This could include conveyor transfers and drops into curing piles if the heights are too high. Lignin binders are therefore not recommended for ambient curing operations as conveyor transfers and drops cannot be avoided. The samples with the starch binder yielded acceptable drop test results and may have enough green strength for rough handling.

Drying of the pellets is an issue that should be considered. The low green strengths of the lignin pellets probably prevent nearly any kind of handling into bunkers or piles. Furthermore, stacking the pellets into piles or bunkers is not conducive to drying which correlates to strength gain. In addition, the chemical reaction of the very fine FeSi dust with water should be considered. Allowing wet materials to sit in piles for prolonged periods may create conditions for the fine grained FeSi materials to react with water causing phosphine gas generation and maybe even spontaneous ignition of the dust in the material.

The results indicate that as the moisture in the pellets drops down to 1% or below then the strength of the pellets increases a great deal. The 8 day cured samples are the best, both in terms or axial compression and drop resistance. The moistures were below 1% for the actively dried samples as well, but the drop resistance and axial compressions were not as good. This may be due to stresses induced from the rapid drying and shrinkage of the pellets in the oven. This rapid shrinkage may actually have induced some cracking. Nonetheless, the strengths of the actively dried pellets are still quite good. A better result could probably be obtained by slowing the drying cycle, lower temperature and slightly more time. The 8 day cured samples can be thought of being tougher than the dried samples. Not only did they withstand compressive forces better but were less brittle than the dried samples.

References

[1] Noskov, V.A., Bolshakov, V.I., Maimur, B.N. et al. 2004. Pilot industrial production of briquettes from screenings of ferroalloys at JSC "NFZ". Metallurgical and Mining Industry. 3: 124–126 (in Russian).

[2] Davey, K.P. 2004. The development of an agglomerate through the use of FeMn waste. Proceedings of Tenth International Ferroalloys Congress, INFACON-X: Trans-Formation through Technology, Cape Town, South Africa, pp. 272–280.

[3] Hycnar, J.J., Borowski, G. and Józefiak, T. 2014. Conditions for the preparation of stable ferrosilicon dust briquettes. Inżynieria Mineralna – Journal of the Polish Mineral Engineering Society 33(1): 155–162.

[4] Electronic resource. http://www.silingen.eu/products/.

8.6 Study of the Kinetics of Solid-Liquid Phase Carbothermal Reduction of Lean Oxidized Nickel Ore

The applicability of the solid-liquid-phase carbothermal reduction process for poor oxidized nickel ore is assessed by the form of the obtained metal phase, its chemical composition, the amount of slag, the degree of nickel extraction from the ore, the temperature of the process and some other factors [1].

For poor oxidized ores, the formation of a single bead of reduced metal can be hindered by an excess amount of solid carbon, as well as a lack of it.

To determine the applicability of the process of solid-liquid-phase carbothermal reduction of nickel from poor oxidized nickel ore, ore-coal briquettes were prepared from the nickel ore of the Buruktal deposit, the composition of the ore is shown in Table 8.21. Three different types were used as a reducing agent - charcoal (was chosen as the most reactive type of carbonaceous reductant), semi-coke (has good reducing ability - high reactivity, cheaper than coke) and Shubarkul coal (cheap reducing agent, large reserves on the territory of the Russian Federation) [2]. The technical analysis of the selected reducing agents is shown in Table 8.22.

Metallurgical limestone was used as a flux introduced into the composition of the briquettes, the composition of which is shown in Table 8.23. The flux was introduced into the composition of the briquette to impart the necessary physical and chemical properties to the slag, in particular the necessary fluidity at low temperatures. As a binder for cold briquetting, molasses (waste of sugar production) was chosen, which proved to be an effective binder for ore-coal mixtures [3–5]. In the experiments, an analogue of molasses was used—a 40% solution of sugar in water.

Table 8.21. The composition of the nickel ore of the Buruktal deposit.

Element	Ni	Fe	Si	Ca	Cr	Mn	Al	Cu
%	1.14	32.10	8.95	0.43	0.17	0.48	0.64	0.13

Table 8.22. Technical analysis of the composition of reducing agents.

Parameter	Charcoal	Shubarkul coal	Semi-coke
W^r (physical moisture)	4.00	11.40	10.80
V^{daf} (volatiles)	10.00	39.80	16.00
A^d (ash content)	1.45	2.60	11.20
P^d (phosphorus)	-	0.01	0.03
S^d (sulfur)	0.04	0.34	0.17
C_{fix} (solid carbon)	84.50	45.90	61.80

Table 8.23. Chemical composition of limestone as delivered.

Oxide	Fe_2O_3	SiO_2	$CaCO_3$	Al_2O_3	MgO	SO_3	P_2O_5	w
%, mass	0.38	1.80	94.59	0.98	0.73	0.20	0.02	2.00

Table 8.24. Composition of ore semi-coke briquettes.

No.	Reducing agent		Ore, %	$CaCO_3$, %	Molasses, %
	Type	%			
1	Charcoal	3	92	5	2
2	Charcoal	5	90	5	2
3	Charcoal	10	85	5	2
4	Shubarkul coal	3	92	5	2
5	Shubarkul coal	5	90	5	2
6	Shubarkul coal	10	85	5	2
7	Semi-coke	3	92	5	2
8	Semi-coke	5	90	5	2
9	Semi-coke	10	85	5	2

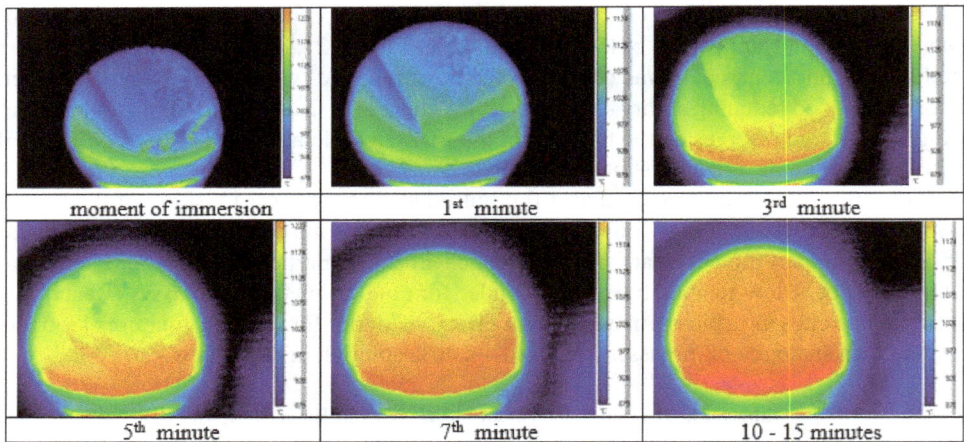

| moment of immersion | 1st minute | 3rd minute |
| 5th minute | 7th minute | 10 - 15 minutes |

Figure 8.33. Thermal images of the ore-carbon briquette reduction process.

The compositions of the investigated briquettes are shown in Table 8.24.

The progress of briquettes reduction was recorded using a Pyrovision M9000 thermal imager, the main stages of recovery are shown in photographs taken with a thermal imager (Fig. 8.33).

The nature of the recovery of all briquettes was the same: when the briquette enters the hot zone of the furnace, the residual volatiles are quickly removed from the briquette within the first minute, the briquette begins to warm up and the oxides are reduced, which manifests itself in the appearance of cracks on the briquette due to a loss of mass with gaseous reaction products. The cylindrical shape of the briquette disappears only due to the accumulation of critical amounts of liquid-solid phases, closer to 10–12 minutes, the formation of a single bead of metal ends.

The appearance of reduction products with different types of reducing agent and different concentrations is shown in Fig. 8.34.

If the content of the reducing agent is at the level of 3%, the metal always has a dispersed character, regardless of the type of reducing agent. This is clearly observed in the case of charcoal, semi-coke and Shubarkul coal, drops of metal are formed, but metal losses are large due to complications with the separation of metal from slag, while the analysis of the obtained metal

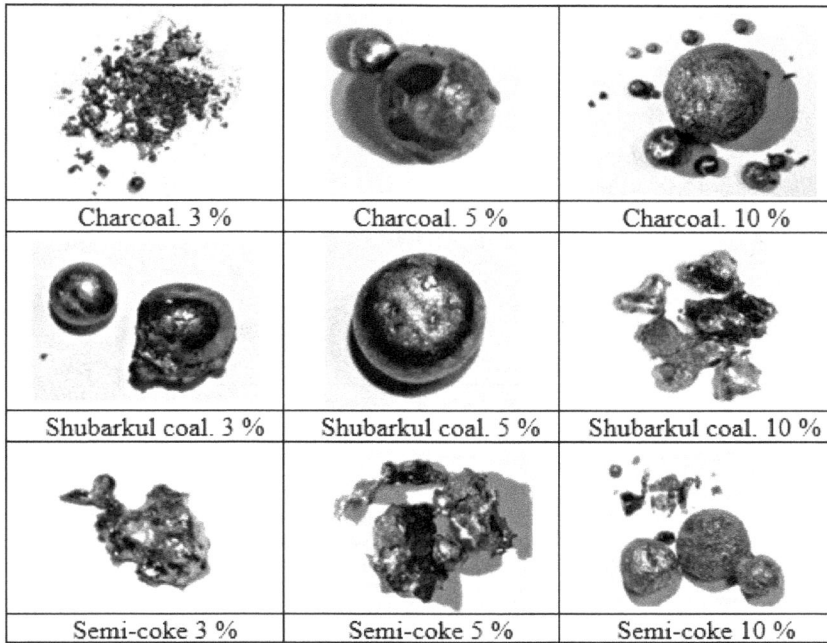

Charcoal. 3 %	Charcoal. 5 %	Charcoal. 10 %
Shubarkul coal. 3 %	Shubarkul coal. 5 %	Shubarkul coal. 10 %
Semi-coke 3 %	Semi-coke 5 %	Semi-coke 10 %

Figure 8.34. Appearance of reduction products.

showed a nickel content of 15–23%. When the content of the reducing agent is at the level of 5%, the metal droplets are already larger and more compact, the analysis of the metal showed a nickel content of 11–14%. At a high content of the reducing agent at the level of 10%, an excess of the reducing agent is observed—an unreacted reducing agent floated on the surface of the melt, which prevented small drops of metal from merging into a single drop of metal, which entailed the loss of metal. Analysis of the metal showed the content of nickel in the metal to the level of 4.6–5.8%. This corresponds to the value of the oxygen affinity of the elements that make up the nickel ore—first, fragile oxides of nickel and cobalt are reduced, then iron oxides. The more the iron is reduced, the more it dilutes nickel and cobalt, therefore, the concentration of nickel in the metal decreases.

Figure 8.35 shows the dependence of the concentration of nickel in the metal and the dependence of the degree of extraction of nickel into the metal on the type and concentration of the reducing agent in the original briquette.

The nature of the reduction and the concentration of the leading element (nickel) in all briquettes, regardless of the type of reducing agent, is the same: the concentration in the final metal at a low content of the reducing agent in the briquette (3%) is high 15–23% (see Fig. 8.39). With an increase in the concentration of the reducing agent in the original briquette, the nickel content decreases and is approximately in the same concentration range—for 5% reductant 11–15% nickel in metal, for 10% reductant 4.5–6%.

The iron content in the resulting metal confirms the earlier stated tendency—with an increase in the concentration of the reducing agent in the original briquette, the iron content in the metal increases: for 3% of the content of the reducing agent, iron contains about 80% in the case of char and charcoal, in the case of Shubarkul coal, the iron concentration decreases to 54%; for 5% of the reducing agent content in the original briquette, in all cases, the iron concentration is at the level of 82–85%; for 10% of the content of the reducing agent, iron is contained in the resulting metal 90–93%.

Thus, from the point of view of the composition of the obtained metal, the type of the reducing agent does not have a strong effect on the reduction. But considering the degree of recovery of nickel from ore, depending on the concentration of the reducing agent and the type of reducing agent in the

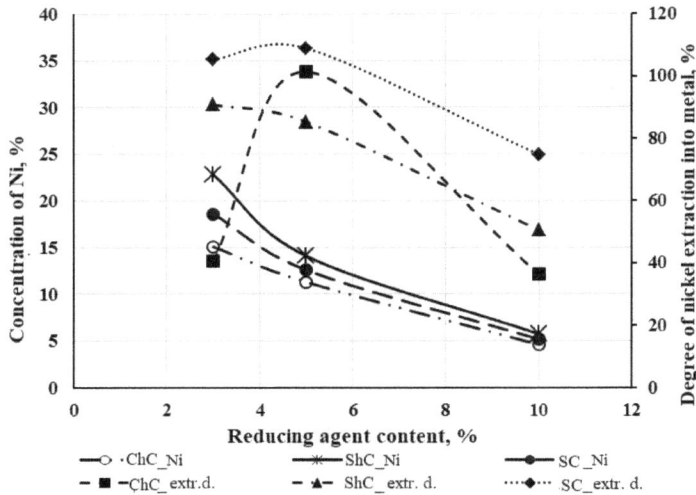

Figure 8.35. Dependence of the nickel content and the degree of recovery on the type and concentration of the reducing agent (ChC – charcoal, ShC – Shubarkol coal, SC – semi-coke).

briquette, and also taking into account the shape and type of the obtained metal (see Fig. 8.34), the advantage of charcoal over other types of reducing agents is clearly visible—the degree of recovery of nickel at low concentrations (at 3 and 5%) is close to 100%, with a reducing agent content of 10%, a decrease in the degree of recovery to 80% is observed, which is explained by the lack of adhesion of drops due to an overabundance of the reducing agent, which floats on the surface of the melt and interferes with the merger of small metal drops into one large, thereby causing an apparent decrease in recovery. Regardless of the concentration of semi-coke in the original briquette, all nickel from the ore is reduced and goes into metal, but at 10% content of the reducing agent, metal losses increase during magnetic separation of the metal from the slag.

A similar behavior of the reduction process is observed in the case of Shubarkul coal, but the degree of nickel recovery is notably lower relative to semi-coke.

In the case of charcoal, there is a marked deviation from the general trend and, as can be seen from the graph (see Fig. 8.35), the recovery varies greatly depending on the concentration of charcoal in the original briquette. Visual analysis of the obtained metal shows that the metal is dispersed, and this causes large losses of metal. The combination of these factors at first consideration indicates that charcoal as a reducing agent for oxidized nickel ore is not entirely suitable.

Accordingly one of the main requirements for assessing the applicability of the method for processing nickel ore, the method of solid-liquid-phase carbothermal reduction is applicable: drops of compact metal are formed, the content of leading elements in the resulting metal and the degree of extraction are satisfactory.

Figure 8.36 shows the primary data of laboratory modeling of the process of carbothermal reduction of ore-coal briquettes in a continuous oxygen reactor. The graph reflects the compositions of waste gases obtained by gas chromatographic analysis with ore-coal briquettes of different composition.

The nature of the recovery of all samples is the same regardless of the reducing agent and its concentration in the original briquette. The differences are in the height and width of the peaks, which is determined by the content of the reducing agent in the original briquette and the type of reducing agent. The highest peaks are observed in charcoal, slightly lower in semi-coke, the lowest in Shubarkul coal at the same concentrations, this is explained by the C_{fix} content in reducing agents (see Table 8.22).

The obtained primary data of gas chromatographic analysis were processed—the degree of conversion for each briquette composition was calculated using equation (14).

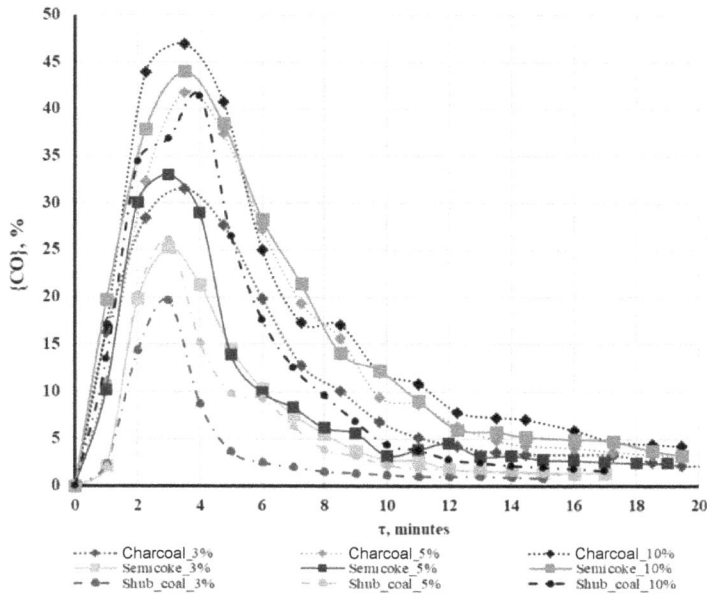

Figure 8.36. Primary experimental results.

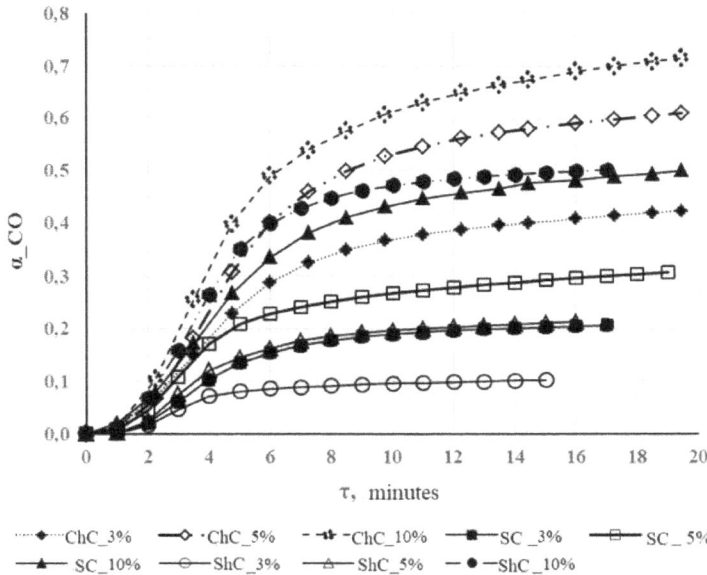

Figure 8.37. The results of calculating the degree of conversion from the time of testing according to the data on the composition of the gas phase (ChC – charcoal, SC – semicoke, ShC – shubarkol coal).

Figure 8.37 presents the calculated data on the degree of conversion α, obtained by integrating the trapezoidal method of the experimental data on the dependence of the gas phase composition on time minus the furnace background and taking into account the corrections of the blank experiment with pure reducing agents (in terms of the carbon monoxide content).

Having calculated the degree of transformation α (O) in two independent ways (using equations (15) and equations (16)) and comparing them (Fig. 8.38), almost complete coincidence (the points lie on the control straight line passing through the origin at an angle of 450) in the case semi-coke and Shubarkul coal was noted. The slight deviation from the straight line is explained by the loss

278 Briquetting in Metallurgy

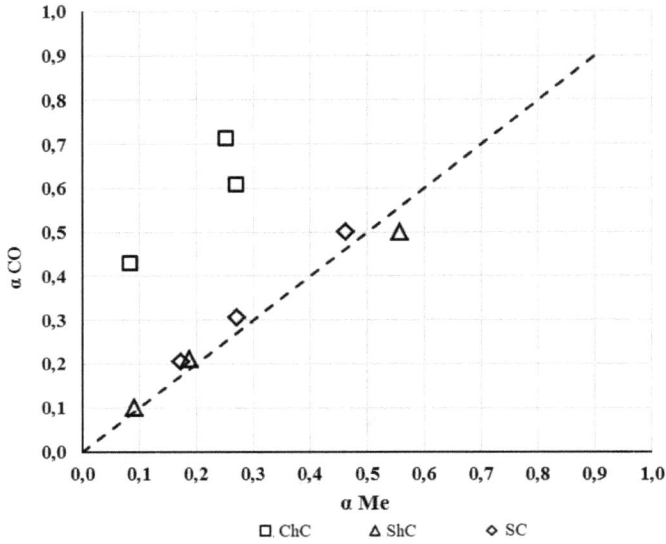

Figure 8.38. Comparison of the results of calculating the degree of conversion by the composition of the gas phase α_CO and by the composition of the obtained metal α_Me (ChC – charcoal, SC – semicoke, ShC – shubarkol coal).

of the metal phase during the separation of the reaction products—metal and slag. In the case of charcoal, there is a significant deviation from the control line, which is also explained by the loss of metal mass, but already significant, since the metal is finely dispersed (see Fig. 8.34). In general, a comparison of the calculated data confirmed the accuracy of the experimental procedure and calculations of the main parameter—the degree of conversion α (O).

In the general case, the process of reduction in a mechanical mixture of oxides with carbon consists of several simple physical and chemical "elementary" processes (phenomena, stages) of movement of particles of a substance, changes in its structure, state of aggregation and chemical composition [6]. Among these phenomena (stages), a chemical reaction, diffusion, sublimation or evaporation, melting and other stages, each of which can be a limiting stage in the entire process, can be of importance [7]. To determine the limiting stage of the process of carbothermal solid-liquid reduction of oxidized nickel ore, it is necessary to determine the kinetic constants of the process. As mentioned earlier, to determine the kinetic constants, a direct differential method was chosen, the use of which made it possible to determine the approximate order of the reduction reaction and the apparent activation energy of the process (E_act).

Using equation (17), the experimental data were converted into a logarithmic form, made it possible to determine the coefficients of the equation necessary for calculating the activation energy, which is shown in Fig. 8.39.

When calculating the apparent activation energy, the earlier found coefficients of the equation were used—the value of E_act was determined for each value of the order of reaction n in the range from 0 to 3 for all three concentrations of charcoal in the original briquette, which is shown in Fig. 8.40.

In the case of charcoal, there is a sharp difference in the values of the activation energy E_act at a charcoal content of 3% in the original briquette from 5 and 10% content. In the case of a coal content of 3% in the entire range of reaction order values from 0 to 3, the activation energy greatly exceeds the value of 500 kJ/mol. In the case of the content of 5 and 10% in the initial briquette at n from 0 to 1, characteristic of the limiting processes in the form of internal or external mass transfer processes, the activation energies are too high in all cases; at n 1.5–2 the activation energy varies from 150 to 500 kJ/mol. In the case of charcoal, the values differ quite a lot, it is likely that many dissimilar processes occur simultaneously, this case requires a deeper study of the ongoing processes.

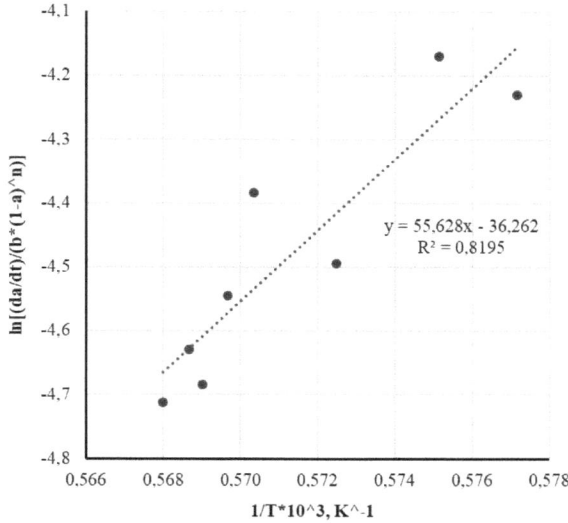

Figure 8.39. Calculation of the apparent activation energy in the case of charcoal.

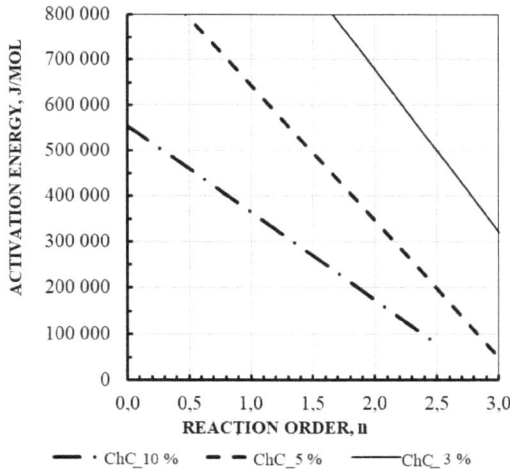

Figure 8.40. Activation energy of the reduction process for the case of ore-coal briquettes with charcoal (ChC – charcoal, ShC – Shubarkol coal, SC – semi-coke).

The obtained experimental data in the case of semi-coke were processed in the same way as the experimental data with charcoal.

Figure 8.41 reflects the method for calculating the apparent activation energy from the coefficients of the equation, calculated using the graph plotted in the coordinates of the logarithmic dependence of the degree of conversion on the inverse temperature of the process.

The calculation of the apparent activation energy was carried out according to the same principle as before for charcoal.

Figure 8.42 shows the dependence of the activation energy E_act on the value of the reaction order in the case of char. For all three variants of the semicoke content in the initial briquette for the reaction order n in the range from 0 to 1, the apparent activation energy is at the level of 380–630 kJ/mol. In the most probable range of the order of the reaction of 1.5–2, the activation energy is at the level of 300–500 kJ/mol, which indicates the chemical reaction as a limiting link in the process.

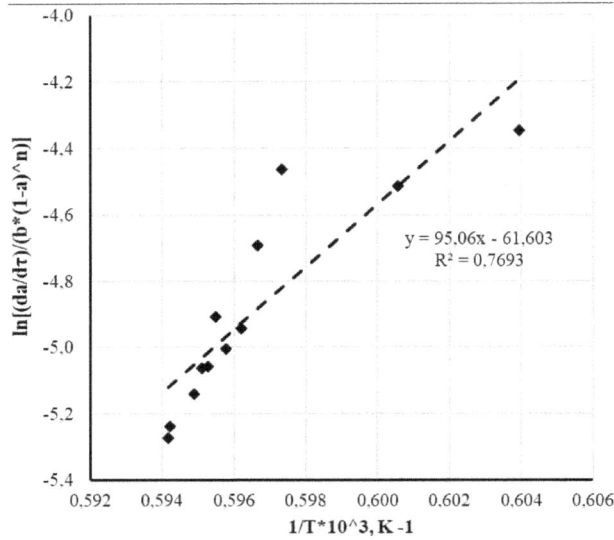

Figure 8.41. Calculation of the apparent activation energy in the case of semi-coke.

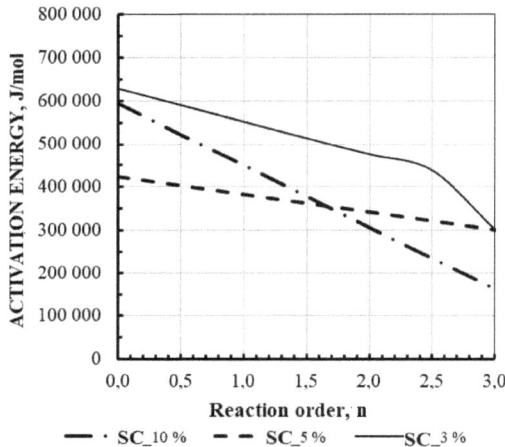

Figure 8.42. Activation energy of the reduction process for the case of ore-coal briquettes with semi-coke (ChC – charcoal, ShC – Shubarkol coal, SC – semi-coke).

To calculate the apparent activation energy of the process of carbothermal reduction with Shubarkul coal, the coefficients of the differential equation were determined according to the graph built in logarithmic coordinates, which is shown in Fig. 8.43.

Using the calculated coefficients of the differential equation in logarithmic form, we switched to calculating the apparent activation energy in the case of Shubarkul coal.

Figure 8.44 shows the results of calculations of the apparent activation energy E_act. There is a sharp difference in the values of the activation energy in the most probable range of the order of the reaction 1.5–2 for 3% in the initial briquette—the activation energy is at the level of 700 kJ/mol, in the case of 5 and 10% coal in the initial briquettes, the activation energy is in the range 170–200 kJ/mol. This difference in the values of the activation energy is most likely due to the high content of volatiles in the Shubarkul coal.

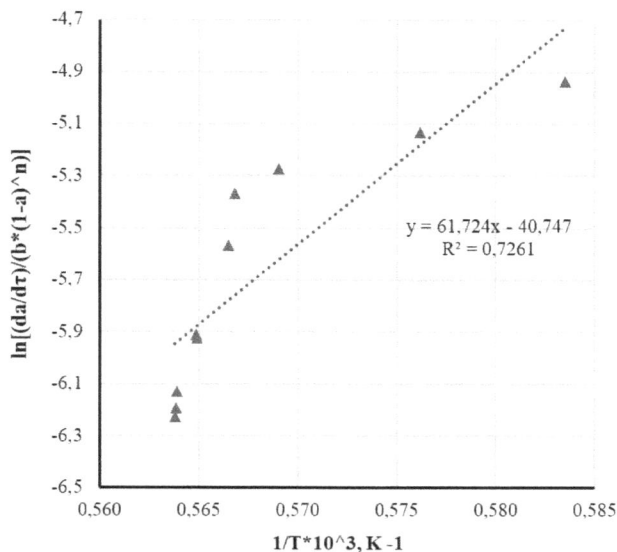

Figure 8.43. Calculation of the apparent activation energy for the case with Shubarkul coal.

$$y = 61.724x - 40.747$$
$$R^2 = 0.7261$$

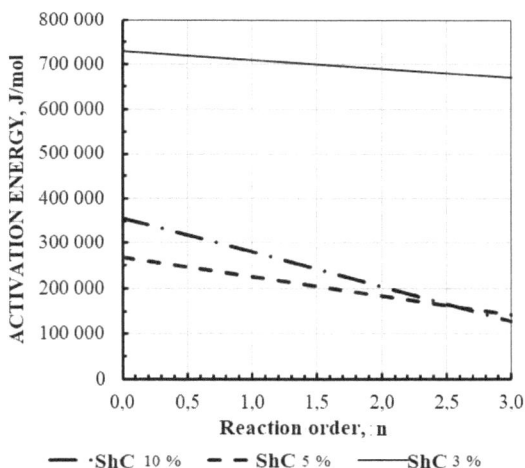

— ·ShC_10 %   – – ShC_5 %   ——ShC_3 %

Figure 8.44. Activation energy of the reduction process for the case of ore-coal briquettes with Shubarkul coal (ChC – charcoal, ShC – Shubarkol coal, SC – semi-coke).

As can be seen from the graphs, the apparent activation energy in all cases, regardless of the type of reducing agent and its concentration in the initial briquette, in the range of reaction order values n from 0 to 1 (characteristic of limiting processes in the form of internal or external mass transfer processes), correspond in all cases too high activation energies E_act (from 300–400 kJ/mol and more). For mass transfer processes, the corresponding values are in the range of 50–150 kJ/mol. The reaction orders of 1.5–2 correspond to activation energies in the range of 200–600 kJ/mol, which corresponds to generally accepted data indicating that the process is limited by a chemical reaction.

It can be concluded that the reduction process proceeds in a mixed mode with simultaneous limitation by internal mass transfer and chemical reaction. The simplest and most universal way to increase the rate of such limiting stages is to increase the temperature at which the process is carried out.

Basic experiments with different types and concentrations of the reducing agent in the briquette have established the following intermediate conclusions:

- the type of reducing agent - semi-coke, hereinafter semi-coke is one of the end products of the industrial process;
- component fraction - less than 1 mm, the limiting stage is close to a chemical reaction;
- the process of solid-liquid-phase carbothermal reduction for oxidized lean nickel ore of the Buruktal deposit is applicable.

To discover the optimal conditions for the carbothermal reduction of oxidized nickel ore—the process temperature, the required holding time of the briquettes in the hot zone of the furnace to maximize the extraction of nickel from the ore, the content of the reducing agent in the original briquette—additional studies were carried out.

The determination of the boundary conditions for the carbothermal reduction of oxidized nickel ore from the Buruktal deposit began with the selection of the optimum process temperature favorable for the rapid and complete reduction of nickel from the ore.

The experimental technique was the same as in the case of the earlier series of experiments, the process temperature varied—four different temperatures were chosen: 1400°C, 1450°C, 1500°C, 1550°C, and the composition of the briquettes—two types of briquettes with a char content of 5% (PC_5%) and 10% (PC_10%) in the original briquette. As a reducing agent, only semi-coke was used as the best reducing agent; in the experiments before, when using semi-coke, the highest degree of reduction of nickel from ore was recorded—about 98% with a high content of it in the metal, and as the second end product of an industrial process under conditions of a continuous oxygen reactor.

The behavior of the briquettes was the same to similar briquettes of the first series of experiments, the products of the experiments are reflected in Fig. 8.45. As can be seen from the photographs, with a char content of 5% in the original briquette and temperatures of 1400°C and 1450°C, the process proceeds only in a solid-phase mode, as it turned out, the temperature at this content of the reducing agent in the original briquette is insufficient for the process to go into the solid-liquid phase mode, while the metallization proceeds completely. Large droplets and small droplets of metal are not formed, which makes it impossible to separate the metal phase by simple and affordable magnetic methods widely used in production. As for the higher temperatures—1500°C and 1550°C, in this case the metal is formed and is easily separated from the slag, the nickel content is noted at the level of 12.5% at T = 1500°C and 16.7% at T = 1550°C, at such temperatures, the process proceeds quickly (up to 15 minutes) and completely.

With a semi-coke content in the initial briquette at the level of 10%, metal formation is observed at all temperatures, but at T = 1400°C and 1450°C, a sharp slowdown of the process was visually observed, which is why the experiment time was increased to 20 minutes. The absence of a large drop of metal is also observed, which is explained by the excess of the reducing agent, which

	T = 1400 °C	T = 1450 °C	T = 1500 °C	T = 1550 °C
Semi-Coke_5 %				
Semi-Coke_10 %				

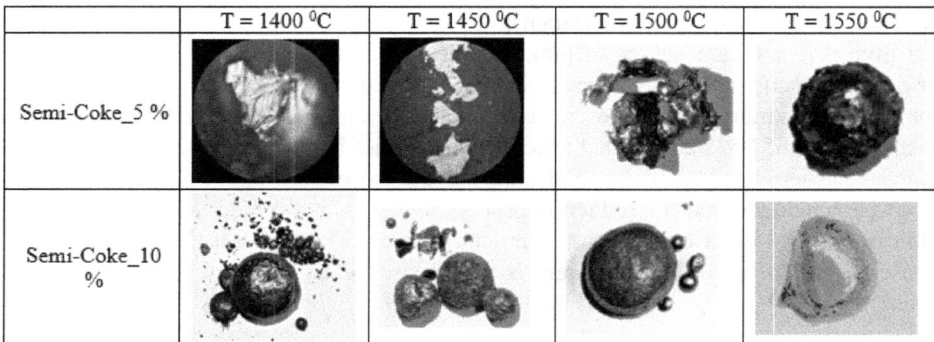

Figure 8.45. Appearance of the resulting metal at different temperatures.

floated on the surface of the melt during the experiment in the entire temperature range and had an uncoupling effect on the metal melt. The content of the leading element nickel in the metal is observed at the level of 4–5.2%, in parallel, iron is reduced from the ore, due to which the resulting metal is diluted with respect to nickel, therefore, the nickel concentration decreases.

Visual analysis of experiments at different process temperatures showed that it is preferable to conduct the recovery process at high temperatures of 1500–1550°C.

This series of experiments was processed in the same way as the basic series of experiments— gas chromatograms were analyzed, the degree of conversion α (O) was calculated by two independent methods. Comparison of the results is shown in Fig. 8.46, which shows that the best coincidence of the degree of conversion and the best course of the process is observed at T = 1500°C for 5% char content in the briquette and at T = 1500°C and T = 1550°C for 10% char, the points practically lie on a straight line passing through the origin at an angle of 450.

To determine the optimal time and temperature for the complete reduction of nickel and its maximum recovery from oxidized nickel ore, we analyzed the change in the degree of conversion of α_CO during experiments with different temperatures and at different times with a step of 2 minutes, the results are shown in Figs. 8.47 and 8.48.

After analyzing the dependence of the degree of conversion α_CO on temperature and time in the case of 5% char content in the original briquette (see Fig. 8.47). At lower temperatures, such as 1400°C and 1450°C, α_CO is higher, but no metal is formed. As for the higher temperatures of 1500°C and 1550°C, here α_CO is lower, but a compact drop of metal has formed. The recovery of nickel from ore to metal is about 80% at 1550°C and more than 98% (that is, complete recovery from ore to metal is observed) at 1500°C. Based on the observations made earlier, we came to the conclusion that for a semi-coke content of 5% in the initial briquette, the most preferable temperature is 1500°C—the process proceeds quickly with the formation of a compact drop of metal, without loss of metal and with a maximum degree of extraction from the ore of more than 98%, therefore, overheat unit up to 1550°C is impractical.

From the analysis of the dependence of the change in α_CO on temperature over time, it was revealed that the recovery process ends at 10 minutes. Figure 8.47 shows that α_CO changes slightly after 10 minutes of the experiment, which also corresponds to visual observation with a thermal imager—the metal formed at 10 minutes. After 10 minutes of holding the sample in the furnace, a deeper reduction of iron from the ore is observed, the concentration of iron in the product increases, due to which the melt is diluted and the concentration of nickel decreases.

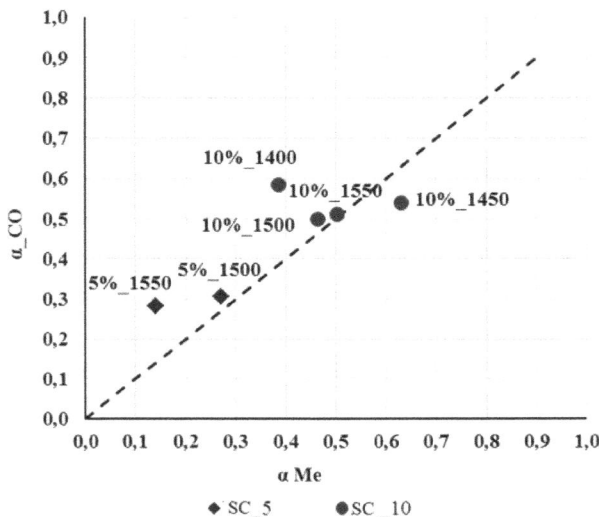

Figure 8.46. Comparison of the degree of conversion at different temperatures (SC – semi-coke).

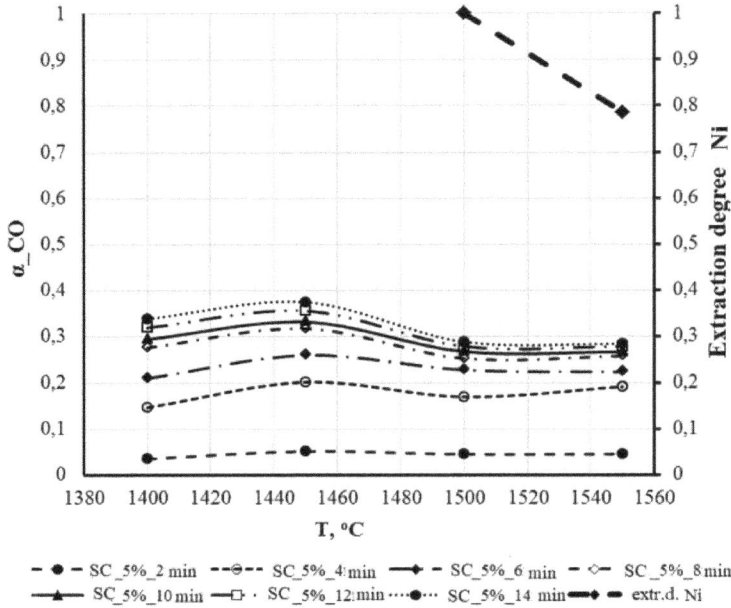

Figure 8.47. Dependence of the degree of conversion on temperature and time in the case of 5% semi-coke content in the original briquette (SC – semi-coke).

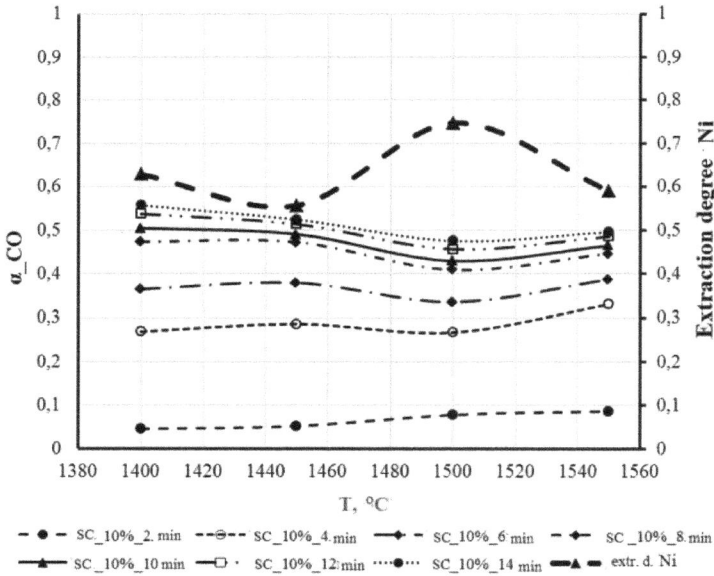

Figure 8.48. Dependence of the degree of conversion on temperature and time in the case of a semi-coke content of 10% in the original briquette.

Thus, for a 5% semi-coke content in the original briquette, the optimal holding time is 10–12 minutes at the required and sufficient furnace space temperature of 1500°C, these conditions ensure the complete extraction of nickel from the oxidized nickel ore of the Buruktal deposit.

Analysis of the dependence of the degree of conversion α_CO on temperature and time at a semi-coke content of 10% in the original briquette (see Fig. 8.48) showed that at all four temperatures, the value of the degree of conversion slows down after 8 minutes of the experiment and increases insignificantly with each subsequent minute, this increase occurs due to the additional reduction of iron from the ore, therefore, with a semi-coke content in the original briquette of 10%, the holding time of the sample in the furnace is 8–12 minutes. The analysis of the degree of recovery of the leading element (nickel) from ore to metal shows the best recovery at T = 1500°C and is about 75%, which is significantly less than at 5% semi-coke in the briquette.

From the analysis of the behavior of ore-carbon briquettes with different contents of semi-coke at various temperatures and a comparison of the degree of extraction of nickel from the ore, its concentration in the obtained metal, the degree of conversion and the time required for complete reduction and formation of a compact metal droplet, it follows that the best option for industrial conditions is a briquette containing 5% semi-coke at a process temperature of 1500°C and a holding time of 10–12 minutes. These conditions provide the most complete recovery of nickel from ore and allow achieving high technology productivity.

In order to clarify the best concentration of char in the original briquette to ensure rapid and complete recovery of nickel from oxidized nickel ore to obtain a commercial product, laboratory studies were carried out with briquettes containing a reducing agent in a wide concentration range from 3 to 14%.

Figure 8.49 demonstrates the dependence of the nickel concentration in the obtained metal on the content of char—the more char in the test sample, the lower the nickel concentration, this phenomenon is explained by the more complete reduction of iron from the oxidized ore due to the excess content of the reducing agent in the briquette, i.e., the resulting metal is diluted with respect to nickel, thereby reducing the concentration of nickel in the metal.

An excess of the reducing agent is observed in the initial briquette at concentrations over 6% of the char in the initial briquette—an unreacted reducing agent floated on the surface of the metal melt, which complicated the formation of a large drop of metal. In the case of an increased content of the reducing agent in the initial sample, the concentration of nickel in the metal notably drops to 10% or less, due to a more complete reduction of iron from the ore as described above.

In the case of a semi-coke content in the original briquette of 4% or less, its deficiency is observed, which manifests itself in the finely dispersed nature of the obtained metal (mainly solid-phase reduction), which is not suitable for industrial conditions, since it complicates the process of separating the metal from the slag.

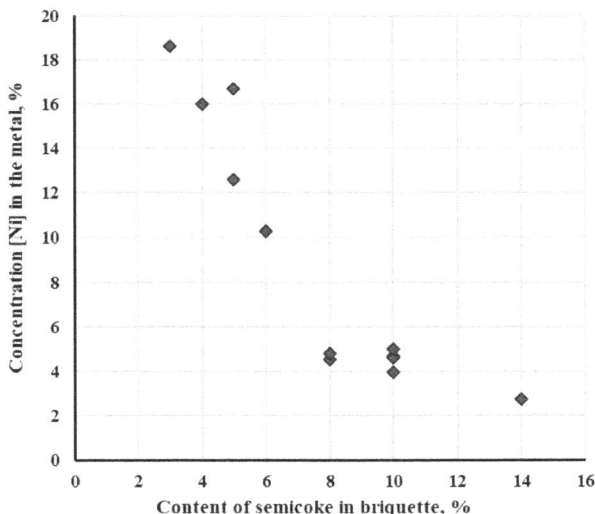

Figure 8.49. Dependence of the concentration of nickel in the metal on the content of semi-coke in the briquette.

Thus, the following conclusions can be drawn:

- the optimum process temperature is not less than 1500°C, which provides a recovery time of less than 12 minutes.

- the optimum concentration of the reducing agent in the briquette is 5%, which leads to a nickel recovery of more than 98% with a simultaneous nickel concentration in the metal of more than 12%.

References

[1] Bizhanov, A.M., Pavlov, A.V. and But, E.A. 2018. The study of the reduction process parameters of ore-coal briquettes for ferroalloys production. The Fifteenth International Ferro-Alloys Congress. Infacon XV (Cape Town, South Africa). pp. 139–151.

[2] Mizin, V.G. and Serov, G.V. 1976. Carbonaceous reducing agents for ferroalloys. - M.: Metallurgy, 272 p. (in Russian).

[3] Mashchenko, V.N., Kniss, V.A., Kobelev, V.A. and Polyansky, L.I. 2005. Preparation of oxidized nickel ores for smelting. - Yekaterinburg: Ural Branch of the Russian Academy of Sciences, 316 p. (in Russian).

[4] Smirnov, V.I., Khudyakov, I.F. and Naboychenko, S.S. 1967. Selection of a method for preparing oxidized nickel ores for mine smelting. Non-ferrous Metallurgy. 3: 24–26 (in Russian).

[5] Mashchenko, V.N. 2007. Improvement of the technology of briquetting oxidized nickel ores of the Serov deposit: Diss. Cand. Tech. Sciences. - Yekaterinburg (in Russian).

[6] Khrushchev, M.S. 1977. On the mechanism of interaction of metal oxides with carbon. Message 1. Ferrous Metallurgy. 2: 13–16 (in Russian).

[7] Khrushchev, M.S. 1977. On the mechanism of interaction of metal oxides with carbon. Message 2. Ferrous Metallurgy. 4: 13–16 (in Russian).

CHAPTER 9
Briquetting in Direct Reduced Iron (DRI) Production

9.1 Application of Briquettes in the Charge of a Midrex Reactor

The process of mini steel mills using oxidized pellets as a feedstock includes production of metallized pellets, which are used further for the production of Direct Reduction Iron (DRI), production of steel in Electric Arc Furnaces (EAF) and its continuous casting and rolling. Companies that are importing large amounts of iron ore pellets annually for DRI production and also during the course of the transportation, stockpiling and charging of the pellets into metallization reactors, as well as the discharging of the metallized pellets, generate tens of thousands of tons of fine materials containing iron every year. Pellets fines and DRI sludge are usually dumped in piles and EAF dust and mill scale are sold to a third-party. The results of a preliminary analysis indicate that the recycling of such materials in the form of briquettes would help to produce additional quantities of steel, and would also free up a large area occupied by dumped wastes. Recovery of these wastes would generate additional revenues, surpassing revenue from direct sales of wastes.

We have studied the possibility of using stiff vacuum extrusion to produce agglomerated products suitable for use as charge components in DRI reactors. The experiments were conducted in an industrial Midrex reactor operated by one of the customers of the J.C. Steele & Sons, Inc. This company imports 3.5 million tons of iron-ore pellets annually to produce 2.35 million tons of Hot-Briquetted Iron (HBI). Tens of thousands of tons of finely dispersed iron-bearing wastes are formed each year during the unloading, storage and subsequent charging of pellets into metallization reactors, and also during the discharging of the metallized pellets and their briquetting.

9.1.1 Tests with brex in a rigid steel basket in the charge of a midrex reactor

Two series of basket tests that entailed the reduction of brex in an industrial Midrex reactor were conducted. In the first series 25–30 strengthened brex were placed inside a rigid steel basket (Fig. 9.1a) that were then charged into the reactor together with pellets. The brex were extricated from the reactor after the tests were concluded in order to be able to visually evaluate the condition of the reduced briquettes, and study their composition and properties. The mechanical strength of the brex did not play a significant role in this case, since they were not subjected to pressure from the column of charge materials.

The stiffness of the basket completely prevented the brex from being subjected to pressure from the layer of pellets and could possibly be deformed or fractured. Figure 9.2 shows images of a steel basket and its contents after being removed from an industrial Midrex reactor at the end of the test. In the second series of tests, brex were placed inside deformable gas-permeable steel packets (Fig. 9.1b), which made it possible to study their behavior under conditions similar to those encountered inside a layer of pellets in a Midrex reactor. The results of the first series of tests are described below.

Figure 9.1. Rigid steel baskets before feeding into the reactor (left) and steel deformable packages (right) for loading the brex into the Midrex reactor.

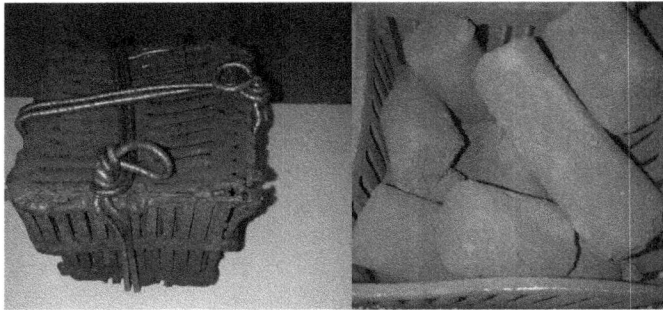

Figure 9.2. Rigid steel basket after removal from the Midrex reactor and its contents.

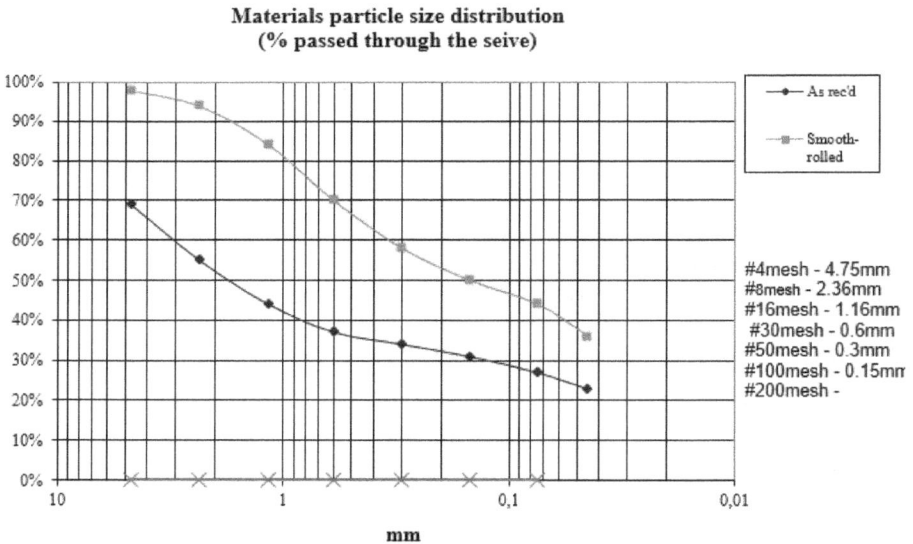

Materials particle size distribution
(% passed through the seive)

Legend:
- As rec'd
- Smooth-rolled

#4mesh - 4.75mm
#8mesh - 2.36mm
#16mesh - 1.16mm
#30mesh - 0.6mm
#50mesh - 0.3mm
#100mesh - 0.15mm
#200mesh -

Figure 9.3. Granulometric composition of the mixture of materials for briquetting: (1) initial mixture; (2) after additional pulverization (the maxima on the curves correspond to the mixture's content of particles smaller than 5 mm).

To conduct the tests, we used brex produced from a mixture of pellets fines – 55.6% (no more than 92% of which were 6.3 mm or smaller in size), metallized DRI sludge – 27.8% (also with a 92% particle content no coarser than 6.3 mm), and mill scale – 16.6% (with 99% no coarser than 10 mm). The main goals of the investigation were to evaluate the reduction properties of brex in the Midrex process and to select a binder that would give the brex the necessary strength and maximize their metallization. The mixture was subjected to additional pulverization (Fig. 9.3) on a

Table 9.1. Chemical composition of the components of the mixture for brex production.

Elements and oxides	Content in the component, wt.%		
	Pellets fines	DRI sludge	Mill scale
Fe	65.0	66.6	70.0
SiO_2 + CaO	3.8	6.52	1.00
CaO	1.3	4.38	0.15
MgO	0.75	0.69	0.10
Al_2O_3	0.95	0.83	0.25
MnO	0.1	0.16	1.20
P	0.055	0.045	0.02
TiO_2	-	0.010	0.020
V_2O_5	-	0.12	0.025
C	-	-	0.30
S	0.015	0.01	0.015
Na_2O + K_2O		0.33	-

Table 9.2. Component-by-component composition of the brex.

Components of the charge	Type of brex			
	01–01	01–02	01–03	01–04
Mixture of pellets fines, DRI sludge and mill scale	95.0	95.0	91.5	92.0
Slaked lime	5.0			6.0
Portland cement		5.0	8.0	
Bentonite			0.5	
Molasses				2.0

roll crusher before it was briquetted. Table 9.1 shows the chemical composition of the components of the mixture.

Four types of experimental brex were tested (Table 9.2). The types differed negligibly from one another in terms of their chemical composition and varied only in the type, and the amount of binder that was used. The brex was made by J.C. Steele & Sons on a laboratory extruder. The mixture was prepared and homogenized by using a Hobart laboratory mixer that simulated the processing of the charge in a pug sealer.

There was a substantial difference (Table 9.3) between the green brex and the brex after strengthening (that is after they had been strengthened over the course of 14 days).

The porosity of brex is one of their most important properties and determines their reducibility. The microstructure of the specimens and the specifics of their porosity were investigated with the use of a Scanning Electron Microscope (SEM).

LEO 1450 VP (Carl Zeiss, Germany) with a guaranteed resolution of 3.5 nm, and STIMAN software [1] to quantitatively analyze SEM images obtained in the back-scattered electron regime. The porosity of briquette specimens 01–01 and 01–03 was studied on material that was freshly cleaved from the specimens' surface. Morphological studies of the microstructure were performed in the secondary-electron regime and allowed us obtain high-quality half-tone images over a broad range of magnifications. The STIMAN method makes it possible to obtain correct images with distinct boundaries between the pores and the particles. The morphological parameters of brex

Table 9.3. The physical and mechanical properties of brex after strengthening.

Properties	Type of brex			
	01–01	01–02	01–03	01–04
Density, 1 g/cm³	3.314	3.458	3.300	3.464
Density, 2 g/cm³	3.085	3.005	2.844	3.038
Compressive strength1, N/mm²	2.2	1.6	1.6	18.4
Moisture content*1, %	11.5	10.5	9.8	8.9
Moisture content*2, %	1.11	1.07	1.06	2.53
1 Green brex. 2 Strengthened brex.				

Table 9.4. The morphological parameters of the pore microstructure in brex specimens.

Type of brex	Parameters	Class of pore, µm					D_{max}	n_{im}, %	K_a, %
		D_1 (< 0.1)	D_2 (0.1–1.0)	D_3 (1.0–10)	D_4 (10–100)	D_5 (> 100)			
	N, %	0.7	13.5	56.2	29.6	-	32.36	21.25	8.50
	K_f			0.42–0.50; 0.58–0.67					
	N, %	1.9	18.2	46.9	33.1	-	32.2	20.88	17.69
	K_f			0.33–0.42; 0.50–0.58					

Notes: 1. N represents pores of different size classes expressed as fractions of the total number of pores n_{im} calculated from the SEM-image; D1–D5 are the different pore-size classes; Dmax is the maximum equivalent diameter of the pores; K_a is the anisotropy coefficient, or the degree of orientation of the solid structural elements; K_f is the shape factor of the pores. 2. The coefficient K_f is calculated as the ratio of the semi-minor of an ellipse inscribed in a pore to its semi-major axis. The coefficient $Kf = 0.66–1.00$ for isometric pores, $K_f = 0.1–0.66$ for anisometric pores, and $K_f < 0.1$ for slit-shaped pores

01–01 and 01–03 that were measured in this way, together with the characteristics of the pores are shown in Table 9.4. This information clearly indicates that the only significant variation in the porosity of the specimens was with respect to the anisotropy coefficient. This difference was probably due to the type of binder.

The distribution of the macropores (with sizes greater than 100 µm) can be determined by computer-assisted x-ray tomography. The images in Fig. 9.4 illustrate the difference in the distributions of the macropores in green brex specimens 01–01 and 01–03. The images were captured with a Yamato TDM-1000 x-ray computer micro tomograph (Japan). The magnification was 32 and the resolution was 11 µm. It is possible to conclude that the 01–01 brex had a higher percentage of pores larger than 100 µm than did the 01–03 brex.

The degree of open porosity determined by the STIMAN method is very consistent with the open porosity measured by liquid saturation in a vacuum in accordance with the standard DIN 51056 (GOST 26450.1–85). The porosity values were within the range 21–24% for the specimens that were examined.

Cracks could clearly be seen on brex specimens 01–02 and 01–03 after they were removed from the reactor (Fig. 9.5).

Both specimens had been prepared with the use of a cement binder. There were more cracks in the specimens with the 01–03 designation, which had higher binder content (Table 9.2). It is possible to conclude that the hot strength of brex is related to their crushing strength. Brex specimens 01–02 and 01–03, which were weaker and less dense in the cold state (see Table 9.3), also had a lower hot strength. The lower value for their hot strength manifested itself in the formation of surface cracks

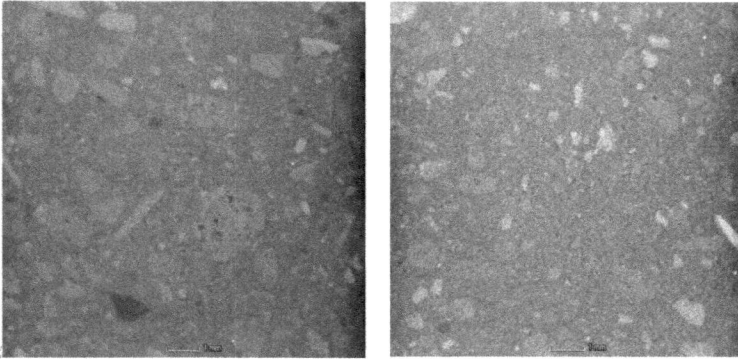

Figure 9.4. Distribution of pores coarser than 100 μm in specimens of briquettes 01–01 (a) and 01–03 (b) (micro-x-ray tomography).

Figure 9.5. The look of the reduced briquettes after their extraction from the rigid steel baskets.

in the brex, which is consistent with the results obtained from studying the hot strength of iron-ore pellets and comparing it to their crushing strength [2]. In the case of the pellets, there was also a negative dependence of hot strength on the ratio Al_2O_3/SiO_2. An increase in this ratio decreased the pellets' hot strength. Brex specimens 01–02 and 01–03, which exhibited a lower hot strength than brex prepared with a lime binder, had the highest value for elastic modulus due to the substantial amount of alumina in the cement and, in particular, in the bentonite (brex 01–03). At the same time, the presence of lime and Hematite (pellet fines and scale) in brex 01–01 and 01–04 helped calcium ferrites form at just 400–500°C during the course of the briquettes' heating in the Midrex reactor. These ferrites, which have low melting and softening points, strengthen the structure of brex and improve their reducibility [3].

After the brex were extracted from the rigid steel baskets, their chemical composition, total iron content, content of metallic iron, and degree of metallization were determined (Table 9.5). The highest degree of metallization was seen for the 01–01 brex with 5% lime content, while the lowest degree of metallization was registered for the 01–03 brex with the maximal (8%) content of cement binder.

Table 9.5. Chemical composition of the green and reduced brex.

Type of brex	Green brex		Reduced brex			
	Fe_t,%	C, %	Fe_t, %	Fe_{met} %	Metallization, %	C, %
01–01	64.53	2.32	80.07	77.39	96.65	0.98
01–02	64.06	1.88	79.22	74.63	94.21	0.84
01–03	62.53	1.72	75.20	68.37	90.92	0.87
01–04	60.46	2.60	76.40	70.17	91.85	1.70

The changes in the porosity of the brex during their reduction were evaluated by using a Phoenix V|tome|XS 240 x-ray micro tomograph with two x-ray tubes: a micro focus tube with a maximum accelerating voltage of 240 kV and a power of 320 W; a nanofocus tube with a maximum accelerating voltage of 180 kV and a power of 15 W. The initial analysis of the data and the construction of a three-dimensional model of the specimens based on the x-ray photographs (projections) were done using the datos|x reconstruction software, while the VGStudioMAX 2.1 and AvizoFire 7.1 software was used to visualize and analyze the data based on elements of the three-dimensional image. The photographs were taken at an accelerating voltage of 100 kV and a current of 200 mA; the resolution during the recordings was 5.5 µm. Table 9.6 shows the results obtained from measuring the volume and number of pores in the brex before and after reduction.

Figure 9.6 depicts the visualization of the pore distribution in the form of a 3D-model.

Table 9.6. Volume and number of pores in brex specimens.

Characteristics	Type of brex (green/reduced)	
	01–01	01–03
Total volume of the pore space, mm³	0.77/0.45	0.69/0.87
Number of pores in the investigated volume of the specimen	186063/93321	198420/241634

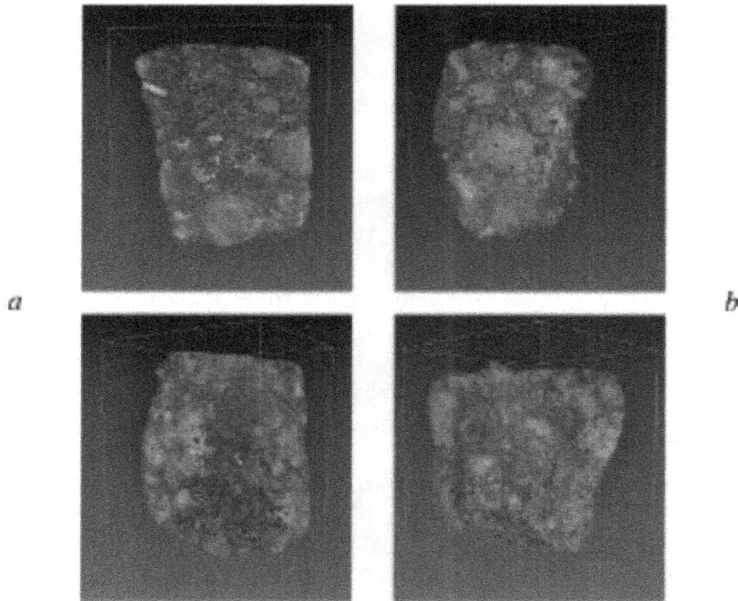

Figure 9.6. 3D-model of the pore space in a 01–01 brex (a) and a 01–03 brex (b).

It is apparent that the porosity of the brex which were examined (and had coarse and ultra-coarse pores with a size of 1–100 µm or larger) changed in different directions. The total volume of the pore space decreased by a factor of 1.6 in the 01–01 brex and increased by 20% in the 01–03 brex. The increase in the porosity of the reduced brex with a cement binder can be attributed to the disintegration of the cement stone at 850°C, and the accompanying decrease in the strength of the brex.

The brex remained intact due to the formation of a metallic matrix composed of reduced iron. Conversely, the reduction of the brex with a lime binder was accompanied by a decrease in their porosity. This can be attributed to the agglutination of fine pores during the formation of the metallic matrix composed of reduced iron. Basket tests of brex made from a mixture of pellet fines, mill scale and finely dispersed wastes from HBI production in an industrial Midrex reactor showed that they can be effectively metalized without loss of integrity. The brex with a lime binder displayed the highest degree of metallization and the greatest strength. While the brex with a cement binder remained intact, cracks were formed on their surface. The number of cracks was found to be greatest in the brex with a higher binder content.

9.1.2 Tests with brex in deformable steel packets in charge of midrex reactor

At the second stage of testing, the experimental brex were placed inside deformable steel packages that were gas permeable for the reducing gas (Fig. 9.1b). This input method allows brex behavior to be simulated adequately in actual reactor conditions, including the mechanical pressure of the surrounding traditional charge. At the end of the process, these packages were drawn out of the reactor, which enabled the condition of the reduced brex to be determined visually, and their reduced chemical composition and properties to be explored.

The prepared mix of pellets fines (52.6%); metallized sludge from DRI production (26.4%), mill scale (15.8%) and EAF dust (5.2%) has been used.

The chemical composition of the mix components is given in Table 9.7.

All test samples were extruded using a 25 mm round pelletizing die. A Hobart laboratory mixer was used to simulate the mixing with water and pugging of the ground feed material in the open tub of the pug sealer. The laboratory extruder simulates the processing of the material through the sealing auger and die, into the vacuum chamber, and then final extrusion. The laboratory extruder consists of two chambers with a sealing die between them. The rear chamber is fitted with a 3-inch diameter sealing auger that pushes material through the sealing die. The second chamber can be subjected to a vacuum. It is also fitted with a 3-inch diameter auger that extrudes material through a pelletizing die. A PC-based data system monitors and records extrusion data. All mixes were extruded immediately after mixing. Three types of the experimental brex were produced with the compositions given in Table 9.8.

Table 9.7. Chemical composition of the mix components.

Chemical compounds	Pellets fine	Mill scale	DRI sludge	EAF dust
Fe_{tot}	65.00	70.0	66.2	29.68
SiO_2	2.50	1.00	2.14	4.25
CaO	1.30	0.15	4.38	19.74
MgO	0.75	0.10	0.69	24.27
Al_2O_3	0.95	0.25	0.83	1.32
MnO	0.10	1.20	0.16	0.96
S	0.015	0.015	0.01	0.13
$Na_2O + K_2O$	0.034	–	0.33	1.42

Table 9.8. Experimental brex compositions, %.

Charge component	Brex No. 1	Brex No. 2	Brex No. 3
Pellets fines	50.0	50.0	50.0
Sludge	25.0	25.0	25.0
Mill scale	15.0	15.0	15.0
EAF dust	5.0	4.75	5.0
Slaked lime	5.0	-	-
Portland cement	-	5.0	-
Magnesium binder	-	-	5.0
Bentonite	-	0.25	-

Table 9.9. Chemical composition of experimental brex before reduction, %.

Elements and oxides	Brex No. 1	Brex No. 2	Brex No. 3
Fe_t	62.47	61.30	62.61
C	1.49	1.40	1.05
CaO	9.24	8.41	4.30
MgO	3.61	3.07	6.24
SiO_2	2.45	4.91	2.76
Al_2O_3	1.29	1.81	0.99
TiO_2	0.10	0.12	0.12
V_2O_5	0.076	0.07	0.08
MnO	0.34	0.36	0.35
P_2O_5	0.07	0.09	0.06
S	0.08	0.08	0.50
$Na_2O + K_2O$	0.19	0.93	0.83
Cl	0.04	0.03	0.02
ZnO	0.36	0.40	0.35

A magnesium sulfate-based binder was composed using the magnesium sulfate heptahydrate ($MgSO_4 \cdot 7H_2O$). The binding properties of magnesium sulphate were first described by Zhuravlev et al. [4]. The change in form that occurs when a dehydrated inorganic salt is converted to the hydrated crystal form is the basis for the recommendation that a partially dehydrated magnesium sulfate be used industrially as a binding material. Setting begins in 3 minutes and is complete in 6 minutes. Chemical composition of the brex is given in Table 9.9.

A calibrated electronic scale with a density measuring attachment was used to determine brex density. A compression tester was used for sample strength measurements, tensile splitting strength was measured in accordance with ASTM (American Society for Testing and Materials) C1006-07, Paragraph 7.1. Porosity was measured in accordance with DIN (Deutsches Institut für Normung) 51056. Moisture content was measured using a moisture balance. The physical and mechanical properties of the raw brex are given in Table 9.10.

Experimental brex with the above compositions were placed into the deformable and gas-permeable steel packages (each type of brex was placed in a separate package) and these packages were added to the traditional charge of the Midrex reactor. The tests were conducted during the course of a night shift from midnight to 8:00 a.m. The reducing gas temperature was around 900°C. At the end of the process, the packages were removed from the reactor due to the presence of so-called cold discharge (with a discharge temperature of 40°C) thus enabling the condition of the

Table 9.10. Physical and mechanical properties of raw brex.

Property	Brex No. 1	Brex No. 2	Brex No. 3
Density, g/cm^3	3.5	3.48	3.66
Compressive strength, N/mm^2	4.8	11.1	4.4
Tensile splitting strength, N/mm^2	1.5	1.4	1.2
Porosity	29.7	24.7	25.1
Moisture content, %	8.4	8.4	8.6

(a)   (b)

(c)   (d)

Figure 9.7. Raw brex (a) and reduced brex after their extraction from the steel packages (b) brex No. 1, (c) brex No. 2, (d) brex No. 3.

reduced brex to be visually determined (Fig. 9.7) and their chemical composition and properties to be examined.

An observation can be made that the brex that used a magnesium sulfate-based binder demonstrated the highest degree of integrity. The cement-bonded brex demonstrated the largest degree of fines generation. The brex that used lime binders were destroyed to a markedly lesser extent.

The mechanical strength values of the reduced and non-destroyed brex were defined in the course of splitting tensile testing (0.4 N/mm^2 for lime-bonded brex and 1.2 N/mm^2 for brex with a magnesium sulfate-based binder). The strength of the brex with a magnesium sulfate binder did not changed compared to its original value (Table 9.10). Thus, the brex with the lowest values for the cold mechanical strength showed the highest values for hot strength. Similarly, the brex with the magnesium sulfate-based binder also showed the highest reducibility. The decrease in the reducibility of the lime- and cement-bonded brex is related to the decrease in the permeability of the steel packages after the particles of the generated fines have blocked the holes. The chemical composition of the reduced brex and their degrees of metallization are presented in Table 9.11. A serious discrepancy in terms of the total iron contents in the reduced brex can be seen. This, in turn, comes from a difference in metallic iron content. The reason for this is related to the different hot strength levels of the brex and to the aforementioned blocking of the holes by fines from partially disintegrated brex (No. 1 and No. 2), which prevents the intake of the reducing gas.

The Midrex reactor, which has been used for full-scale testing, produces DRI that has the following guaranteed parameters: a degree of metallization of 94%, total iron content—91%, metallic

Table 9.11. Chemical composition and metallization of the reduced brex.

Elements and oxides	Brex No. 1	Brex No. 2	Brex No. 3
Fe_{tot}	74.86	69.02	86.86
Fe_{met}	49.11	18.66	84.00
Metallization, %	65.60	26.96	96.71
C	1.75	0.89	1.02
CaO	9.11	6.71	5.07
MgO	2.93	2.24	7.81
SiO_2	3.41	4.35	4.04
Al_2O_3	1.57	1.45	2.13
TiO_2	0.11	0.12	0.12
V_2O_5	0.07	0.08	0.07
MnO	0.34	0.43	0.39
P_2O_5	0.08	0.07	0.01
S	0.03	0.09	0.26
$Na_2O + K_2O$	0.48	0.51	0.74
Cl	0.09	0.01	-
ZnO	0.18	0.30	0.05

iron content—85%, carbon content—1.3%, sulfur—0.005%. It follows from the testing results that the brex with the magnesium sulfate-based binder reached a higher degree of metallization (96.71%), and almost the same level of metallic iron and carbon content, but this was smaller than in the DRI total iron content. The sulfur content is significantly higher in this brex. This is evidently a limiting factor for the utilization of brex with a magnesium sulfate-based binder in Midrex reactors. Sulfur will partially go to the flue-gas and in, the absence of the flue-gas desulfurization, it could spoil the quality of the catalyst. A lime-bonded brex has much lower sulfur content.

These results with those obtained in the Midrex reactor of the ArcelorMittal in Hamburg with roller briquettes of the following composition: 37.6% DRI filter cake, 47% oxide fines, 9.4% return fines and 6% of hydrated lime and molasses as binders can be compared [5]. The total iron content in the raw briquettes amounted to 59%, in the reduced—77%. The degree of metallization of the briquettes extracted from deformable steel packages was 86% [6]. The brex with a magnesium sulfate-based binder revealed a greater degree of metallization (96.71%) with a lower initial value of the total iron content (62.61%) in the raw brex and with a smaller binder content (5%).

9.1.3 Mineralogical study of the reduced brex

The structure of the reduced samples of brex has been studied using light microscopy methods with the "Nikon" equipment (a polarized ECLIPSE LV100 POL microscope equipped with a digital photomicrography system DS-5M-L1) (Nikon, Tokyo, Japan). Analysis of mineral phases in polished sections of briquettes was conducted by MLA 650 (FEI Company, Hillsboro, OR, USA), including an FEI Quanta 650 SEM scanning electron microscope. The main mineral phases were determined by a diffraction analysis using an analytical complex ARL 9900 Workstation IP3600 (combined X-ray fluorescence spectrometer) (Thermo Fisher Scientific, Waltham, MA, USA). Reduced brex with a magnesium sulfate-based binder and retained integrity samples of the reduced brex with lime and cement binders were tested.

Fines from pellets with a low basicity (CaO/SiO_2 = 0.3–0.5) are represented by fragments with a different mineral composition: cores in the form of residual magnetite with a glass phase and shells consisting of splices of Hematite with calcium ferrites. The mill scale has a magnetite composition—a solid solution $(Fe,Mn)O \cdot Fe_2O_3$. A special role in the composition of the brex is played by EAF dust, which consists of the main melt generating components: CaO, MgO, MnO, SiO_2 and Na_2O.

In the reduced and non-destroyed cement-bonded brex, iron is represented by partly recovered iron containing minerals made from pellets fines and mill scale.

In the main body of the cement-bonded brex extracted from the steel mesh packages, there are areas with the phases that are close in terms of their composition to the binary metasilicates $2Na_2O \cdot CaO \cdot 3SiO_2$ (N_2CS_3) and $Na_2O \cdot 2CaO \cdot 3SiO_2$ (NC_2S_3) without any presence of iron (Fig. 9.8).

N_2CS_3 melts incongruently [7] (melting starts at 1141°C with the creation of 78.5% of the melt and 21.5% of NC_2S_3, and ends at 1203°C). NC_2S_3 melts congruently with an extensive field of primary crystallization. In the Midrex process temperature range only two eutectics exist (Fig. 9.9): 755°C—N_3S_8 + NCS_5 +S (N_2O—22%; CaO—3.8%; SiO_2—74.2%); 827°C—N_2CS_3 + NC_2S_3 + NS_2 (N_2O—36.6%; CaO—1.8%; SiO_2—60.7%).

In the phases detected by the X-ray spectral microprobe analysis, CaO content exceeds 10%, indicating that limited melt formation takes place in these areas. The phase detected at temperatures

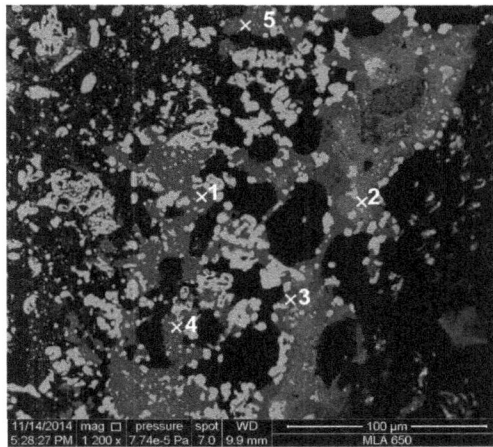

Figure 9.8. SEM-image of the reduced cement-bonded brex. 1, 3, 4—binary metasilicates; 2—silica containing calcium ferrite; 5—dicalcium ferrite.

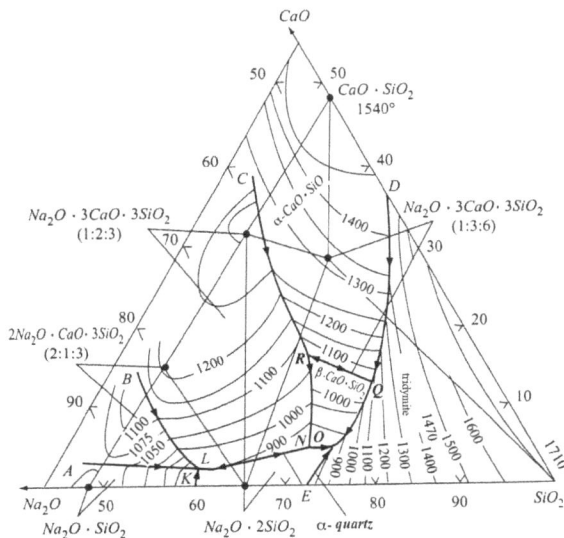

Figure 9.9. Diagram of the system Na_2O-CaO-SiO_2.

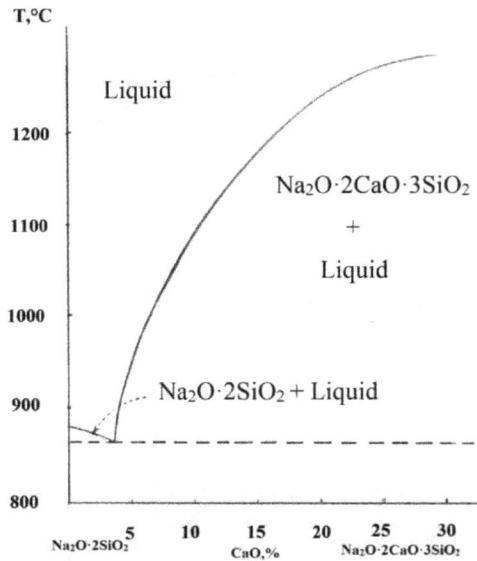

Figure 9.10. Diagram of system $Na_2O \cdot 2SiO_2$–$Na_2O \cdot 2CaO \cdot 3SiO_2$.

ranging from 870–950°C can be in an amorphous state with low amounts of liquid phases, as illustrated in Fig. 9.10.

In the cement-bonded brex, the alkali-silicate phase is predominantly associated with the smallest dusty fractions, which are contained in the alkali. The silicate bonding is not fully developed in the sample's body. Optical study of the samples of the reduced brex shows the pieces of dicalcium silicate (Ca_2SiO_4) crystals.

Lime-bonded brex have a higher degree of metallization and a deeper interaction between charge components. Such samples are characterized by a variation in the composition of iron-silicate melts in different parts of the body of the brex. The most prevailing is the combination of the primary melt with iron. In this case, creation of the melt is related to the enveloping of the pellets fines particles by the dust fractions. The primary iron-silicate melt composition is close to olivine phases. The melt has a high basicity ($CaO/SiO_2 = 0.8$–0.9), Na_2O content in different parts of the sample ranges from 1.0 to 10%, magnesium and manganese oxides content does not exceed 1.0–2.0 mass %. Under the conditions of the Midrex process as the pellet fines contact the melt, metallic whiskers are generated (Fig. 9.11).

Rarely are mineral formations with a high content of calcium oxide (up to 40%), silicon oxide content (25–26%) and the same amounts of bivalent iron observed in brex with lime as a binder. No other oxides are found in the composition of the amorphous iron-calcium phases with high calcium content. The composition of such phases is close to the Melilite phase ($Ca_2Fe_3 + Si_2O_7$).

The predomination of the fine fraction within the brex is of particular interest. It consists of a combination of metal with calcium ferrites. Apparently, in pellets fines, consisting of the combination of Hematite with calcium ferrite, the Hematite phase was the first to reduce and such reduction took place without breaking contact with the ferrite.

Brex with magnesium sulfate binder exhibited the highest degree of metallization and a well-developed melt generation process.

For the first time in this study, it was found that during the metallization, the original microstructure of brex does not persist. Brex is close to a two-phase system: a metal and a silicate phase. However, the composition of the silicate part remains non-uniform in the adjacent volumes of the reacted fines. According to X-ray spectral microprobe analysis, among the studied melts two mineral species are dominant—close to the olivine and Melilite structures, but each with a different

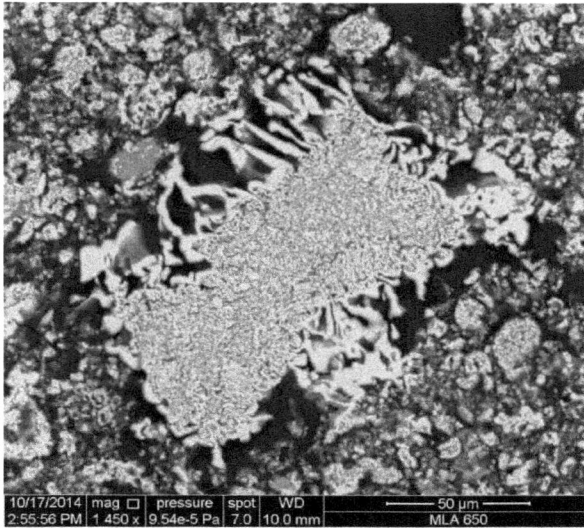

Figure 9.11. Generation of the metallic "whiskers" at the contacts of the pellet fines and melts.

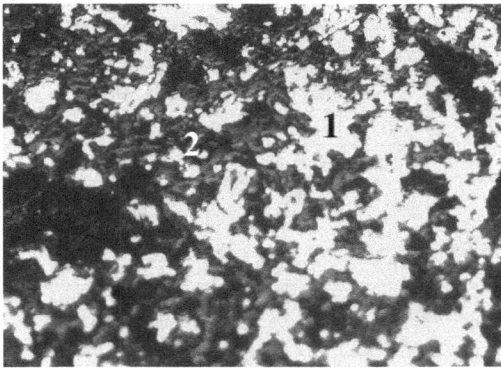

Figure 9.12. Optical microscopy of the reduced lime-bonded brex (1—iron, 2—Mellilite; magnification 200).

content of iron oxides, magnesium and manganese. This conclusion has also been confirmed by the results of an optical microscopy of the brex samples (Fig. 9.12).

The areas of the melt, which have reached equilibrium, are of particular interest. The mineral phase in combination with residual melt is observed in these areas (Fig. 9.13).

Thus, it is clear that the mechanism of the strengthening of brex samples in each case has its own specific characteristics. For cement-bonded brex, a lack of hot strength during heating in the reducing atmosphere is associated with their swelling and with the limited formation of silicate bonding. The strength of lime-bonded brex during the reduction process is provided by formation in the solid phase of the iron-calcium silicates, similar in composition to the olivine phase with a low melting temperature. Brex with the magnesium sulfate-based binder has the highest degree of generation of the two-phase metal-silicate system at the end of the metallization process.

During the process of reduction of the brex placed in deformable steel packages their porosity changed in a different way. With the aim of studying the nature of such changes, we investigated the structure of the porosity of the original and reduced brex. The microstructure of samples was studied using a scanning electronic microscope (SEM) 1450 LEO VP (Carl Zeiss, Jena, Germany) with a resolution of 3.5 nm. For morphological studies of the microstructure of the samples the secondary electrons mode has been used, which allows high-quality halftone images to be obtained

Figure 9.13. Structure of the reduced brex (1, 2—mineral phase; 3, 4—residual melt).

Table 9.12. Morphological parameters of the original and reduced brex.*

Brex	Porosity SEM/SEM + CT	Contribution of Pores of Different Dimension Categories to Total Porosity (p_{tot}), %. Diameter in Microns.					Maximum Diameter, µm
	p_{tot}, %	D_1	D_2	D_3	D_4	D_5	D_{max}
		< 0.1	0.1–1.0	1.0–10	10–100	> 100	
Brex No. 1	31.6	0.6	8.6	25.6	65.2	0.0	57.6
	37.3	0.5	7.2	21.4	66.8	4.1	407.2
Brex No. 1 (reduced)	32.9	0.2	9.0	38.4	52.4	0.0	70.94
	38.3	0.1	7.7	32.4	49.8	10.0	504.49
Brex No. 2	31.2	0.6	8.0	30.1	61.3	0.0	45.78
	35.7	0.5	7.0	26.2	61.3	5.0	299.96
Brex No. 2 (reduced)	37.9	0.2	9.3	31.1	59.1	0.3	100.51
	40.4	0.2	8.9	29.5	55.2	6.2	583.77
Brex No. 3	31.8	0.6	5.8	28.4	65.2	0.0	62.92
	32.9	0.5	5.5	27.2	65.3	1.5	191.89
Brex No. 3 (reduced)	32.8	0.2	3.2	35.7	60.9	0.0	71.75
	39.3	0.2	2.6	28.9	58.0	10.3	736.41

* p_{tot}—total porosity calculated by the analysis of SEM image; D_1–D_5—different dimension categories of pores, D_{max}—maximum pore diameter; CT—computer tomography.

across a wide range of magnifications. Quantitative analysis of microstructure was conducted using STIMAN software by two different methods: The first—is based on a series of SEM images obtained in back-scattered electrons, which enables the correct picture to be obtained with clearer boundaries between the pores and particles; the second—by a complex analysis based on a set of SEM images, and images captured using Yamato TDM-1000 (Yamato Scientific, Tokyo, Japan) computed tomography. The results of the morphological studies are given in Table 9.12.

During the process of metallization, the total porosity of brex with a magnesium sulfate binder increased by 19.5%, cement-bonded brex by 13.15% and brex with lime as a binder only by 2.6%. The porosity increased due to large and mega pores (larger than 100 µm): for brex with magnesium

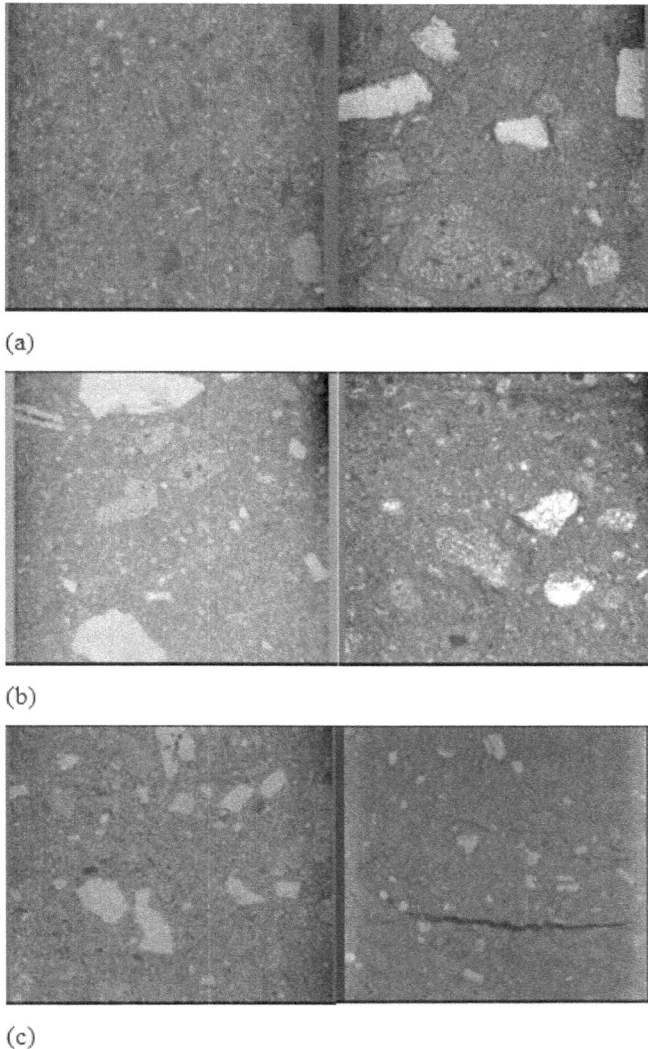

(a)

(b)

(c)

Figure 9.14. Distribution of pores in the original (left) and reduced (right) brex; (a) lime-bonded; (b) cement-bonded brex; (c) brex with a magnesium sulfate-based binder, magnification 64.

sulfate binder—they increased by 6.9 times, and for lime-bonded brex—increased by 2.43 times. In cement-bonded brex, the growth of share of large share was 1.24 times. The maximum diameter of pores in brex with a magnesium sulfate binder has increased by almost three times and in brex with a lime—by 1.23 times. The difference in the nature of changes of porosity between the original and reduced brex can also be clearly seen on images captured by computer tomography (Fig. 9.14). In brex with magnesium sulfate- and lime-based binders, the development of internal cracks took place. This contributed to the formation of coherent porous systems and to the increase in the reduction rate.

It can be assumed that the driving mechanism for the development of porosity in the brex was the crystal lattice type that changes during the Hematite-magnetite phase transformations, which results in an increase in the volume of the brex. This increase in volume is accompanied by mechanical stresses, which may cause the partial destruction of brex, which leads to an increase in the proportion of large and mega-pores and to the formation of internal cracks. The stronger the brex is mechanically, the less it is prone to such partial destruction.

The brex with the magnesium binder were further manufactured at the industrial SVE briquetting line belonging to Suraj PL (Rourkela, India). Phase analysis of the industrial brex was done in Rikagu Miniflex600 x-ray diffraction system using copper target and nickel filter. Figures 9.15–9.19 show the phase composition of the brex after reduction at different temperatures. The testing has been performed with the help of the experimental facility described in 5.2 (Figs. 5.19 and 5.20) and adjusted for the conditions simulating blast furnace process. Heating of the brex samples in a reducing atmosphere was carried out to temperatures of 1000°C, 1100°C, 1200°C, 1300°C and 1400°C.

Mineralogical study of the reduced samples was done on polished section under reflected light in Leitz Universal microscope with the image analyzer. Figures 9.20–9.24 show the different stages of the brex metallization.

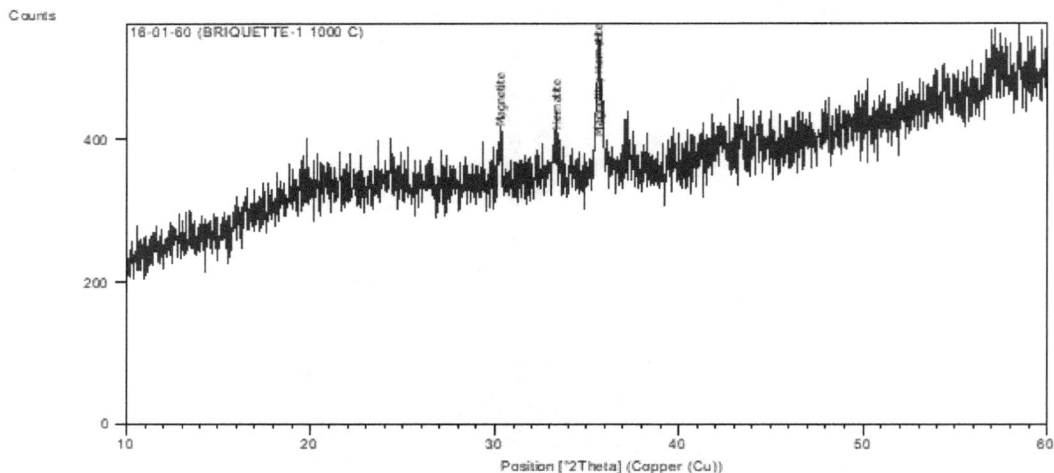

Figure 9.15. Phase composition of brex sample after reduction at 1000°C. Main phases: magnetite, Hematite.

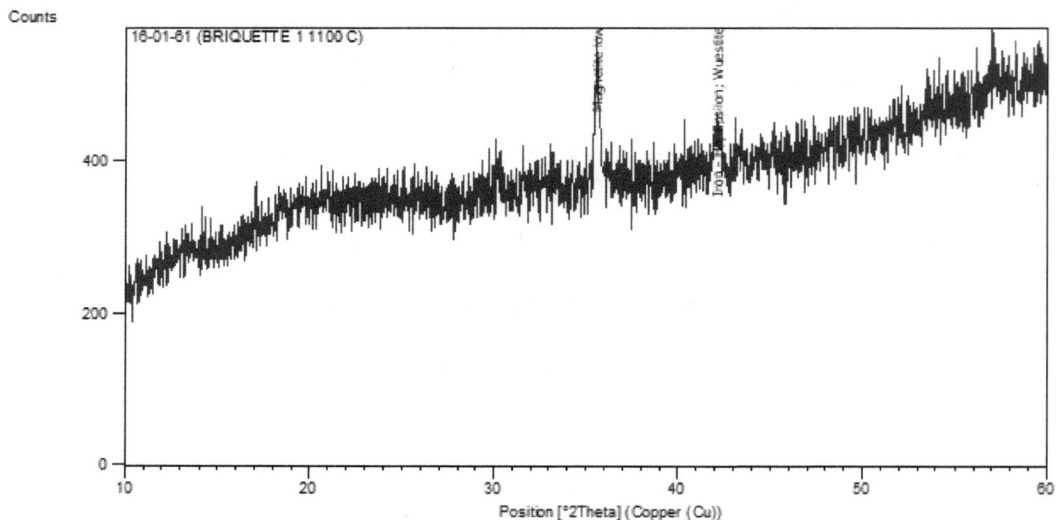

Figure 9.16. Phase composition of brex sample after reduction at 1100°C. Main phases: magnetite, wustite, iron.

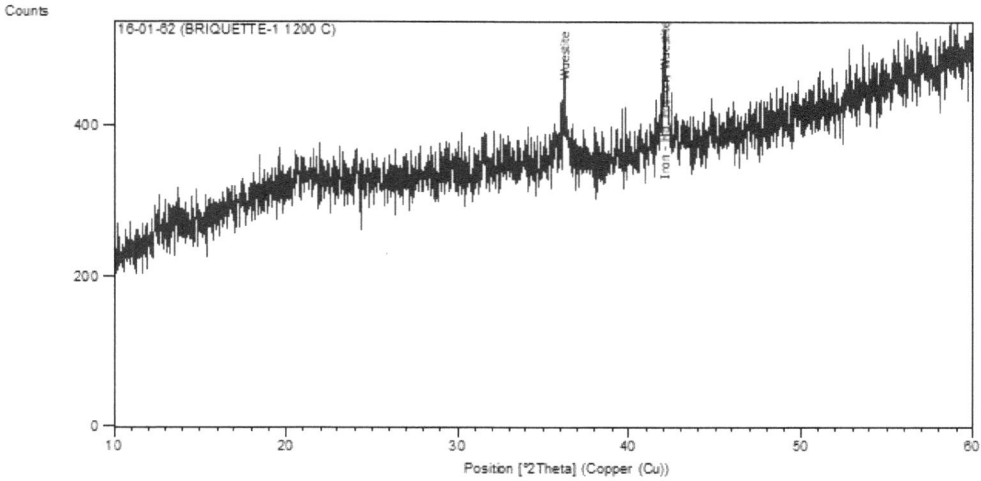

Figure 9.17. Phase composition of brex sample after reduction at 1200°C. Main phases: iron, wustite.

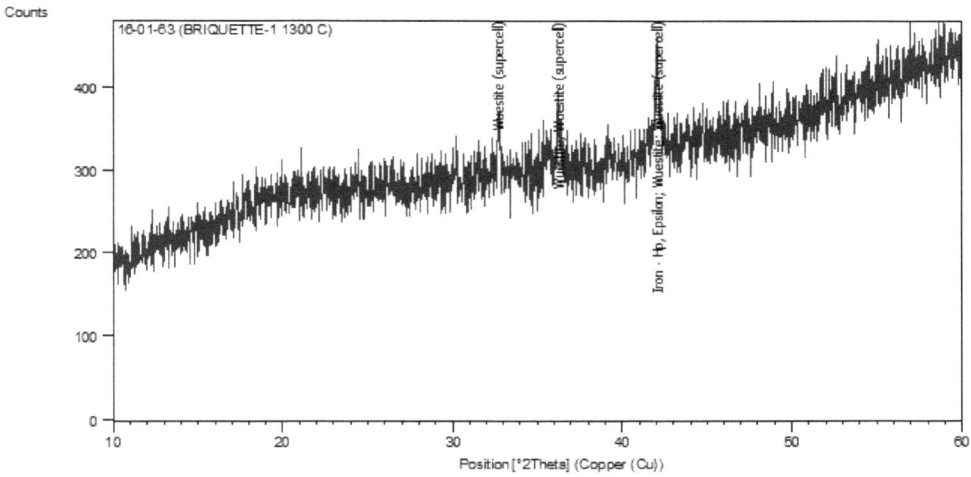

Figure 9.18. Phase composition of brex sample after reduction at 1300°C. Main phases: α-iron, wustite.

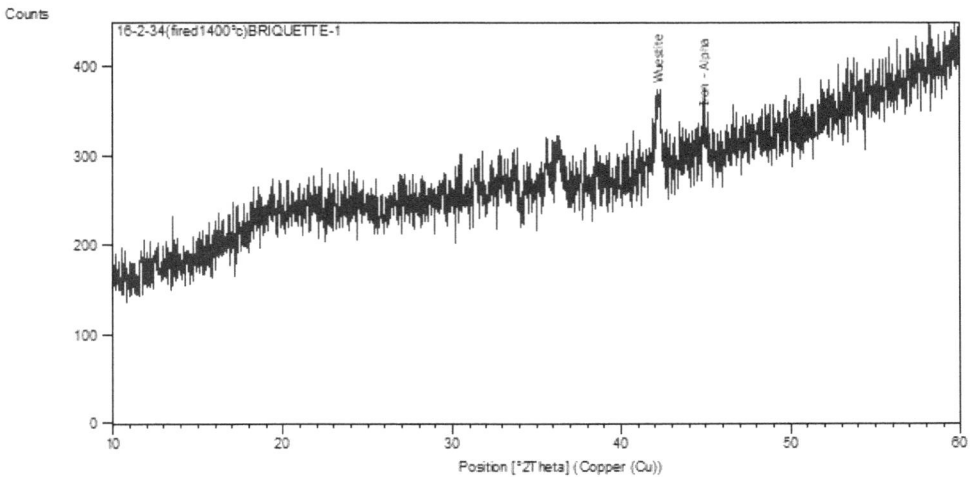

Figure 9.19. Phase composition of brex sample after reduction at 1400°C. Main phases: α-iron, wustite.

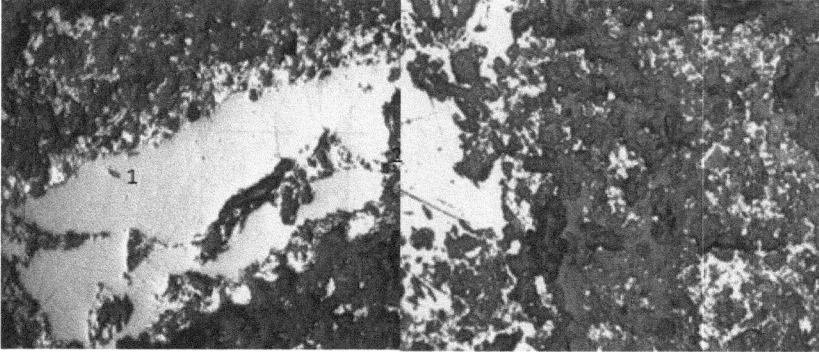

Figure 9.20. Photomicrograph of Briquette-1 fired at 1000°C. 1 - magnetite; 2 - Hematite; magnification x200.

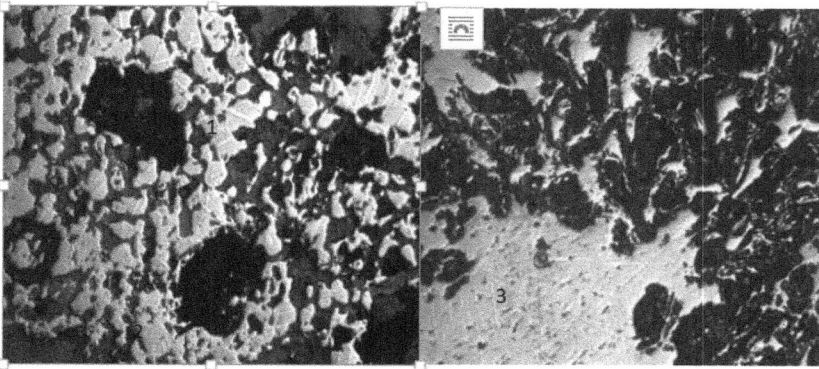

Figure 9.21. Photomicrograph of Briquette-1 fired at 1100°C. 1 - iron; 2 - wustite; 3 - magnetite; magnification x200.

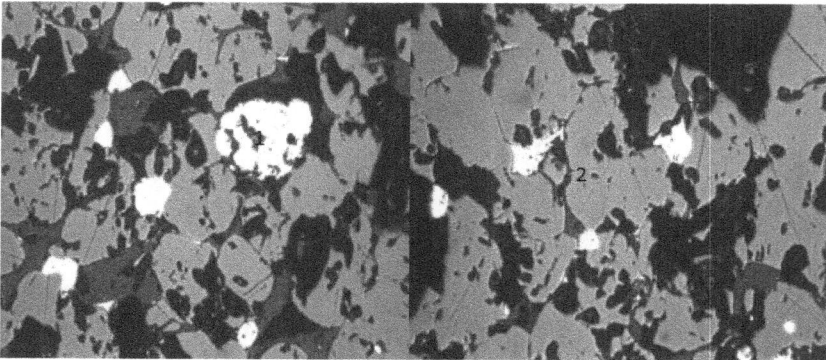

Figure 9.22. Photomicrograph of Briquette-1 fired at 1200°C. 1 - iron; 2 - wustite; magnification x200.

The results of this study show clearly that magnesium bonded brex could be considered as candidates for the BF charge.

As an appendix to the study of the properties of the brex as a charge component in Midrex reactors we studied the properties of brex made from direct reduced iron (DRI) fines, that is formed the in course of the classification process designed to classify the pellets, and the hot briquetted iron that are manufactured at the Stock Company «Lebedinsky GOK» (Belgorod region, Russia). The entire metallized product is subjected to screening in terms of its classification: less than 4 mm, less than 25 mm. In order to obtain a competitive, and economically viable product the brex must be able

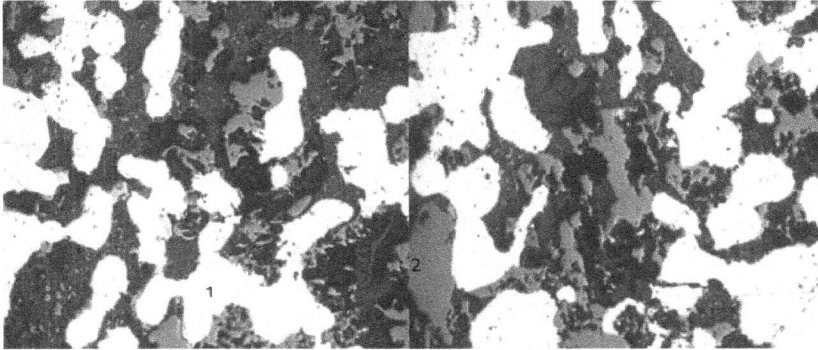

Figure 9.23. Photomicrograph of Briquette-1 fired at 1300°C. 1 - iron; 2 - wustite; magnification x200.

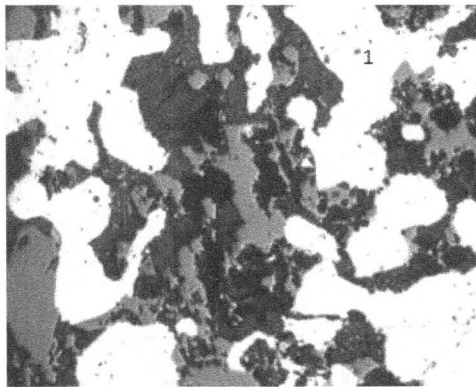

Figure 9.24. Photomicrograph of Briquette-1 fired at 1400°C. 1 - iron; magnification x200.

Figure 9.25. Brex made from DRI fines (at the «Lebedinsky GOK»).

to be used as a charge component in an arc furnace, and correspondingly its size, density and binding agent should be such that the briquette does not float in the slag, but descends down to the melt and does not input any components that are harmful to the steel. The durability should be sufficient to prevent any destruction during the course of transportation (either by vibration or friction), and during its storage in the stockpile or in the hold.

From the DRI fines formed at the «Lebedinsky GOK» brex with the following composition was prepared and tested by ourselves (Fig. 9.25): DRI fines 92%, molasses - 4%, lime - 4%, bentonite - 0.5% (in excess of 100%).

The value for the crushing strength of the brex was on average 122 kg/cm². The porosity of the brex was 27.36%.

Thermal testing in the Tamman furnace during the melting of the brex at a temperature of 1513°C demonstrated their high thermal stability. The brex melted smoothly, and was immersed fully into the melt. The duration of the melt of a single brex was 10 minutes.

References

[1] Sokolov, V.N., Yurkovets, D.I. and Razgulina, O.V. 1997. Determination of tortuosity coefficient of pore channels by computer analysis of SEM images. Proceedings of Russian Academy of Sciences, Physical Series. 61(10): 1898–1902 (in Russian).

[2] Dwarapud, S. and Ranjan, M. 2010. Influence of oxide silicate melt phases on the RDI of iron ore pellets suitable for shaft furnace of direct reduction process. ISIJ Int. 50(11): 1581–1589.

[3] Vegman, E.F., Zherebin, B.N., Pokhvisnev, A.N. et al. 2004. The Metallurgy of Iron: Textbook, IKTs Akademkniga, Moscow, 774 p. (in Russian).

[4] Zhuravlev, V.F. and Zhitomirskaya, V.I. 1950. Binding properties of crystal hydrates of the sulfate type. J. Appl. Chem. USSR 23: 115–119.

[5] Harald, H.-K., Alexander, F. and Stefan, H. 2017. Briquetting of ferrous fines—saving resources, creating value. pp. 973–978. *In*: AISTech 2017 Conference Proceedings, vol. II.

[6] Bizhanov, A.M., Kurunov, I.F. and Wakeel, A.Kh. 2016. Behavior of extrusion briquettes (brex) in Midrex reactors. Part 2. Metallurgist (3): 112.

[7] Berezhnoy, A.S. 1988. Multicomponent Alkaline Oxide Systems, 200p. Kiev: Naukova Dumka (in Russian).

9.2 High Temperature Reduction of Brex

We investigated the behavior of the iron ore and coal brex in a neutral atmosphere at a temperature range of 1350–1370°C partially simulating the conditions of the reduction of the iron ore and coal pellets in the ITmk3 process developed by «Kobe Steel» [1–2]. This process is implemented in the annular chamber Rotary Hearth Furnace (RHF). Green (unburned) iron ore and coal pellets are loaded in one or two layers onto a rotating hearth covered by a refractory. The heating of pellets and reduction of iron oxides is performed per single revolution of the hearth over the course of 10–15 minutes at a maximum temperature that is higher than 1350°C. Iron oxides are reduced by carbon in the iron ore and coal pellets or briquettes. The reduced iron in the pellets or briquettes is carbonized with its particles being coagulated and melted to produce bean shaped cast iron drops (3–12 mm) inside the softened slag shell. When cooled in a neutral atmosphere these drops harden forming cast iron pellets (or "nuggets"). The metallurgical properties of pellets or briquettes must maintain their integrity during heating, and reduction and the slag shell that is produced should prevent the spread of liquid droplets of iron.

For the production of the experimental brex we used the Hematite iron ore concentrate (63.13% of iron content; Fe_2O_3 = 90.18%, SiO_2 = 4.63%, Al_2O_3 – 3.11%, MnO – 0.807%, CaO – 0.346%), the Globe coal (bulk density 768 kg/m³; moisture content 7.8%; ash 4%; volatiles 37.7%; and where 73% of particle's are less than 0.6 mm in size) and the Shubarkul coal (bulk density 800 kg/m³; moisture content 10%; ash 7.1%; volatiles 38.5%; S 0.40%; P 0.021%.; and where 99% of the particles are less than 0.3 mm in size). The Type I Portland cement (general use) has been used as the binder. Volclay DC-2 Western (Sodium) Bentonite was used as the binder and plasticizer. We manufactured brex with both round, and oval cross-sections (with diameters of ½" and 1"; and a length equal to 1.5–2.0 diameters). The laboratory extruder simulates the processing of the material through the feeder to the pug mill and then to the sealing auger and die, into the vacuum chamber, and then final extrusion. Moisture content was measured using a moisture balance. A calibrated electronic scale with a density measuring attachment was used to determine brex density (Mettler MS603S and Mettler MS-DNY-43).

The results of the sieve analysis of Globe coal and Shubarkul coal are shown in Fig. 9.26.

Table 9.13 shows the chemical composition of iron ore concentrate (w-content).

Globe

Shubarkul

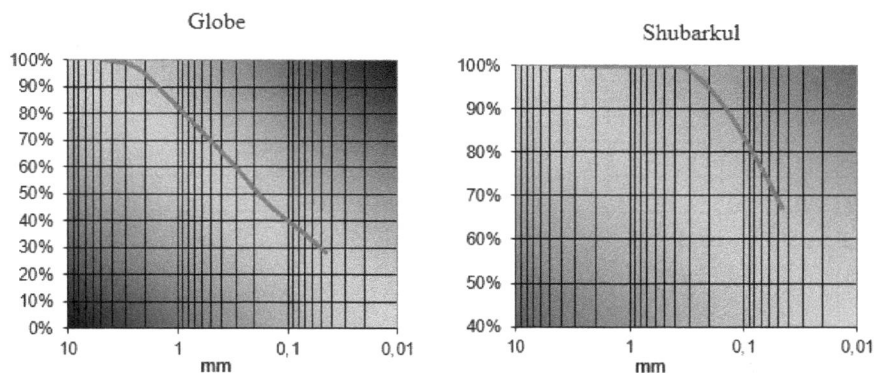

Figure 9.26. Sieve analysis: Left–Globe coal (73% particles less than 0.6 mm in size); Right – Shubarkul coal (99% particles less than 0.3 mm in size).

Table 9.13. Chemical composition of iron ore concentrate.

Component	w, %
Fe_2O_3	90.18
SiO_2	4.63
Al_2O_3	3.11
MnO	0.807
CaO	0.346
MgO	0.337
Na_2O	0.163
P_2O_5	0.129
TiO_2	0.091
S	0.050
BaO	0.026
CuO	0.022
ZnO	0.021
NiO	0.017
Cl	0.016
Co_3O_4	0.015

Table 9.14 gives the brex compositions, shapes and their physical properties. Wet base moisture of the brex No. 1 and No. 2 was 18.5%. For the brex No. 3 moisture content was equal to 15.2%. The vacuum applied for the production of the brex No. 1 and No. 2 was at the level of 48.26 mm Hg and for the brex No. 3 the value was 38.10 mm Hg.

For the production of the brex No. 3 we applied a 24 hour souring using Bentonite. This procedure enhances the plasticity of the mix and can lead to an increase in the mechanical strength of the brex (or to the corresponding decrease of the binder consumption).

We measured the compressive strength of the cured brex. These values were measured based on the Instron 3345 device (USA). For the measurement procedure we prepared the required amount of the specially shaped brex samples and subjected them to compression. The values of the compressive strength for the brex were as follows (kgF/cm²): sample No. 1 – 59.0; sample No. 2 – 24.0; sample No. 3 – 49.0. The influence of the shape of the brex on the mechanical strength were observed. For brex No. 1 and No. 2 with the same chemical composition we attained a significant increase in the compressive strength for the oval shaped brex when compared with the regular cylindrical form of

Table 9.14. Brex components, shape and physical properties.

Brex sample	Component, %				Density, g/cm³	Porosity %	CCS, kgF/cm²
	Iron ore	Coal	Portland cement	Bentonite			
No. 1*	63	32	5	–	2.072	34.8	59.0
No. 2	63	32	5	–	2.034	36.9	24.0
No. 3	63.5	31.3**	4.7	0.5	2.207	36.4	49.0

* oval shape (1.25 cm × 2.5 cm);
** Globe Coal

the brex. As far as the values for the tensile splitting strength are concerned this increase takes place when the applied pressure is perpendicular to the main axis of the oval. When the applied pressure is perpendicular to the short axis of the oval, we observed a corresponding decrease in the strength value. It is clear, however, that the most probable orientation for the oval shaped brex lying on its side is with its long axis parallel to the flat surface. It is also worth mentioning that the transition to the oval shape increases the value of the external brex surface when compared with the round brex of the same length and volume.

We expected that the increased density of the brex might prevent their carbothermic reduction. However, the results of the porosity study show that for all brex compositions that were considered the value of the porosity favors their reducibility. We used an LEO 1450 VP (Carl Zeiss, Germany) Scanning Electronic Microscope with a resolution of 3.5 nm together with the X-ray computed high resolution computed tomography system Phoenix V|tome|X S 240 (General Electric, USA). Computed tomography has been used to detect the share of the macro pores (larger than 100 μm in size). For the smaller size pores the SEM has been applied. The approach based on STIMAN [3] computer software has been used to calculate the porosity detected by SEM. The results of these measurements showed that the total porosity of the brex No. 1 was equal to 34.8% (10.9% measured by X-ray computed tomography and 23.9% by SEM) brex No. 2 was equal to 36.9% (12.3% measured by X-ray computed tomography and 24.6% by SEM). For brex No. 3 the same value was equal to 36.4% (10.8 and 25.6% correspondingly). The spatial distribution of the pores detected by the X-ray computed tomography is illustrated by the image given in Fig. 9.27.

The reduction tests of the iron ore and coal brex following the conditions of the ITmk3 process were conducted by us in laboratory resistance electric furnaces with a controlled atmosphere based on the Electric Furnace with a vertical arrangement of the graphite heater and with an inside diameter equal to 65 mm. We positioned a working removable Alumina Crucible (98% Al_2O_3) in the

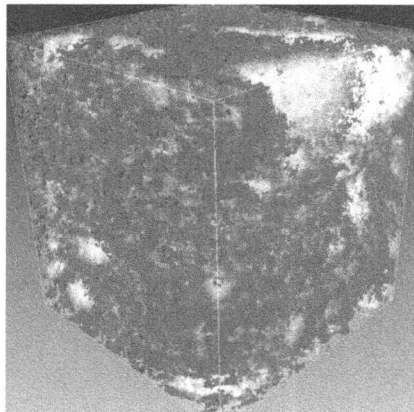

Figure 9.27. Spatial distribution of the pores in brex No. 2 (X-ray computed tomography; dark and black- pore space; light and transparent – brex body; resolution 6 μm).

isothermal zone of the heater. The temperature in the furnace was regulated and stabilized by us with the help of the thermocouple BP (A) 5/20 located in the isothermal area of the furnace on the outside of the heater. The true value of the temperature was measured using the thermocouple 2 BP (A) 5/20, located inside the heater and lowered from above into the working Alumina crucible. According to available data on the process of ITmk3 we selected 1360°C as the working temperature, which we maintained with an accuracy of ±10°C (1350–1370°C).

Before the test the furnace with the working replaceable Alumina Crucible was pumped out using the forevacuum pump until the residual pressure reached a value of 10^{-1} Pa and was then filled with a high-clean grade argon. Then we opened the gas discharge to atmosphere and kept the argon consumption through the furnace at the level of 0.5 l/min. The heating of the furnace was switched on, reached the desired temperature level and stabilized it, and then dropped the brex sample through the dosing gateway into the crucible. This moment was considered as the test start of the test. Thus, we subjected the sample brex to the rapid heating from the ambient temperature to the working temperature of the furnace. During every single test we recorded the process on a video using thermal imaging infrared camera "Pyrovision M9000" ("Micron", USA). This device records the film and also shows the temperature at each point of each frame. Sample exposure time was around 15 minutes; if necessary it could be extended it up to 20 minutes. After that we depressurized the furnace, extracted the crucible with the products and tempered them in air; installed a new working crucible into the furnace and the new sample brex into the dosing gateway. We closed the furnace, washed it out using argon, and after setting the required temperature we repeated the cycle.

Figures 9.28 and 9.29 show the photos of the consequent stages of the brex melting (volatiles release, solid-state reduction, melting and hot metal and slag drop creation) based on the visual data obtained by «PyrovisionM9000».

In all cases, the high density of briquette and large amounts of volatile coal did not interfere with the solid-state reduction of iron. During the heating the cracks at the brex surface appeared however

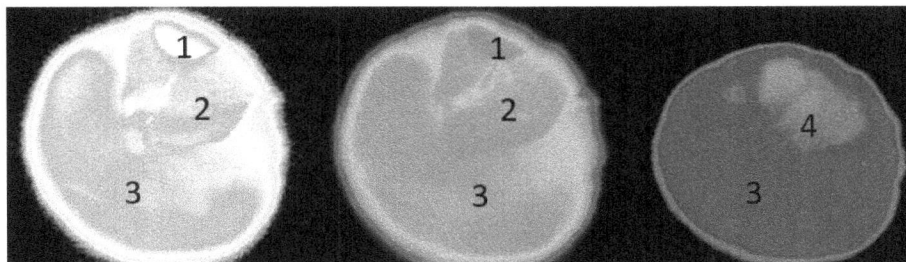

Figure 9.28. Brex No. 1 (from left to right: volatiles release, t = 1 min, T = 1000°C; beginning of the melting, t = 6 minutes, T = 1200°C; before complete melting, t = 7 minutes, T = 1 360°C); 1 - thermocouple; 2 - brex; 3 - carbonaceous bed zone; 4 - melting brex (solid and liquid).

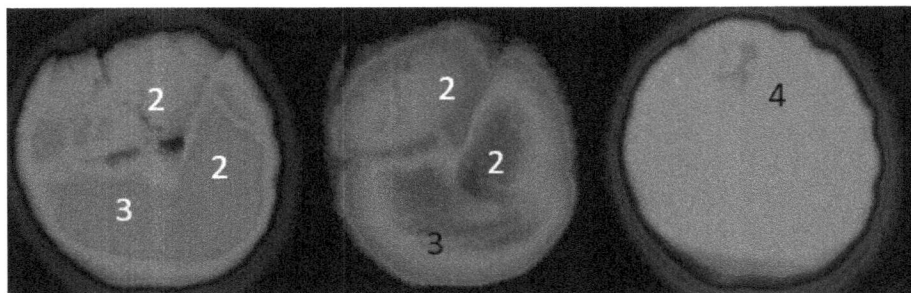

Figure 9.29. Brex No. 2 (from left to right: the end of the coal volatiles release, t = 1 min 40 seconds, T = 1000°C; solid state reduction, t = 6 minutes, 37 seconds, T=1200 °C; creation of the nuggets, t = 13 minutes, 20 seconds, T = 1360°C); 2 - brex; 3 - carbonaceous bed zone; 4 - liquid nugget surrounded by slag.

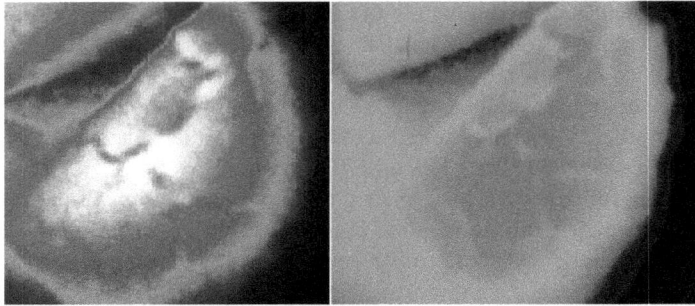

Figure 9.30. Brex No. 3 at temperatures T = 1100°C, left; and T=1250°C; right.

Figure 9.31. Products of the brex reduction. From left to right: brex No. 1, brex No. 2, brex No. 3.

Table 9.15. Chemical composition of metal (nuggets and shell).

Element %	Sample No. 1	Sample No. 2	Sample No. 3
[Fe]*	95.53	96.05	94.08
[C]	3.740	3.333	4.800
[Si]	0.252	0.154	0.605
[Mn]	0.076	0.121	0.136
[S]	0.107	0.118	0.185
[P]	0.135	0.058	0.085
[Cu]	0.0030	0.0074	0.0090
[Co]	0.0905	0.1030	0.0343
[Ti]	0.0580	0.0656	0.0570
[Zn]	0.0060	0.0005	0.0050

the brex did not decompose and kept its integrity (Fig. 9.30). Finally, the brex was transformed into liquid metal (nuggets) and slag drops. In the event of an excess of reducing agents the formation of metallized shells mixed with drops of slag were observed.

Figure 9.31 Shows the appearance of the products of the brex reduction.

Chemical composition of the reduced products (nuggets and shells) is given in Table 9.15. Due to the high content of coal in the brex the carbon concentration in the metal was at the level of 3.3–4.8%. Metal samples (nuggets and shells) were studied using a Jeol JSM 6490 LV Scanning Electron Microscope. All samples except No.3 have a similar microstructure: small island-type non-metallic inclusions in a metal matrix. The smallest size of inclusions is observed in the sample No. 2 (10–20 μm), in the sample No. 1 the inclusions are 50–60 μm in size. There are no oxide inclusions in the

No. 2 sample. Sample No. 3 is characterized by nonhomogeneous inclusions of a linear form with a width of around 5–10 μm and a length of 200–300 μm. As can be seen from Table 9.15 the C and Si content are high in sample No. 2. This shows that the reduction material in sample No. 2 is excessive compared to samples No. 1 and No. 3. The decrease of the amount of the reduction material in sample No. 2 could help to improve the strength of the brex. Synchronized thermal analysis methods and Mössbauer spectroscopy has been applied for the study of the phase composition of the brex (STA 449 C, Germany). A comparison of the peaks of the metallic iron formation shows that the reduction of the metal iron takes place to a larger extent in the brex samples No. 1 and No. 2 (with the Shubarkul coal) rather than in the brex No. 3 which contains Globe coal. The degree of fineness of the Shubarkul coal particles might have influenced their enhanced reducibility. For a definitive conclusion on the suitability of the brex, it is necessary to conduct further research on optimization of their size and on the composition of the fluxing oxides in the brex.

Metal samples (nuggets and shells) were studied using a Jeol JSM 6490 LV Scanning Electron Microscope. Figure 9.32 shows the images of the samples in the backscattered electrons.

Figure 9.32. Images of metal samples in backscattered electrons (from the left to the right - No. 1; No. 2; No. 3).

Conclusions

1. The research into the properties of brex made from dispersal wastes from mini-plants, producing rolled products, and using oxide pellets as a raw material, and the study into the behavior of this brex as they are heated in a reducing atmosphere in an industrial Midrex process reactor, enables in concluding—that SVE technology makes full and effective recycling of these above wastes possible.

2. The difference in the nature of the changes in the porosity of both original and reduced brex depending on the type of binder used, has been established. In brex that used a magnesium binder and which retained their tensile strength, and did not form any fines, the metallization process is accompanied by the formation of a metallic carcass, and the small pores disappear, while the larger pores increase in size.

3. Brex that use a magnesium binder have a high degree of metallization after reduction, in the course of this process the microstructure in the original brex is not retained, and begins to resemble the condition under the two-phase «metal-silicate phase» system.

4. One limitation on the use of brex that use a magnesium binder is the increased sulfur content in the binder, which is partially transformed into a flue gas, and in the absence of desulfurization of the flue gas this reduces the quality of the catalyst. It is this same parameter that will also define the permitted content of these brex in the charge of a metallization furnace.

5. SVE can be used effectively to agglomerate HBI fines.

6. Brex made from coal and ore could be considered as a possible alternative to the coal and ore pellets for use in the ITmk3® process.

7. Finally concluding on the suitability of these brex, research needs to be undertaken into the optimization of the dimensions, and composition of slag forming oxides in the brex.

References

[1] Kurunov, I.F. and Savchuk, N.A. 2002. Status and prospects of non blast furnace iron metallurgy. Ferrous Metallurgy Information. p. 198 (in Russian).

[2] Kikuchi, S., Ito, S., Kobayashi, I. et al. 2010. KOBELCO Technology Review. 29: 77–84.

[3] Sokolov, V.N., Yurkovets, D.I., Razgulina, O.V. and Melnik, V.N. 2004. Study of characteristics of solids microstructure by computer analysis of SEM images Proceedings of Russian Academy of Sciences. Physical Series. 68(9): 1332–1337.

Index

About the Author

Aitber Bizhanov was born on October 6, 1956 in Buinaksk, Russia. He graduated from the Moscow Institute of Physics and Technology in 1979. For eleven years, Aitber worked at the Institute of High Temperatures of the USSR Academy of Sciences. In 1992, he left academia and joined EVRAZ, a large vertically integrated metallurgical and vanadium enterprise with global assets. In 2004, Aitber went to the Kosaya Iron Works as a commercial director, having introduced metallurgical waste briquetting technology. Since 2006, he has been working on agglomeration and briquetting of metallurgical waste as a waste management expert at HARSCO Metals. Today Aitber is an independent representative of J.C. Steele & Sons, Inc., Direxa Engineering and Haendle in Russia and the CIS, Eastern Europe and Turkey. He holds a PhD in agglomeration of natural and anthropogenic materials in metallurgy. Author of over 70 publications (including six books), owner and co-author of 14 Russian patents in this area. Author and owner of the BREX trademark. With his personal participation, projects for briquetting natural and anthropogenic raw materials in ferrous metallurgy have been successfully implemented in a number of countries. Aitber has been a member of the Institute for Briquetting and Agglomeration (IBA) since 2011.

For Product Safety Concerns and Information please contact our EU
representative GPSR@taylorandfrancis.com
Taylor & Francis Verlag GmbH, Kaufingerstraße 24, 80331 München, Germany